Handbook of Research on Engineering Education in a Global Context

Elena V. Smirnova
Bauman Moscow State Technical University, Russia

Robin P. Clark
University of Warwick, UK

Volume I

A volume in the Advances in Higher Education and Professional Development (AHEPD) Book Series

IGI Global
DISSEMINATOR OF KNOWLEDGE

Published in the United States of America by
IGI Global
Information Science Reference (an imprint of IGI Global)
701 E. Chocolate Avenue
Hershey PA, USA 17033
Tel: 717-533-8845
Fax: 717-533-8661
E-mail: cust@igi-global.com
Web site: http://www.igi-global.com

Library of Congress Cataloging-in-Publication Data

Names: Smirnova, Elena V., 1954- editor. | Clark, Robin P., 1964- editor.
Title: Handbook of research on engineering education in a global context / Elena V. Smirnova and Robin P. Clark, editors.
Description: Hershey, PA : Information Science Reference, [2017] | Includes bibliographical references.
Identifiers: LCCN 2017016128| ISBN 9781522533955 (hardcover) | ISBN 9781522533962 (ebook)
Subjects: LCSH: Engineering--Study and teaching.
Classification: LCC T65 .H234 2017 | DDC 620.0071--dc23 LC record available at https://lccn.loc.gov/2017016128

This book is published in the IGI Global book series Advances in Higher Education and Professional Development (AHEPD) (ISSN: 2327-6983; eISSN: 2327-6991)

British Cataloguing in Publication Data
A Cataloguing in Publication record for this book is available from the British Library.

All work contributed to this book is new, previously-unpublished material. The views expressed in this book are those of the authors, but not necessarily of the publisher.

For electronic access to this publication, please contact: eresources@igi-global.com.

Advances in Higher Education and Professional Development (AHEPD) Book Series

Jared Keengwe
University of North Dakota, USA

ISSN:2327-6983
EISSN:2327-6991

Mission

As world economies continue to shift and change in response to global financial situations, job markets have begun to demand a more highly-skilled workforce. In many industries a college degree is the minimum requirement and further educational development is expected to advance. With these current trends in mind, the **Advances in Higher Education & Professional Development (AHEPD) Book Series** provides an outlet for researchers and academics to publish their research in these areas and to distribute these works to practitioners and other researchers.

AHEPD encompasses all research dealing with higher education pedagogy, development, and curriculum design, as well as all areas of professional development, regardless of focus.

Coverage

- Adult Education
- Assessment in Higher Education
- Career Training
- Coaching and Mentoring
- Continuing Professional Development
- Governance in Higher Education
- Higher Education Policy
- Pedagogy of Teaching Higher Education
- Vocational Education

IGI Global is currently accepting manuscripts for publication within this series. To submit a proposal for a volume in this series, please contact our Acquisition Editors at Acquisitions@igi-global.com or visit: http://www.igi-global.com/publish/.

The Advances in Higher Education and Professional Development (AHEPD) Book Series (ISSN 2327-6983) is published by IGI Global, 701 E. Chocolate Avenue, Hershey, PA 17033-1240, USA, www.igi-global.com. This series is composed of titles available for purchase individually; each title is edited to be contextually exclusive from any other title within the series. For pricing and ordering information please visit http://www.igi-global.com/book-series/advances-higher-education-professional-development/73681. Postmaster: Send all address changes to above address.

Titles in this Series

For a list of additional titles in this series, please visit: www.igi-global.com/book-series

Handbook of Research on Media Literacy in Higher Education Environments
Jayne Cubbage (Bowie State University, USA)
Information Science Reference • copyright 2018 • 396pp • H/C (ISBN: 9781522540595) • US $255.00 (our price)

Changing Urban Landscapes Through Public Higher Education
Anika Spratley Burtin (University of the District of Columbia, USA) Jeffery S. Fleming (University of the District of Columbia, USA) and Pamela Hampton-Garland (University of the District of Columbia, USA)
Information Science Reference • copyright 2018 • 301pp • H/C (ISBN: 9781522534549) • US $175.00 (our price)

Preparing the Next Generation of Teachers for 21st Century Education
Siew Fun Tang (Curtin University, Malaysia) and Chee Leong Lim (Taylor's University, Malaysia)
Information Science Reference • copyright 2018 • 377pp • H/C (ISBN: 9781522540809) • US $210.00 (our price)

Promoting Ethnic Diversity and Multiculturalism in Higher Education
Barbara Blummer (Center for Computing Sciences, USA) Jeffrey M. Kenton (Towson University, USA) and Michael Wiatrowski (Independent Researcher, USA)
Information Science Reference • copyright 2018 • 309pp • H/C (ISBN: 9781522540977) • US $185.00 (our price)

Innovative Practices in Teacher Preparation and Graduate-Level Teacher Education Programs
Drew Polly (University of North Carolina - Charlotte, USA) Michael Putman (University of North Carolina - Charlotte, USA) Teresa M. Petty (University of North Carolina - Charlotte, USA) and Amy J. Good (University of North Carolina - Charlotte, USA)
Information Science Reference • copyright 2018 • 720pp • H/C (ISBN: 9781522530688) • US $275.00 (our price)

Critical Assessment and Strategies for Increased Student Retention
Ruth Claire Black (Imperial College London, UK)
Information Science Reference • copyright 2018 • 352pp • H/C (ISBN: 9781522529989) • US $195.00 (our price)

Strategic Learning Ideologies in Prison Education Programs
Idowu Biao (University of Botswana, Botswana)
Information Science Reference • copyright 2018 • 307pp • H/C (ISBN: 9781522529095) • US $195.00 (our price)

Advocacy in Academia and the Role of Teacher Preparation Programs
Ursula Thomas (Georgia State University, USA)
Information Science Reference • copyright 2018 • 375pp • H/C (ISBN: 9781522529064) • US $195.00 (our price)

IGI Global
DISSEMINATOR OF KNOWLEDGE

701 East Chocolate Avenue, Hershey, PA 17033, USA
Tel: 717-533-8845 x100 • Fax: 717-533-8661
E-Mail: cust@igi-global.com • www.igi-global.com

List of Contributors

Table of Contents

Detailed Table of Contents

for the 150th Anniversary of Hokkaido University," a blueprint for drastically reforming the university. In the 2014 fiscal year, a strategy to further internationalize education, "Hokkaido Universal Campus Initiative" was chosen by MEXT for the "Top Global University Project." The author analyzes Hokkaido University's internationalization progress, focusing on the strengths and activities of major projects and the changes in the overall management.

This chapter describes how the government of Uruguay believes that all children have the right to have technology at their fingertips and that all children have the right to connectivity and access to the internet, that it is as important to have electricity and running water as to have access to the internet, and that it would have a high impact on the country's technological deployment and, obviously, on education and teaching. The parts of the chapter are concerned with the technology and pedagogy relationship: how to improve pedagogy through technology, the importance of teaching English and math online with help of education inspectors and the teachers using modern platform—virtual classrooms, books in digital format, digital technology laboratories—that allows collaborative work, work on projects, logical thinking, an online assessment system. All these integrated tools transfer to the biggest investments which the author calls "global learning network."

The need for a reliable supply of engineering talent is accepted globally, but in many parts of the world the many challenges mean that this is not easily achieved. Even if the graduate supply is a reality, often there are concerns about the quality of the engineers entering the workforce. This chapter will explore this landscape, and after identifying the many challenges, explore solutions and potential ideas for the future of engineering education and the university/industry collaboration.

On the basis of its own hypotheses and analysis of foreign studies, the author determines how the informational and motivational functions of rankings have impact on Russian universities' student choice. The analysis of the rankings position of universities and some of their key performance indicators (the foreign students number, the quality of applicants) showed there is not any direct correlation between them. According to the author, in order to maximize the effectiveness of rankings, it is necessary to improve the ranking methodology, to develop universities' decision-making processes based on the rankings results, and to create customer culture on the rankings results using.

Academic and professional recognition of engineering degrees is an important problem in higher education and human resources mobility. The chapter presents a review of academic and professional recognition systems features in Cyprus and Russia. Both Russia (non-EU-member country) and Cyprus (EU-member country) are Bologna countries, use similar education curricula, and will potentially follow the qualification framework in the European Higher Education Area. However, national qualification frameworks are different. The chapter discusses the academic and professional recognition systems features in Cyprus. Then the Russian system of engineering degrees is explained and the academic and professional recognition approach is clarified. Case studies for both countries are outlined. A comparison of the academic and professional recognition systems features in Cyprus and Russia is given.

This chapter presents prognostic analysis results concerning the changes in the engineering competence requirements. It is noted that professional competences of future experts in this field are undergoing certain changes related to the need for operating complex systems and working in a team in uncertain contexts in order to support and ensure good management throughout the entire high-tech systems life-cycle. It has been established that certain technological areas of the National Technology Initiative (NTI), which is being implemented now, are not provided with the educational training programs by the adopted Federal National Educational Standards (FNES). This chapter also focuses on the role of Worldwide CDIO Initiative international engineering standards of education in the development of new engineering competence assessment tools to enhance the national system of educational standards and includes National Research Nuclear University MEPhI's own educational standards in higher education as an example.

Russian experience in the implementation of CDIO (conceive, design, implement, operate) standards for modernization of BEng programs focused on graduate training for complex engineering activity are considered. The CPD program "Applying CDIO Standards in Engineering Education" for managers and faculty staff at Russian HEIs is described. Further development of the CDIO concept for MSc and PhD engineering programs design are discussed taking into account the priorities of innovative and research engineering activities. The FCDI (forecast, conceive, design, implement) standards focused on MSc program graduate training for innovative engineering activity and FFCD (foresight, forecast, conceive, design) standards focused on PhD program graduate training for research engineering activity are presented.

As known to all, we are living in an information society today. Learners can acquire knowledge, skill, and ability by participating in teaching, training and learning systems. As the basis of teaching, training, and learning system, the management system is significantly important in the management of online materials and intercommunion between students and teachers. In order to construct more effective systems of management, virtual organizations should be considered with latticed structure, which can easily adapt to external environment changing and may even inspire more innovations in the fields of education.

The chapter deals with the concept of training for technicians, able to adjust to ongoing changes in the manufacturing environment. It is proposed to apply the systematic approach with a targeted use of collected methods and acmeological planning in a learning process when extra time is found owing to interdisciplinary integration and increased self-learning. Such planning is done with crosscutting design and problem-modular training with cross-rated instruction materials and learning outcomes. Education information systems are recommended as tools (like those based on modular object-oriented dynamic learning environment according to shareable content object reference model standard) together with automated information systems for self-learning, business and role plays with methods to rate career guidance and multimedia aids as learning tools.

This chapter proposes a new model of managing educational institutions' activities to provide staffing needs: customer synchronized resource planning (CSRP). It describes a technology that rebuilds the learning process in order to reduce the time needed to prepare staff adequately with the competencies required by employers as requested by the economy sector. At present the development of an open system for the educational institution is being carried out. This system is able not only to create an educational program dynamically, which allows us to get the right number of specialists with the desired competencies in the minimum period, but also to rebuild the agency's management system for new tasks: to generate the necessary training materials, make changes in the timetable, and rebuild the educational portal by adding new data.

Section 3
Innovation in Engineering Education

The second section presents network traffic analysis as an example of recent effective research topics and provides a set of open source tools to perform the research's practical steps. The research example with the suggested tools is valid as practical lab work for telecommunication and IT-related majors.

This chapter proposes the concept of hybrid intelligent information system (HIIS) as a "glue" concept that helps to unite disparate sections of a course on intelligent information systems. The chapter discusses a generalized structure of HIIS based on modules of consciousness and subconsciousness. The authors show that a HIIS may be implemented using a multiagent approach based on holonic organization. They provide a formalized model of metagraph and a review of methods to describe holonic agents based on the metagraph approach. Thus, a HIIS allows one to combine different approaches which are taught in a course on intelligent information systems.

The chapter considers a task of teaching undergraduate students practical skills using artificial neural networks to solve problems of information processing and control systems. It represents and proves the methods of teaching, based on the gradual increase in the complexity of tasks to be solved by students. The developed complex of laboratory works includes classical problems and methods of their solutions, as well as original methods for solving problems of automatic control. The technology base of the laboratory works are both well-known programs and software package developed by the authors. In addition to the practical experience in the use of software packages, students obtain experience in conducting comparative studies of traditional and neural network methods for solving control problems.

Teaching for creativity is one of the most challenging problems in engineering education. Two approaches are mostly applied in teaching creative skills: using creative problem-solving exercises and emerging people into a creative environment for stimulating their creativity. One of the most important requirements to creative digital environment is creativity of its non-player characters (NPC). The chapter discusses the advantages of applying a multi-agent (MA) approach to achieve creative behavior of the NPCs. The agent architecture is based on a behavior tree model, extended with three additional classes of nodes, implementing agent reactions and adaptive action planning according to agent priorities. The proposed agent architecture is implemented in a typical survival action game where all players, represented as agents, should explore the world to find resources. The assessment of the quality of agents' behavior shows that all the agents successfully demonstrate rational and adaptive behavior in the complex dynamical environment.

The chapter reveals the issue of implementing snapshots for maintaining virtual machines used for students' lab stands. Hence, implementing that backup/restore method means significant reduction of the amount of effort required for lab stands maintenance. The actuality of this chapter is in the increasing appliance of virtualization methods in the educational process, as it currently is not very developed due to the lack of a systematic approach to the development and application of new technologies. The object of study is the practical part of the course "Network Software" of the Department IU5 in BMSTU. The subject of research is the process of preparing a virtual stand for lab works. The purpose of research is to prove the significance of applying virtualization technologies such as using snapshots in the educational process.

This chapter considers the current state, types, and relevance of modern virtual laboratories, virtual practices, and training in higher engineering education. The four types of virtual laboratories are considering. This work also offers examples of virtual scientific and engineering processes simulation laboratories and virtual remote laboratories, virtual practices, and internship. It analyzes the experience of universities and companies in the virtual laboratories, virtual practices, and internship. Particularly interesting for online learning platforms are the virtual laboratories of edX and the National Platform for Open Education. Finally, the chapter provides recommendations on the development of shared knowledge centers for collective use of virtual installations and laboratories, on ways of remote participation in collaborative work with real unique installations, and on participation in the distributed research unique installations and data processing tools. The authors also indicate directions of the development of supportive virtual internship programs for students.

axiomatic basis. The core categories of the basis are activity classes. The whole conceptual framework and fundamental laws are represented as consequences of initial axioms and postulates of the basis. This approach allows the higher education pedagogy to overcome the conceptual crisis, which resulted from the variety of existing conceptual frameworks.

This chapter raises the issues of development of continuous education of managerial engineering personnel in industrial companies. The case of designing implementation of the additional education program for the course "Innovation Management" with flexible learning paths was considered. This case was created on the basis of the Foresight study's results of current and future development challenges facing LLC Lihoslavl Factory "Svetotechnika." The results of the survey of the audience—heads of structural enterprise's subdivisions—were shown. They demonstrated the need for organizational learning, as well as the most appropriate forms and instruments of its implementation.

Leading organizations of the national nanotechnology network (NNN) monitor staffing and develop mechanisms for coordination of educational processes of enterprises of the nanotechnology industry. To estimate the current state of training for nanotechnology industry in the leading universities of the Russian Federation, a study of their publications indexed in the Scopus database in 2012, 2014, and 2015 years in the subject area of "nano" was made. As a result of analysis, the universities, which form the background for the production of highly qualified specialists in the field of nanotechnology, were determined.

Long-term practice to employ the university graduates to work in industrial enterprises as well as the analysis of the "adaptation" process of a young specialist to the production process show that during the first two years he has to learn new areas of expertise. Teaching of these disciplines within the frames

Foreword

Dear Readers,

Handbook of Research on Engineering Education in a Global Context is written by participants of the International Scientific Congress "Science and Engineering Education (SEE2016)" held in 2016 within the walls of one of the oldest engineering institutions in Russia, the Bauman Moscow State Technical University. I was privileged to greet the members of the SEE2016 Congress together with my colleague, Cosmonaut Alexey Ovchina, when we were both on Board the International Space Station (ISS).

I know that the participants of the Congress discussed the need to explore changing the approaches used to prepare engineers for their future roles in society. This affects all countries of the world, so the focus was on concrete and practical steps for improving the system of engineering education globally. Speakers from Russia, Great Britain, Norway, Japan, China, Sri Lanka, Uruguay, Germany, Ecuador, and France noted how rapidly the demands on the engineering profession are changing and looking for approaches to help employers achieve global technological breakthroughs in the engineering field that will enhance the economy, industry and wider society.

The SEE2016 Scientific Congress provided an important opportunity for colleagues from across the globe to discuss systemic, cross-cutting issues and innovative ideas contributing to the preparation of the engineering personnel for the future. In our time the technological way of life is changing so quickly that it is sometimes difficult to look beyond the horizon and see where the jumps may occur in the next 10 - 20 - 50 years. This is perhaps best understood when it comes to space. A miracle of science and engineering that saw the first flight of Yuri Gagarin into space. These days the participants of the Congress are communicating directly with astronauts who are aboard the International Space Station – an event that is both stunning but already quite familiar.

On the ISS all things are created through engineering work and thought. I graduated from Bauman University and I am sincerely glad that the demand for engineering degrees is again high and rising. I am glad that both in Russia and globally, universities are preparing students for both the traditional and the most advanced engineering professions. After all, engineers are not only changing the world, they are creating the future – new homes and buildings, smart phones and computers, unique materials and medical technology, aircraft, space stations....

It is so important that we meet the current demand for young people to enter the engineering professions, professions which will directly affect our tomorrow. Among the authors of the chapters in this book there are those who teach young people, preparing them to enter the engineering professions, there are the representatives of industrial companies and there are those who determine public policy. All of them share a passion for the engineering profession and are, to varying degrees, people who are creative, talented, enthusiastic and seeking new solutions in science, engineering and technology education.

This is an important book – a collection of chapters concerning latest thinking, conceived by people involved in the training of engineers. I wish the authors good luck on their journey of discovery and understanding and with this book – may there be more. For good readers like us, please enjoy and be stimulated. Finally to all, many long years of life!

Oleg Ivanovich Skripochka
Independent Researcher, Russia

Oleg Ivanovich Skripochka *is a cosmonaut, engineer, member of the Group of Cosmonauts of the Russian Scientific Center "Energia," hero of the Russian Federation.*

Preface

Engineering is receiving increased attention in all corners of the world. Governments and industry see engineering as the mechanism to address the global challenges faced by our planet. It is also the means to ensure prosperity and the well-being of the people inhabiting the Earth. As we constantly hear, these challenges are many, but then so too are the opportunities.

To realize the opportunities and address the challenges, it is important that we develop a passionate and knowledgeable workforce, at ease with engineering and science thinking, a workforce that can be creative and then turn that creativity into practical and tangible action. Underpinning the development of this workforce is the need to explore the approaches and thinking we adopt when considering engineering education.

Globally, engineering education has become the subject of much activity, whether considering practice or research. Meetings such as the SEE2016 Scientific Congress held in Moscow in the summer of 2016, that stimulated this book, are occurring across the world on a regular basis – ASEE in the USA, SEFI in Europe and A^2E^2 in Australia are national events that sit alongside global conferences such as WEEF, Frontiers in Engineering Education and the research focused REES meeting. Ideas abound, yet often these events and the associated publications in the field are not always as 'joined up' as perhaps they should be. To that end, this volume tries to acknowledge these forerunners in the field, but aims to add to the dialogue and the innovative thinking by presenting a group of papers that are new to the existing community. In presenting this new material, we intend for the volume to be a form of outreach in that the authors hope to share and connect in a way that will lead to future collaboration.

There is still so much to explore in this field, so uncovering new approaches and thinking is always welcome. The book is organized in 6 sections, each of which will be familiar to readers at the top level, although what is explored in individual chapters will hopefully offer some new material for readers to consider.

The 6 sections are as follows:

1. The Global Context
2. Quality and Standards
3. Innovation in Engineering Education
4. Communication in Learning
5. Technology-Enhanced Learning
6. Employability and Entrepreneurship

The opening section focuses on the Global Context with four chapters, each of which considers one of the four subsequent sections, with Communication in Learning being common to all. The first paper from Norway explores the approach to Quality Assurance in Norwegian Higher Education. The emphasis of the paper is the need to embed quality in a way that is empowering and based around consultation rather than imposition. Quality needs support and nurturing rather than a 'big stick' approach if real and sustainable results are to be realized. With a robust quality environment in place, innovation in programme design and delivery can thrive and be encouraged. To that end, the second paper from Japan looks at innovation from a holistic viewpoint. In the demanding market that is Global Higher Education, exploring new ways to enter and become established in that market place are essential for all universities. The globalization approach of Hokkaido University presents a unique case study of how a clear strategy has been implemented with targets identified and an ambition for long term sustainability central to management thinking.

Introducing and making available technology at the earliest stages of a child's education is the focus of the third paper from Uruguay. The author discusses a wide ranging project whose aim was to make IT and the Internet available to as many young people as possible. Today, this access has almost become a 'right' if a community or country is to consider itself truly inclusive and able to providing the best educational starting point for its children. Moving to the other end of the educational spectrum, the final paper in the initial section from the UK explores what is needed to be successful in the engineering profession and how universities can best prepare their students for a successful career after study. Again the theme of inclusion, accepting diversity and the global nature of work are implicit within the ideas explored and evidence presented.

The five sections that follow explore each of these areas in greater detail. The range of chapters explores each section from various different perspectives – some are very much rooted in practice, whilst others are exploring novel ideas about how to gain greater understanding that can then impact learning and the wider student experience. At the start of each section, the chapters will be summarized such that the reader can see how they fit together to tell a coherent story.

In viewing the author listing, the reader will be very aware that a large proportion of the contributions are from authors within Russia. This is perhaps to be expected given that the book was stimulated by the SEE2016 Scientific Congress. This is perhaps though, more importantly, a real benefit of this book. For too long many practitioners, researchers and authors in the field of Engineering Education within the Russian Federation have been unknown to the wider Engineering Education community. Their work, approaches and thinking have not been visible. That is what makes this book so important as it starts the process of making the work of these people more available to the global community. Another important feature of this volume is the amount of high level input in the submissions, Rectors and Vice Rectors are represented throughout the contributions in a way that is not perhaps as common in other Engineering Education books available today.

As editors we sincerely hope that this volume will help to create and foster some new relationships and innovation in the field of Engineering Education. We also hope that this will be a starting point for further global collaboration and sharing. Russia has long had a reputation for the creation of capable and technically excellent engineers, so it would seem sensible to explore the ways in which this process takes place such that others can learn. At the same time, new approaches to engineering pedagogy being implemented in other parts of the world has the potential to enrich the Russian approach to Engineering

Education. One of the best and most visible examples of this would be the enthusiastic way in which Russian universities have explored and adopted the CDIO (Conceive, Design, Implement, Operate) Framework. This particular example is explored in a couple of the contributions in this volume.

In compiling these volumes our target readership is very much those who are teaching and researching in the area of engineering education, those in industry and policy making roles interested in the latest thinking around how universities prepare engineering students for subsequent employment and in particular those worth a passion for engineering education who wish to gain a truly global perspective. We hope you find something of value and interest in the book, and we strongly encourage you to reach out beyond the book to form new connections that will take this field, we know is important and want to see develop, to new heights.

SECTION 1: THE GLOBAL CONTEXT

This section serves as an opening to the book by setting the scene for the sections that follow.

The Engineering Education space comprises many different facets. The opening chapter from Gynnild explores the challenges of implementing a robust quality system. With increasing attempts to measure the student learning and quality of the student experience in universities across the globe, having an effective and organic quality assurance system to guide programme development and operation is essential. The chapter is most helpful in that it not only explores the features of a good system but also the ways in which the implementation and operation of that system can be comprised if the key stakeholders in the process are not involved and consulted.

Innovation is implicit in much Engineering Education today. This is often most obvious at the practitioner level, something that is explored more fully in Section 3 of the book. In the second chapter in this section, Nawa takes a more overarching viewpoint and considers the innovative challenge of developing a globalization strategy for Hokkaido University in Japan. It presents the different levels of complexity in producing such a strategy and then taking it forward into an implementable form. Key to the success of the work has been the development and adherence to clear goals and their associated measures. By adopting this approach, Nawa suggests that the model for change becomes sustainable and can promote further growth and innovation.

STEM (Science, Technology, Engineering and Maths) education as a whole is increasingly developed on a technological foundation. In early years education the exposure to and familiarization with technology is an essential feature that prepares students for their subsequent learning and ultimately for the workplace. Brechner presents a case study from Uruguay, Project Ceibal, in Chapter 3 that demonstrates that, with a strong commitment, the introduction of IT and Internet access in schools on a wide scale can have important benefits for both students and the wider society. These are most often seen in the educational abilities of the students but also in the way it acts as a way to promote inclusiveness across the whole of society.

The final paper in this section ‘closes the loop’. By that we mean it looks at what happens within the university and beyond that ensures engineering graduates are able to enter the workforce and become productive members of that community. The arguments put forward by Clark and Andrews suggest that engineering degree courses need to be designed and delivered with a very clear eye on what industry and society need. The chapter also reflects on the changing nature of learners – a diverse, technology literate and demanding group of people that require engineering educators to be creative, engaging yet thorough in their learning and teaching practice. With scholarship and research to guide them, these practitioners are constantly introducing new thinking into the field.

Together these papers set the scene for the subsequent sections. Taken as a whole the volume thus captures the vibrant and progressive state of the engineering education field.

SECTION 2: QUALITY AND STANDARDS

Quality and Standards are the guiding lights when it comes to the implementation of credible Engineering programmes. On the one hand universities strive for accreditation to ensure that the learning meets the demands of the profession, yet equally in this modern era, the attractiveness of the educational approach is also a key consideration. The opening chapter in this section by Yagudina explores the role of university rankings in aiding student choice as to where to study. The paper argues that although of value and recognized by both universities and potential students, the understanding of what they are conveying and the methodology used to generate them needs further consideration.

The recognition of engineering degree programmes is the focus of the next chapter in this section, in particular a comparison of case studies from Cyprus and Russia. Kassinopoulos and Zinchenko explore the EU / non EU differences. In particular the authors note that despite sharing similar guiding principles and standards, the application in each case is different. This highlights that despite the desire to promote global mobility amongst engineering students and graduates, barriers still exist.

Fundamental to the development of the guiding standards is a clear definition and understanding of the professional engineering competence requirements. Berestov et al take a forward looking approach to what industry needs and will need in the future. They suggest that the CDIO framework in Engineering Education is a valuable starting point, something that is developed further in the subsequent chapter by Chuchalin. The extension of the CDIO thinking to Masters and PhD education in engineering is an interesting development explored in Chuchalin's paper. Until now the main focus across the world has been on Bachelor's education, so this chapter adds a new level of experience to the CDIO conversation.

When considering innovation in Engineering Education, a book would be remiss if it did not explore the value and importance of Information Technology (IT) in the facilitation of effective learning approaches and the management of student records. Considering a more structured approach to this element of Engineering Education is the subject of the chapter by Aleksandrov et al. Their 'scientific' approach suggests that efficiencies can be gained when compared to the less rigorous approaches that are often adopted in universities.

It is not just IT that needs consideration, the training and preparation of staff is an often overlooked aspect of learning and teaching. Vlasov et al look at a specific area by considering technician training and how technology can be a helpful component of developing the needed vocational skills. Martynov et al take this approach to a higher level by considering the overarching management information system that drives the interconnection of staff training, educational design and resource planning.

These preceding three chapters have been included in this section as they all take a higher level, institutional view of the use of technology in Engineering Education. A subsequent section of the book (Section 4) looks at the application and technology of technology in learning in much greater depth.

Quality assurance and quality enhancement are both equally important in engineering education. Kolsenkov and Taganov propose a 'scientific' approach to quality assurance that makes use of GIS (Geographical Information System) data and has the potential to facilitate improved decision making by managers when considering the operation of courses and of the wider university. Quality enhancement is the theme of the chapter by Sukhobokov et al when they consider the improvement of Engineering Masters and Doctoral programmes. The key driver for the improvement is the fast pace of change within industry that necessitates a proactive and future-looking approach to change within the university.

The concluding chapter in this section explores the challenge of helping students to transition from the school environment into that of the Technical University. Tsibizova explores the various elements that if addressed proactively can enable students to settle into the university and start to develop as novice engineers from the earliest point in their tertiary educational journey. Although not a quality process or standard explicitly, ensuring a consistently high level of understanding and practice in this sensitive area of university operations can ensure students progress and the problem of student retention is overcome.

SECTION 3: INNOVATION IN ENGINEERING EDUCATION

As engineers and engineering educators we are by our very nature creative individuals and teams. It is thus not surprising to find, in a volume such as this, a large number of case studies showcasing exciting and innovative developments in Engineering Education. Innovation can take many forms so the diversity of material in this section of the book explores a range of possibilities.

The section opens with an exploration of a challenge for engineering educators across the world, that of the mathematics competencies of engineering students. Mathematics is often cited as a key cause of student drop out from engineering courses. Exploring the findings of a European Union funded project, Soldatenko et al offer suggestions for the 'modernisation' of the engineering curriculum with respect to the mathematics requirements.

The opening chapter is followed by two examples of Problem and Project Oriented Learning approaches that cover mathematics (Dotsenko et al) and software engineering (Vishnevskaya et al). In each case the authors explain not only the approach itself but also the supporting requirements for success and the challenges that can be presented. Accepting and actioning these requirements successfully has the potential to promote learning and help student motivation.

A key feature of the Problem and Project Approaches to Learning is student interaction in the learning process rather than passive engagement. The third chapter in this section explores a range of interaction possibilities, often making use of technology (Baryshev et al). Returning to an earlier argument, CDIO is considered to be a fundamental driver of the changes taking place. Disciplinary innovations in avionics (Proletarsky et al), economics (Ponkratov et al) and biotechnology (Suyatinov) teaching make use of some of these ideas promoting student interaction in the learning process. A common theme in each case study is the relevance of learning and the efforts that go into defining authentic problems and involving industry in the education.

Interdisciplinarity is explored in the contribution from Buldakova and Suyatinov. In the modern era, whether working in industry or within the academic environment, the ability to work across disciplines is an essential element of an engineering student's education. The authors propose active 'learning through research' approaches as a way to realise the necessary skills and awareness in this space.

The final chapter in this section explores the value of practical classes and how the activity design is best approached when developing tasks that are most beneficial when considering student independent working. The particular example relates to a course in Digital Signal Processing and Syuzev et al introduce an analytical approach that is both novel and thought provoking.

SECTION 4: COMMUNICATION IN LEARNING

The chapters in this section focus on language proficiency and communication skills. Often the ability of engineers to communicate with colleagues, customers, suppliers and authorities in different parts of the world is vitally important in order to convey technical information or an engineering opinion. Miscommunication can have serious consequences in terms of misunderstanding, incorrect actions or putting people's safety at risk.

English is a global language in the technical domain and, as such, being able to converse confidently in English in an engineering role is something that universities take very seriously in many parts of the world. The examples in the two chapters are from Russia, but they could be contextualized to other countries quite easily. Margaryan and Alyavdina explore new approaches to English language teaching in a technical university, whereas Kirsanova and Lazarev focus on the embedding of strong technical communication skills in particular engineering courses. The development of both competence and confidence are stated aims in each of the scenarios presented.

The concluding chapter in this short section focuses on foreign language learning in engineering education. In many countries, foreign language learning is not considered an important feature of engineering education, so is not included. In this example the inclusion is, quite rightly in the view of the Editors, required. This particular work by Kurovskaja explores the use and value of foreign language technical textbooks in helping to develop confidence in the 'new' language. With confidence and competence, the students can bring a versatility and new dimension to whatever role they undertake on graduation.

SECTION 5: TECHNOLOGY-ENHANCED LEARNING

The pace of change when considering educational technologies is rapid. It is important though that the value to learning is appreciated and understood by both the practitioner and the university. In this section that focuses on Technology Enhanced Learning (TEL), the opening chapter examines approaches that can be taken when considering the monitoring and forecasting of educational technology development and implementation (Kozlov et al). The power of this work is not only in the understanding it offers but also the enhanced decision making potential for management within the higher education environment.

The complexity and diversity of the TEL landscape is captured in the chapters that follow. Jammoul et al explore the use of Open Source Software in both the teaching and research settings. A particular feature of the work that the authors identify is the way in which OSS can promote collaboration and sharing in learning.

As intelligence becomes more embedded within software solutions in all walks of life, it would seem beneficial to enable student learning using such 'intelligent' approaches. An approach to teaching Intelligent Information Systems is presented by Chernenkiy et al in which they highlight the need to connect the different concepts rather than have the student view them as separate. This is achieved using a technology based solution.

Borodkin et al offer a methodology that uses neural networks as the base tool for teaching the solution of real world problems in the laboratory. In a more refined example, Alimov et al use the digital game environment as way to engage students in learning and in particular the development of creative thinking skills.

Virtual laboratories (Kalistratov et al) and broader thinking about the role of virtual working in Engineering Education (Alekseev et al) present both the challenges and opportunities of the virtual space. The potential for overcoming the most basic of issues, such as a lack of money for equipment for a 'real' lab, demonstrates the power of considering virtual solutions. Teachers have the opportunity to set up a host of scenarios for students to interrogate with minimal additional work or cost.

The use of technology to facilitate student assessment is explored by Pesoshin et al. The chapter explores an approach to both the design and implementation of automated testing approaches. A very different challenge is discussed in the following paper. The use of Virtual Learning Environments (VLEs) in universities is commonplace as they have become the foundational technology platform to facilitate student learning. In the contribution by Chernikov et al, the ability to link one such platform, Moodle, with the wider university IT systems is explored. The benefits being the accuracy and consistency of the information made available to both students and university staff at all times.

The final chapter in this section brings the earlier examples together by considering the synthesis of the professional competences of the engineer with the range of virtual and technological learning opportunities they experience during their education (Chistyakova). This analytical paper offers an opportunity for Programme Leaders to ensure that students are gaining the appropriate experience from their TEL in order to be ready for their future careers.

SECTION 6: EMPLOYABILITY AND ENTREPRENEURSHIP

In this final section, the lens is firmly on what happens after graduation. In the first section, Clark explored the range of approaches being taken within the UK in order to develop the needed employability skills in graduates. Here the authors go deeper and present detailed ideas for how to make this happen. We then extend this into the area of entrepreneurship and the authors discuss how this can be promoted within engineering education.

Serebraykova et al present a case study of a model employed to ensure the development of professional capabilities in graduating students. The model called the 'Elite Engineering Education Programme', promotes both personal and interpersonal capabilities as well as those most readily associated with the needs of an engineering job. The idea of competences is developed further in the chapter by Malakhov et al where the consideration is that of meta-competences and how they can best be developed in the laboratory environment.

In the personal space, Nesterenko proposes an approach that considers the individual's responsibility for and awareness of their development. Understanding how they evolve as professional engineers is a feature that can help individuals 'stand out' in a competitive job market. Another differentiating factor can be a level of competence and ability in management thinking and skills. Maltseva explores this opportunity with a particular focus on innovation management, an area that is particularly relevant to engineers. The following two chapters present case studies in particular disciplines – the nano industry

(Grekhov et al) and the aerospace industry (Dubkov et al). The holistic view of the first chapter contrasts with a more analytical exploration of the computer vision systems segment of the modern aircraft industry in the latter.

Technology entrepreneurship is the focus of the paper by Vlasova and Pilyugina. The authors identify the need to properly define 'technology entrepreneurship' such that the education for engineers can be developed appropriately. Having said this, the importance of a business and management element within the learning is strongly emphasized, with all being developed in an engaging and active learning environment. Borisov considers how the 'technical entrepreneurs' discussed in the preceding chapter can then engage with Small and Medium Sized Enterprises (SME's) to promote regional growth and prosperity.

Although diversity has not been explicitly discussed throughout the book, it is implicit in much of the work presented. The authors state at various points their awareness of student cohort diversity and the focus on developing inclusive learning and teaching strategies. The final paper in the book though discusses the issue directly with an exploration of the characteristic of disability and how more needs to be done to overcome the underrepresentation of disabled people in STEM (Science, Technology, Engineering, Maths) careers (Fell et al). The authors argue that by simply discussing the issue and raising awareness, positive steps can be taken in moving forward.

Elena Smirnova
Bauman Moscow State Technical University, Russia

Robin Clark
University of Warwick, UK

Acknowledgment

The editors would like to acknowledge the help of all the people involved in this project and, more specifically, the authors and reviewers who took part in the review process. Without their support, this book would not have become a reality.

First, the editors would like to thank each one of the authors for their contributions. Our sincere gratitude goes to the chapter's authors who contributed their time and expertise to this book.

Second, the editors wish to acknowledge the valuable contributions of the reviewers regarding the improvement of quality, coherence, and content presentation of chapters. Most of the authors also served as referees; we highly appreciate their double task.

Section 1
The Global Context

Chapter 1
Quality Assurance in Norwegian Higher Education: A Case Study

Vidar Gynnild
Norwegian University of Science and Technology, Norway

ABSTRACT

This chapter starts out by exploring why five higher education institutions failed to meet nationally agreed criteria for the approval of their quality systems. In order to achieve this, the review panels' reports were examined with a particular view to data analysis and data application. The reports were readily available online and represented an excellent data source for research purposes. Panels found that institutional quality reports were descriptive rather than analytical, that quality procedures were unsystematic and conclusions often missing. Unfortunately, the panels failed to come up with radically new approaches that could potentially alter that situation. Rather, the recipe seems to be more of the same, in particular student evaluation of teaching. With this as a backdrop, this study provides conceptual tools that might change the actors' approaches and thus empower those undertaking quality reviews locally. The alternative could otherwise be sustained frustration and wasted efforts.

INTRODUCTION

The Universities and Colleges Act (§ 1-6) stipulates that all higher education institutions in Norway should have a satisfactory internal quality assurance system. Within the framework of national regulations, local institutions enjoy freedom to tailor their system according to professional profile and a range of other factors. To help ensure that systems meet set quality standards, the Norwegian Agency for Quality Assurance in Education (NOKUT) is mandated to review institutions' quality systems using specified criteria (Universities and Colleges Act, § 2-1). NOKUT is a professional, public body, which in addition to its supervision mission, aims to stimulate quality efforts more broadly. It therefore serves as both a supervisory body and resource unit for higher education.

DOI: 10.4018/978-1-5225-3395-5.ch001

To carry out the statutory mandate of the Ministry, NOKUT makes use of expert committees to evaluate institutional quality systems. This started in 2003, while NOKUT in its second phase of evaluation from the spring of 2009 changed its practice by placing greater emphasis on the institutions' use and benefit of quality assurance systems. However, the responsibility of the expert committees is still to judge whether or not systems may be approved based on criteria set by NOKUT. The external panel review consists of an evaluation of the quality system and the institution's active use of it. This is a challenging because there is a widespread distrust to such undertakings among academic staff. Furthermore, there are concerns over the bureaucratic nature of some of the current quality initiatives, sometimes with little evidence of real improvements in student learning, as exhibited in the following quote: "While forces of accountability are strong, those devoted to improvement, including the promotion of innovation, are fragmented" (Newton, 2001, p. 222).

Given the fact that institutional quality initiatives occupy significant resources, it is surprising to note that this domain often remain both under-researched and under-theorized. While there is an abundancy of good intentions around, there is less of evidence to inform policy decisions and practical interventions; "... in reality, we can point to very little research into how 'quality policy', or other areas of strategy designed to improve learning and teaching have been used, how this has impacted on academic practice (Newton, 2002, p. 1-2). This study aims to bring evidence of the application of data for quality enhancement purposes. The Norwegian quality framework features five criteria to guide review panels' examination of quality systems; however, this study places an emphasis exclusively on the last two of them:

- Stimulus to establish a quality culture, if commitment to quality is encouraged by the institution;
- Whether the objectives, responsibilities, processes and actors that are part of the quality system are clearly described, and how the quality assurance system is tailored to meet the institution's needs;
- Securing and evaluating the quality of study programs based on data from multiple sources, and whether there are quality assurance procedures for new study programs;
- Analysis, assessment and reporting: whether data are being analyzed, assessed and presented to the responsible boards and management level;
- Use of knowledge for improvement: if improvement measures are rooted in a proper quality analysis.

To ensure reasonably professional judgments, each review panel recruited people with varied backgrounds and competencies. The following items exhibit competency requirements:

1. Experience with quality work or evaluation;
2. At least one should have experience at managerial level of higher education;
3. At least one should be linked to a relevant institution abroad;
4. There should be one student with experience from governing bodies or student unions;
5. There should be at least one panel member with professorial competence.

DATA AND RESEARCH QUESTIONS

This is a qualitative study of external review panels' assessments of five higher education institutions' self-reporting on own quality systems. These institutions differ in terms of student enrolment and disciplinary focus; however, they are all tertiary institutions and as such exposed to regular reviews of their quality systems. I do not elaborate on the nature of the various institutions since this falls outside the focus of the study. The five external review reports varied in length from 25-42 pages and formed the exclusive data source of the study. The following research questions guided the research:

- Which issues relating to item d and e referred to above caused the committees' concern?
- What advice did the committees provide to enhance institutions' educational quality?
- What conceptual contributions might help in promoting educational development?

In analyzing the data, I utilized a dual approach. First, I wanted to capture key themes in conclusions, including arguments underpinning them. Second, I wanted to check for potential patterns that might be inherent in data gathered from the institutions. Are there unresolved educational issues that keep reoccurring across institutions, and how could they possibly be dealt with conceptually to remedy the situation? In terms of research findings, conclusions of this study are fully valid for the sample; however, given the limited number of cases, we are not in a position to generalize findings uncritically. The author still does not believe this reduces the educational value of the study, as the fact that knowledge cannot be formally generalized "… does not mean that it cannot enter into the collective process of knowledge accumulation in a given field or in a society" (Flyvbjerg, 2010, p. 227).

Analysis and Discussion

A major concern for the review panels appears to be whether or not institutions have made use of the evaluations, while other means of data collection are not mentioned. Yet another salient feature is that reports are *descriptive* rather than *analytical* and the authors do not seem capable of bringing new perspectives and helpful conceptual tools. Recommendations appear to be more of the same, implying "fine tuning" of current practices rather than "radical change" to spark motivation for more productive ways of learning and teaching. Table 1 provides an overview of the panels' judgment of two aspects of quality for five institutions.

A key issue is the need for conceptual tools and expertise to enrich and improve conceptual approaches to quality enhancement. Our data indicate that institutions involved in this study need a more scholarly approach to basic steps in undertaking successful quality enhancement projects. There is a strikingly poor match between the rhetoric and capabilities demonstrated on the ground in order to turn ideals into realities. First, there are apparent challenges of method. Evaluations may or may not be useful depending on the skills of those designing them. Standards and criteria inherent in such studies require methodological insights as well as disciplinary knowledge. In an ideal world, there would be a clear connection between data collection, data analysis and implementation of measures to change the situation to the better:

Table 1. Aspects of quality as presented in five expert panel reviews

Inst.	Analysis, Reporting, Evaluation	Application of Data for Improvement
1	The committee believes that the screening process at the institutional level weakens the flow of accumulated and analytical data to the Board.	There is no comprehensive analysis at the institutional level to securely implement additional measures.
2	Many of the indicators are not based on analysis or discussion of potential measures to be implemented.	When there is insufficient quality of the reporting system, decisions on quality issues are simply not possible.
3	The reports from faculties to Rector contain little, if any, analysis of teaching and learning, but rather indicate challenges to obtain such data. This hampers any further analysis.	It emerged that the system generates some information that is not communicated. … It is not clear what has been done to follow up on matters that were seen as issues last year.
4	The reports draw few conclusions and point to few measures that could be implemented to improve quality. Reports, analysis and assessment appear as sporadic and unsystematic. There is no indication that the reports have made a difference whatsoever.	The main impression of the review panel is that it is unclear how data generated by the system is used to improve the quality of the education. The college needs to improve this area in the future.
5	The system generates an abundance of data, but satisfactory analysis is missing. Reports contain at best inadequate analysis of the quality. Reports are mostly duplicates of faculty reports.	This institution has not brought evidence that any measure has been implemented. An informal quality culture is in itself positive ... but the quality system is not implemented at all levels.

Readers get the impression that the distinction between description and analysis is not sufficiently present, and that institutions have been unable to help academics to gather insights into essential relationships in the learning-teaching environment. The reporting appears descriptive with little, if any, analysis and therefore less interesting conclusions (Review Panel's Report, Institution 2).

In this study, a series of unfortunate circumstances in combination prevent productive use of resources. Expectations to the effects of the quality systems did not materialize, and there is an urgent need to come up with valid explanations in order to remedy the situation. One obvious issue to start with is not to confuse means and ends in the system. Teaching is always a means towards an end, never an end in itself, and there is no one-to-one relationship between teaching and learning. A corollary would be to adopt a focus on learning rather than on teaching, which would have implications for data collection and analysis. Furthermore, as long as there is no evidence of impacts on learning, teachers cannot be sure if implemented measures have had any positive effect. Student satisfaction studies messes up the issue by making "happiness" rather than learning gain the success criterion.

The worst scenario would be just adopting a purely instrumental approach to comply with externally imposed quality requirements; however, with little if any positive outcomes. In such circumstances, frustration and anger may pave the way for negligence or even outright resistance. Anecdotal evidence suggests that lack of trust in quality systems as well as missing evidence of observable effects make it hard to pursue the same track. A more technical approach attending to "deviations" from guidelines may not remedy the situation. This may avoid serious errors, but does not necessarily need to be effective to improve learning, if that is an issue. Finally, it is surprising to see that the panel reports frequently make use of the construct of "quality" without further explanation. This gives the impression that the construct refers to something attractive and valuable, yet still being unable to explain in further depth. It is therefore of interest to examine the committees' advice on how quality systems should be remedied to meet criteria. Table 2 gives an overview for each institution.

Table 2. Review panels' advice to the respective institutions

Institution	Selected Quotes to Describe the Nature of Recommendations
1	Re-introduce the learning environment committee; make explicit the quality assurance system; student involvement should be strengthened; consider the introduction of a deviance system; the QA system should be made more transparent.
2	Examine current QA system; introduce student evaluations of education; provide information on the QA system, including expectations; continue work to produce annual quality reports; consider evaluation of entire study programs; consider indicators of workplace relevance.
3	System description is unsatisfactory; better routines for student evaluations are required; mandatory course evaluations should be implemented and reported.
4	Quality reports should be improved; clear roles and responsibilities of QA are missing; evaluations, analysis and measures need more attention; must clarify the purpose of extensive reporting; must link QA efforts to more explicit follow-up measures.
5	Analysis of quality is fragmented and unsatisfactory; too much responsibility of QA is assigned to one person.

Table 2 exhibits major advice from the review panels to each institution. The main message is that diverse issues are in need of attention. Unfortunately, the panels provide no conceptual models to aid thinking and action at the institutions. Due to the word limit, I will only provide a few approaches that might be helpful in cases such as those demonstrated in Table 2. The point is that basic conceptual approaches of phenomena fundamentally impact analysis and ensuring actions. It is therefore crucial to get the process right from the very beginning.

The construct of "teaching" is often taken for granted based on tacit knowledge; however, it is vital to remember that different interpretations have implications for the execution of roles and responsibilities in quality programs. A conceptual distinction is made between "teaching in a narrow sense", for example lecturing, and "teaching in a broad sense" which would imply a radically different teacher role, primarily as a designer of learning methods and environments (Barr & Tagg, 1995). Not only do different interpretations of the same construct have practical implications, they also build on different theories of learning. While "lecturing" is based on transmission theories of learning, teaching in a "broad" sense relies on constructivist theories of learning. It is the design of learning environments that makes the difference to students, not the teacher's lecturing skills.

Many quality programs are designed as if their creators have been unaware of this distinction. Teaching in a "narrow" sense appears as self-evident, and consequently, data gathering instruments are designed to suit this model, while teaching in a "broad" sense represents a far greater potential for improved learning. The two routes also mark the difference between an individualistic and a systemic approach to learning and teaching. Even self-proclaimed outcomes-based institutions in reality often focus on "teaching" rather than on "learning" in their quality programs, with severe implications for the design of data gathering instruments, and subsequently on data analysis and implemented measures. If the targeted area is "teaching", student satisfaction studies of offerings sounds like the right thing to do. If the targeted area is "learning", a sensible measure would be to conduct a gap analysis in order to identify the correspondence between intended learning outcomes and students' real achievements.

FUTURE RESEARCH DIRECTION

Yet another conceptual tool might be helpful in cases such as those discussed in this study. This is presented in a matrix featuring key dimensions of quality assurance and quality enhancement programs. The four quadrants, each with unique requirements focus on "learning" or "teaching" in combination with a specified purpose as either "quality assurance" or "quality enhancement". The matrix could be used to check the focus and purpose of current instruments as well as to devise instruments for future purposes. The interpretation of dimensions suggested in the matrix is not entirely straight forward and calls for a scholarly rather than a managerial approach. As already stated, everyday terms such as "teaching" and "learning" may take on different meanings and do not make sense unless people agree on their interpretations in local settings. One just needs to be conceptually aware of focus and purpose by clarifying issues of "why", "what", "how" and "when".

A different intervention model would be to start with the identification of a problem (Simmons & Gregory, 2003). The approach here is explanation rather than evaluation. The purpose is to establish an understanding of why certain issues keep re-occurring. Awareness is the first step after which an explanatory understanding of an observed issue has to be established. In theory, a broad range of variables in the teaching-learning environment could be taken into consideration; however, some are of course more influential likely than others. The point is to address key explanatory variables and thereby pave the way for targeted interventions for specified purposes. Unfortunately, the extensive use of student evaluation instruments based on Biggs' theory two of teaching (Biggs,1999) confuses means and ends. What counts is what the students do, not their degree of satisfaction with teaching.

CONCLUSION

This study demonstrates the importance of sound conceptual frameworks capable of identifying targeted areas in relation to "focus" and "purpose". Conceptual shortcomings have important consequences for data analysis and follow-up initiatives. It is noteworthy that, even though the review panels bluntly disapproved the institutions' quality programs, they were less able to help failing institutions out by suggesting innovative approaches to better handle issues. The general recipe seems to be more of the same that did not work in the first place.

A generic challenge in evaluation relates to "standards" and "criteria". These terms are sometimes used interchangeably but refer to different realities. If quality programs are imposed by external decisions without proper involvement of key players internally, institutions run the risk of failing support from academic staff and students. In the worst case, quality initiatives with the best of intentions inadvertently may do a disfavor to their institution. One way to avoid this is for academic staff to become more reflective and grounded in the selection of conceptual approaches. This requires key players to become more scholarly in their approaches than was the case in this study. This applies to senior managers as well as those serving on external review panels.

ACKNOWLEDGMENT

This is a revised version of an article first presented at the 8th International Conference of Education, Research and Innovation, 16-18 November 2015, Seville, Spain.

REFERENCES

Barr, R. B., & Tagg, J. (1995). From Teaching to Learning: A New Paradigm for Undergraduate Education. *Change*, *27*(6), 12–25. doi:10.1080/00091383.1995.10544672

Biggs, J. (1999). What the Student Does: Teaching for enhanced learning. *Higher Education Research & Development*, *18*(1), 57–75. doi:10.1080/0729436990180105

Flyvbjerg, B. (2010). Femmis for staelser om case-studiet. In L. Tanggaard & S. Brinkmann (Eds.), *Kvalitative metoder*. København: Hans Reitzels Forlag.

Gynnild, V. (2014, August 25). *From 'quality assurance' to 'quality enhancement': Addressing the transition from 'teaching' to 'learning' in engineering education*. Paper presented at the iCEER2014 - International Conference on Engineering Education and Research, Hamilton, Ontario, Canada.

Newton, J. (2002). Views from Below: Academics coping with quality. *Quality in Higher Education*, *8*(1), 39–61. doi:10.1080/13538320220127434

Simmons, O., & Gregory, T. (2003). Grounded Action: Achieving Optimal and Sustainable Change. *Forum Qualitative Social Research*, *4*(3).

Universities and Colleges Act. (2005). Retrieved from http://lovdata.no/

ADDITIONAL READING

Aleksandrov, A. A., Neusipin, K. A., Proletarsky, A. V., & Fang, K. (2012). Innovation development trends of modern management systems of educational organizations, In Proc. *2012 International Conference on Information Management, Innovation Management and Industrial Engineering* (pp.187-189).

EU Commission project. "Changing Pedagogical Landscape" Retrieved from http://www.changingpedagogicallandscapes.eu/ (Access date 06.01.2018)

International Education support platform Retrieved from http://www.icef.com/en/educator/overview/ (Access date 06.01.2018)

Matzdorf, F., & Greenwood, J. (2015). *Student choice, league tables and university facilities*. Retrieved May 2, 2017, Retrieved from http://shura.shu.ac.uk/10423/1/EuroFM2015_StudentChoicesLeagueTables%26Facilities_final.pdf

Trifan, L. Project Management for the Implementation of an Internet of Things Platform, Edulearn15, 7th International Conference on Education and New Learning Technologies, 4-8 Iulie 2015, Barcelona, Spania, ISBN 978-84-606-8243-1, ISSN: 2340-1117, pp. 6024-6030, [ISI], indexed IATED Digital Library, Google Scholars, [Vis]

Zabotkina, V. I., & Makolov, V. I. (2016, October-December). Implementation of ESG for international joint education programmes. *Integra Educativa*, *20*(4), 446–455.

KEY TERMS AND DEFINITIONS

Conceptual Approaches: Conceptual approaches to quality enhancement means more scholarly approach to basic steps in undertaking successful quality enhancement projects.

Higher Education: An optional final stage of formal learning that occurs after completion of secondary education.

Quality Enhancement: One of four quadrants in a matrix featuring key dimensions of quality assurance and quality enhancement programs.

Chapter 2
Globalization of Japanese Higher Education and the Case of Hokkaido University

Toyoharu Nawa
Hokkaido University, Japan

ABSTRACT

Institutions of higher education all over the world are facing the pressure to internationalize their operations and academic programs, to enhance its competitiveness in an international education market. The first part of this chapter presents a review of national policy to incentivize the internationalization of higher education in Japan since 1980s. The second part introduces internalization initiatives of Hokkaido University in the last decade. Under the initiative of the president, university formulated its vision of "Hokkaido University, contributing to the resolution of global issues" in the "Future Strategy for the 150th Anniversary of Hokkaido University," a blueprint for drastically reforming the university. In the 2014 fiscal year, a strategy to further internationalize education, "Hokkaido Universal Campus Initiative" was chosen by MEXT for the "Top Global University Project." The author analyzes Hokkaido University's internationalization progress, focusing on the strengths and activities of major projects and the changes in the overall management.

INTRODUCTION

Our world is undergoing a sea change due to the rapid advance of globalization and growth of the internet, enabling people in any corners of the world to instantly connect with others and obtain information. This is growing the importance of the knowledge economy and technology, and thus have always challenged in all countries over the world in the past two and three decades. The idea of internationalization is considered as a positive phenomenon for many people yet questionable for many others, but no one doubts that this new situation is somehow linked to new forms of technology and economy. Many higher education institutions have been tried to internationalize their operations and their academic offerings to enhance its competitiveness in an international education market, by ensuring the delivery of a culturally-enriched educational experience.

DOI: 10.4018/978-1-5225-3395-5.ch002

The first part of this article focuses on how Japanese universities have been reformed to internationalize higher education in Japan. A national policy is also reviewed to examine how the government tried to initiative internationalization.

The last part of this article introduces a recent initiative of Hokkaido University since 2008, which aims to internationalize higher education. The author analyzes the Hokkaido University's internationalization progress with a focus on the strengths and issues of major projects and activities and changes in the overall management.

THE INTERNATIONALIZATION OF HIGHER EDUCATION IN JAPAN

Internationalization in higher education in Japan started in the 1980s (Horie, 2002). One of the measures to progress internationalization and multiculturalism in higher education institutions is to promote the international mobility of students. In 1983, Japanese Ministry of Education, Science and Culture (*Monbusho*, later MEXT) announced a plan to have 100,000 international students studying at higher education institutions in Japan by 2000 (MEXT, 2008). The objective of this plan was to accelerate mutual understandings and deepen friendship between Japan and other countries, strengthen intellectual power over the global society, and contribute to internationalization of economic and social systems. The plan was successful in terms of increasing the number of international students (from 10,428 in 1983 to 64,011 in 2000). However, the plan paused due to the limited capability and attractiveness of Japanese education. One of the main causes is the difficulty of learning Japanese. Furthermore, many stakeholders have started to point out that improvement in the quality of education by internationalization is more important than increase in the number of international students.

In 2008 the Japanese government set a goal of attracting 300,000 international university students by 2020, following the 'Global 30' plan in 2009 aimed at transforming 30 universities into world-class institutions of higher education (Yonezawa, 2011). The '300,000 international students plan' and the 'Global 30' plan have focused more closely on supporting universities to expand their English-taught degree programs.

From 2011, in order to rather enhance the international compatibility and competitiveness of higher education in Japan and to provide prioritized support for the world-class and innovative universities, MEXT launched three projects: the Re-Inventing Japan Project was launched in 2011 to establish collaborative relationship initially and mainly with Chinese, Korean and North American universities; the Go Global Japan Project, originally titled the 'Project for Promotion of Global Human Resource Development', was launched in 2012 to develop 42 target universities' international education offerings and to increase the number of Japanese students studying abroad (MEXT, 2012); 'Top Global University Project (TGUP)', titled in Japanese as the Super Global University Project, launched in 2014 is a large-investment initiative designed to enhance the international compatibility and competitiveness of higher education in Japan and to provide prioritized support for the world-class and innovative universities that lead the internationalization of Japanese universities (MEXT, 2014). The TGUP has provided funding for 13 'Type A: Top Type' universities, which have been identified as having the potential to become top 100 ranked world universities and 24 'Type B: Global Traction Type' universities, viewed as innovative universities that can lead the internationalization of Japanese society. MEXT has indicated that all policies are interconnected; the Go Global Japan Project has been directly linked to the TGUP. Hokkaido University was chosen as one of the 13 Type A universities.

INTERNATIONALIZATION OF HOKKAIDO UNIVERSITY AFTER 2008

In the 2009 fiscal year, Hokkaido University cooperated with the International Association of Universities (IAU, headquartered in Paris), which studied Hokkaido University's internationalization strategy and sent a final report detailing the IAU's advice: 'FINAL REPORT: Collaborative Review of Hokkaido University's Internationalization Strategy' (hereafter, the IAU-ISAS Report 2009). As a result, the IAU began its 'Internationalization Strategy Advisory Service' (hereafter, ISAS), and, by April 2016, had performed ISAS for sixteen universities in nine countries.

The University has incorporated many of the useful suggestions detailed in the IAU-ISAS Report 2009, and during the second period of its 6-year mid-term management plan, which began in the 2010 fiscal year, accelerated its internationalization efforts.

In preparation for its 150th anniversary in 2026, Hokkaido University set itself the goal of achieving the vision 'Hokkaido University, contributing to the resolution of global issues'. Additionally, in March 2014, under the initiative of the president, it formulated the 'Future Strategy for the 150th Anniversary of Hokkaido University', a blueprint for drastically reforming the University.

In the 2014 fiscal year, while working on internationalization, a strategy to further internationalize education, 'Hokkaido Universal Campus Initiative' (hereafter, HUCI) was chosen by MEXT for the 'Top Global University Project'. The amount of money to be given to the University by MEXT from the 2014 to 2023 fiscal year is approximately 2.4 billion yen in total.

HUCI is an implementation strategy for achieving internationalization and education reform, which the Future Strategy outlines. Under HUCI, the University is embarking upon a 10-year university reform up until 2023. Based on their strengths and characteristics, HUCI will establish 'Universal Campus', a framework to offer education in various parts of the world in collaboration with universities and research institutions that have achieved remarkable outcomes for the resolution of global issues as well as people who have demonstrated outstanding leadership in such activities.

The '1-4-4 Reform Plans' are the backbone of HUCI.

1. One plan to reinforce governance: establishment of the Office of Institutional Research to support quick decision-making by the President as well as a framework to ensure various decisions;
2. Four educational reform plans: development and implementation of degree programs and short-term programs to provide education in collaboration with universities across the world; and
3. Four system reform plans: functional enhancement to effectively achieve educational reform while introducing

MAJOR CHANGES IN POLICIES AND IMPLEMENTED STRUCTURES

Expansion of the President's Mandates and the Strengthened Governance of the University's Administration

The president's budgetary and hiring discretion were expanded to promote inter-disciplinary and international research and education that highlights the strengths and distinct qualities of the University. Furthermore, the entities of the central governing body, such as the Institute for the Advancement of Higher Education, the Office of International Affairs, and the Office of Promotion of Industry-Academia

Figure 1. The '1-4-4 Reform Plans'

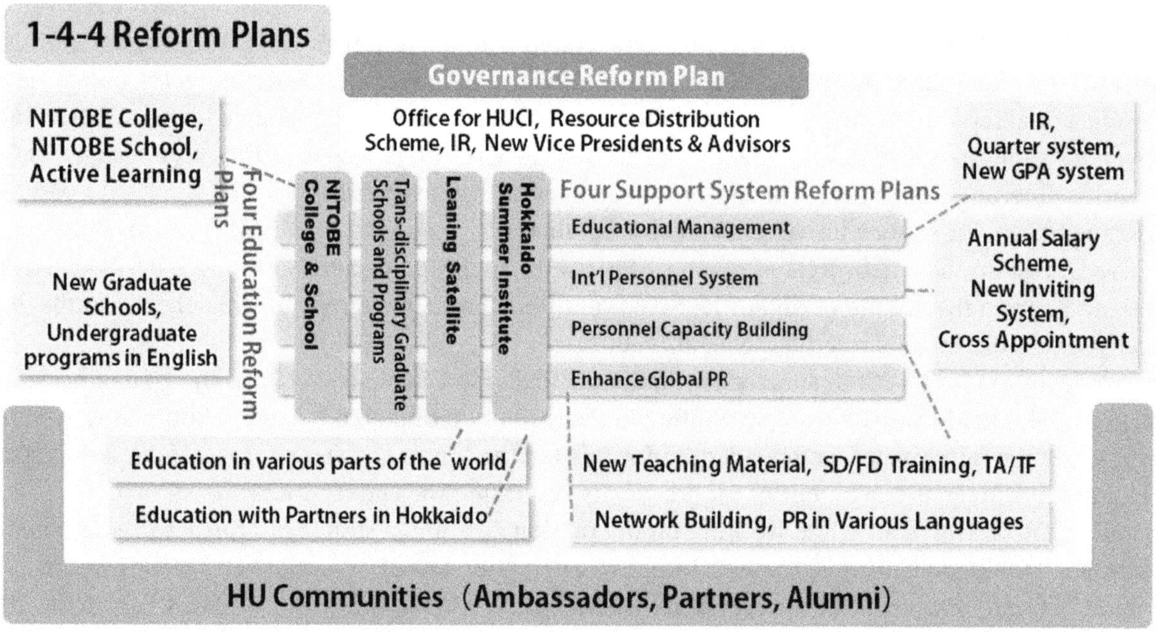

Collaboration, were strengthened. This was a part of a plan to conduct a university-wide initiative and provide services to undergraduate/graduate schools and faculties. To ensure the execution of the Future Strategy, the 'Headquarters for Enhancing Institutional Capacity' was created in 2014.

The first two research themes selected under the framework of the Global Institution for Collaborative Research and Education (hereafter, GI-CoRE) were 'radiation therapy' and 'zoonosis control'. With regard to the former, the 'Global Station for Quantum Medical Science and Engineering" was established in collaboration with Stanford University (USA) in 2014 and, regarding the latter, the 'Global Station for Zoonosis Control' was established in collaboration with the University of Melbourne (Australia), University College Dublin (Ireland), and King Abdullah University of Science & Technology (Saudi Arabia) in the same year. As the third initiative, the 'Global Station for Food, Land and Water Resources' was established in 2015.

In 2014, the GI-CoRE was established directly under the control of the president. Within the GI-CoRE, a team is created for each interdisciplinary research theme, with a system for attracting global top-level research units to work with faculty members of the oversea universities. Each team is named a 'Global Station'. The University will institute a new educational entity based on achievements in research and education as well as professional connections obtained through the activities of each Global Station.

Strengthening of Functions That Support International Activities

In July 2010, the Executive Office for International Exchange, which was mandated to plan international exchanges, was reorganized as the 'Office of International Affairs' (hereafter, OIA). It has two characteristics:

Figure 2. Implementation system

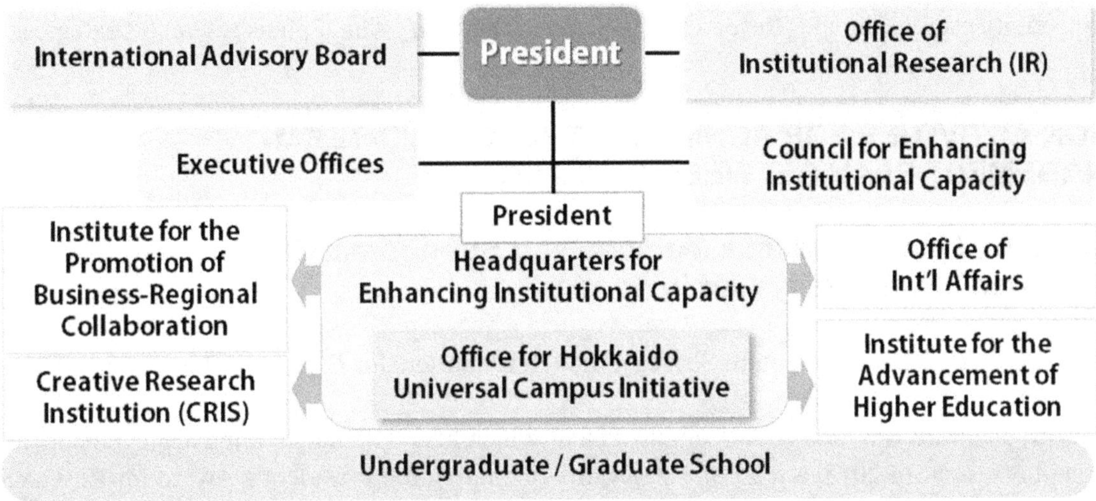

- It not only plans, but also carries out plans for campus internationalization and international exchange activities.
- It consists of both faculty members and administrative staff and facilitates collaborative work between them.

The Office of International Affairs standardized and consolidated the duplicated international administrative duties that existed across faculties, such as a help-desk for international students and the English translation of the University's rules and regulations. Concurrently, it plans and conducts multiple cross-faculty educational programs, as study-abroad programs. They rapidly increased the number of international exchange agreements.

Increasing the Number of Japanese Students Studying Abroad

The Japanese government's policy focused on not only accepting more international students, but also sending more Japanese students abroad. Requests from industries to produce more graduates who can succeed on a global stage had an impact on the transition seen upon this policy.

In response, the University set a target of expanding the number of Japanese students studying abroad. It also began to build a structure to cooperate with alumni, who are active on the global stage, and provide career education.

Internationalization of Faculties

The internationalization initiatives introduced by the University's central governing body, such as the OIA and the Executive Office for Education Reform, affected the faculties. The Faculties/Graduate Schools of Engineering, Fishery Science, Letters (Literature), Medicine, Science, and Veterinary Medicine have all created their own departmental offices for international affairs and hired full-time staff to support the international activities of their faculty members and students.

Research Development

The University established the 'Research Development Section', which plans research strategies.

MAJOR ACTIVITIES FOR ACHIEVING THE 2ND MID-TERM MANAGEMENT PLAN BETWEEN 2010 AND 2015

Expansion of the President's Mandate and Strengthened Governance of University Administration

To ensure the execution of the Future Strategy, the 'Headquarters for Enhancing Institutional Capacity' was created in 2014.

In 2011, an award was created to recognize faculty members with outstanding achievements in education and research. In 2015, a total of 100 awards (including the President's Award for Research and the President's Award for Education), amounting to a total prize money of 33 million yen, were allotted. These awards are expected to function as incentives for faculty members to pursue international activities. In particular, the President's Award for Research gives an incentive to publish research papers in prestigious journals.

Staffing

The University developed a new human resource system, hired international faculty members using the University's financial resources, and increased incentives for faculties that hire international faculty members.

The University proactively translated an information handbook for new faculty members as well as basic university-wide documents, such as rules and regulations related to the employment of international faculty members, into English, and published them on its website.

It also established 'international officer' positions that specialize in international affairs, and hired seven persons at the OIA. Among them, two were granted tenure, including a psychological counselor who is bilingual in Japanese and English.

Education System

Quality Assurance of Education

The University upholds the four basic philosophies of 'frontier spirit', 'global perspective', 'all-round education', and 'practical learning'. Based on these, in 2013 the university clarified the diploma policy and the curriculum policy for undergraduate and graduate programs. Additionally, the admission policy for the undergraduate programs was made clear. The three policies for the undergraduate programs show that the University nurtures internationally minded individuals who will contribute to the development of society.

The University introduced a new GPA system that meets international standards. Moreover, a graduation accreditation system using either the GPA or the common achievement test was established.

The undergraduate and graduate schools introduced an academic quarter system to provide an environment that enables the students to study abroad without repeating a year.

'Manual of Credit Transfer from Overseas Universities', was created newly, which is actively promoting the use of the University Mobility in Asia and the Pacific Credit Transfer Scheme and the European Credit Transfer and Accumulation System. Consequently all students in undergraduate and graduate schools could transfer the credit between HU and overseas universities.

Nitobe College

Nitobe College, which is a program that requires undergraduate students to study abroad, was established in line with the University's new policy to rapidly increase the number of outbound Japanese students. This is a minor program for leadership training named after Inazo Nitobe, who was among the second cohort from Sapporo Agricultural College and served as an Under-Secretary-General of the League of Nations.

The OIA opened two courses that lead Japanese undergraduate students, including students of Nitobe College, to study overseas. One is 'Fellow Study Program' that offers a practical training program for global leader by alumni that have active careers in international society, which is held four times a year, and the other is 'Short-term Study Abroad Special Program', which is held twelve times a year.

Modern Japanese Studies Program

An undergraduate program, 'Modern Japanese Studies Program' (hereafter, MJSP), was established for international students who wish to learn the Japanese language and modern culture. The MJSP is the University's first undergraduate program that does not set Japanese language ability as a requirement for taking the entrance examination.

Adopting the 'Central English Program Unit' (hereafter, CEPU) method proposed in the IAU-ISAS Report 2009, in 2013 the University established a CEPU of ten native English-speaking faculty members. The CEPU largely contributed to the establishment and the initiation of the MJSP.

Along with the MJSP, many of the graduate schools have implemented the following admission systems for their internationalization process:

- Online application service
- Enrollment in April and October
- An entrance examination that does not require travel to Japan
- Application guidebooks translated into English

Student Exchange

The 'Hokkaido University Short-term Exchange Program' (HUSTEP) and the 'Japanese Language and Culture Studies Program' established a six-month program, in addition to the existing year-long program, with the aim of increasing the number of incoming exchange students from partner universities. Moreover, these two programs introduced a system in which the exchange students can earn credits from the University.

Some undergraduate and graduate schools of the University established double degree programs and joint certificate programs that enable students of the University and the partner universities to engage in short-term learning of specialized subjects at overseas counterpart universities.

Internationalization of Research

The Research Development Section improved the database of the University's researchers in regard to their research performance and analyzed the quality of their research. As the result, an evidence-based research management and planning system was established.

Various data on international research competitiveness were provided to the heads of faculties. The data include:

- Time-series data on numbers of published papers, international joint co-authorships, and citation indices for each research faculty
- Numbers of published papers, international co-authorships, and a citation index for each academic discipline in a faculty
- World ranking by academic discipline
- The following initiatives were taken to enhance the dissemination of research works in English:
- Seminars to improve research paper writing in English
- Financial support for participating in international academic conferences, and for conducting international joint research
- Financial support for the use of proof-reading services for paper submission to international academic journals above a certain level

THE MAJOR RESULTS AND IMPACTS OF HUCI ACTIVITIES

Staffing

The number of international faculty members has increased from 66 on May 1, 2009 to 117 (4.8% of all 2,427 full-time faculty members) on the same day in 2015.

Education System

Courses offered in English accounted for 3.1% of courses in undergraduate schools in the 2013 fiscal year and 17.2% in graduate schools. In the 2015 fiscal year, the percentages increased to 4.3% and 20.9%, respectively.

Number of Students Studying Abroad

The number of Japanese students who availed of study abroad programs to earn course credits in the 2013 fiscal year was 276 in undergraduate schools (2.4% of all undergraduate students) and 115 in graduate schools (2.2% of all graduate students). In the 2015 fiscal year, the number increased slightly to 369 (3.3%) and 142 (2.8%), respectively.

Number of Incoming Students

The number of international students reached 1,875 (10.4%) as of November 1, 2015, surpassing the 2015 target of constituting 10% of the overall student population.

The numbers of incoming international students were 1,840 in the 2013 fiscal year, 2,069 in 2014, and 2,412 in 2015, demonstrating an annual increase.

Internationalization of Research

The number of research papers that were recorded in the Web of Science Core Collection as being published in academic journals reached 3,082 in 2015, demonstrating a 10% increase on the number for 2009.

The number of research papers that were published in the past five years, from 2010 to 2015, in academic journals with an impact factor score of 8 or more reached 886, demonstrating a 21% increase on the first period of the mid-term management plan, which ran from 2004 to 2009.

ISSUES AND OF FUTURE ACTIONS AND STRATEGIES TO THE SUCCESS OF HUCI

In 2010, the International Association of Universities (IAU) conducted a review of internationalization programs and strategies at Hokkaido University as part of its Internationalization Advisory Strategy Service (ISAS). Hokkaido University was the first university to participate in ISAS and between 2010 and 2016, approximately a dozen universities from all over the world partnered with the IAU to conduct similar reviews. In 2016, the IAU revised and expanded the ISAS program, creating ISAS (2.0). ISAS (2.0) is a more diversified service, enabling institutions to focus their review and to earn learning badges for different aspects of advancing strategic internationalization.

Having launched its Future Strategy in 2014 and having succeeded in securing one of the Top Global University grants from MEXT, Hokkaido University invited the IAU acknowledged the effectiveness of HU's measures towards its internationalization. Enhancing HU's global visibility and stature will be both a result and a driver of the HUCI. It's ability to attract partners and program participants will, to a large extent, depend on its visibility and reputation. As such, HU has recognized the need to market itself internationally and to ensure that its faculty and staff make international publications and attend international conferences. Hosting international events is also a means to increase the university's global presence. Successful programs build visibility both on a national and international level over time.

However, the Expert Panel also pointed out various issues of the HUCI, of which I would now like to briefly introduce and discuss how to solve them.

Changing Financial Environment

The Expert Panel reported that since HU was awarded funding by MEXT as a Top Global University, the financial circumstances of the University have changed considerably. The HUCI would be jeopardized because of competing priorities and the resulting reduction in the investment of HU's own resources for the initiative. Furthermore, since funding for the HUCI will annually decline over the life of the initiative, financial constraints will undoubtedly have an impact on the overall morale of University personnel and attempts at reform.

HU furthermore faces the major challenge to sustain activities started by the HUCI beyond the period of funding provided by MEXT. This will require faculty and staff to pay for their own activities, and for the university to fund HUCI programs from new sources of revenue. Thus, we are currently searching for new possibilities and funding resources from collaborations between the university and industries, in particular from private enterprises.

To further increase international connectivity, we should consider making greater use of technology. Classrooms can be connected with those of partner institutions fairly inexpensively using synchronous or asynchronous media for joint projects or discussion boards. Skype and other platforms can also be used to connect classrooms in real-time. Similarly, HU may explore how technology can help amplify student experiences before, during and after attending our summer schools at HU (e.g. the Hokkaido Summer Institute) and overseas universities (via our Learning Satellites).

Communication and Consultation

Following the competitive nature of the Top Global University Project, the proposal for the HUCI was developed in a short time frame and leadership at HU moved quickly towards the implementation of measures outlined in the initiative, seeing the rapid launch of new programs. However, the Expert Panel pointed out that faculty members and administrators perceived the initiative as 'top down', without sufficient consultation before the launch about issues related to both design and implementation, and that unrealistic targets were set by the top without consultation as to their achievability by the areas responsible for their delivery.

We believe that executive leadership should consider conducting a series of meetings at each faculty and with selected administrative units to discuss future directions of the HUCI. These might take the form of regularly scheduled meetings or as a special series of events attended by deans, faculty members, and staff. The purpose of these sessions would be the two-way exchange of ideas by senior leaders to review the HUCI, its underlying rationale, philosophy and activities, and to provide an update on its progress. At the same time, the senior leaders would emphasize the HUCI as a means of enhancing the quality and visibility of HU rather than as a goal in itself. Documents and presentations might include a description of the rationale behind the HUCI, how it furthers the accomplishment of the Future Strategy, and its most important lines of attack.

Given the significance of the faculties in advancing internationalization, the senior leaders should ask each faculty to develop its own plan in alignment with the HUCI. Although it is expected that not all faculties will progress in all areas, they will still have different strengths and interests that can contribute to the HUCI.

It is particularly important for those individuals in charge of internationalization efforts to share information - their experiences, successes, and challenges - regularly perhaps in scheduled (e.g. quarterly) sessions. The Faculty Development and Staff Development Sessions could also be utilized to focus on these same issues. Another idea would be to have a venue in which people could meet for coffee or drinks to create interconnectedness among those facing similar issues and challenges.

Integration, Complexity and Synergy

Per the report issued by the Expert Panel, the HUCI sets very ambitious targets. To achieve them, HU will need to step up its activities, going from 'boutique' programs and initiatives tailored to a small group of participants to larger more extensive ones. This will require a thorough examination motivating participation yet addressing obstacles in addition to developing future strategies. Simply increasing the number of actions and initiatives may advance internationalization, but unless they mutually reinforce each other and are a part of a larger strategy, their impact will remain limited and possibly unsustainable. Moreover, such actions contribute to the HUCI being seen as a separate project burdening faculty and staff workloads and thus contributing to their reluctance to fully participate in the initiative. Thus, the university must make great efforts to integrate HUCI programs with currently ongoing work of the faculties.

The HUCI also includes a number of different activities involving international partners and specialized academic fields. The proliferation of topics and partners may reduce the visibility and impact of the HUCI. Thus, we will narrow down the programs and partnerships associated with the HUCI to align with the University's distinctive strengths. This will lead to the university building an even stronger reputation in these fields of expertise and allow for the concentration of resources.

System Reforms for Education

Nitobe College, a distinctive honors program offered by the university, has the potential to be quite successful and to grow. However the Expert Panel pointed that some students drop out of Nitobe College because they cannot fit their participation in the College with the requirements of their undergraduate school. To make Nitobe College a success, we need to consider how to align the requirements of undergraduate schools with the elements of Nitobe College. In other words, adjustments need to come from both sides - the faculties and Nitobe College.

The Expert Panel also drew attention to the fact that some faculty members voiced concerns that the quality of courses, and thus the quality of education programs, has been diminished by being taught in English because some students have a low English proficiency level. Furthermore, many Japanese students are reluctant to take courses offered in English. We therefore should encourage faculty members, staff, and students to continue their efforts to gain a higher English proficiency.

Since a small minority of students study abroad, HU needs to focus on amplifying its on-campus experience to accomplish the goal of producing "globally competent" students. To achieve this, the university should also identify the skills, knowledge, and learning outcomes they would like every HU graduate to process, and make the necessary reforms towards that goal.

CONCLUSION

This article presents an overview of internationalization strategies within Japan's institutions of higher education, and outlines the changes occurring in Japan's higher education since the 1980s.

Hokkaido University has accomplished a great deal towards achieving its goals outlined in the Future Strategy in the two years since the launch of the HUCI. Its current challenges are to consolidate its gains, widen its circle of supporters and participants, and make the necessary adjustments to ensure the sustainability of the HUCI. This is why Hokkaido University was awarded the IAU Learning Badge for Assessing Strategy and Monitoring Achievements from the IAU. Hokkaido University has developed a sound strategy for internationalization, has had targets and indicators assigned for monitoring purposes, and has been reassessing its progress and tactics as and when necessary.

The issues surrounding the globalization of the higher education system in Japan serve as a useful case study on the attempts and challenges faced by many regions in designing and implementing policies to further their national agenda.

REFERENCES

Horie, M. (2002). The internationalization of higher education in Japan in the 1990s: Aeconsideration. *Higher Education, 43*(1), 65–84. doi:10.1023/A:1012920215615

MEXT. (2012). *Selection for the FY2012 Re-Inventing Japan Project.* Retrieved from http://www. mext. go.jp/en/policy/education/highered/title02/detail02/sdetail02/sdetail02/1374092.htm

MEXT. (2014). *Selection for the FY2014 Top Global University Project.* Retrieved from http:// docplayer. net/16379288-Selection-for-the-fy-2014-top-global-university-project-we-hereby-nnounce-the-selectionof-universities-for-the-top-global-university-project.html

Yonezawa, A. (2011). The "global 30" and the consequences of selecting "world-class universities" in Japan. In N. C. Liu, Q. Wang, & Y. Cheng (Eds.), *Paths to a world-class university: Lessons from practices and experiences.* Rotterdam: Sense Publishers. doi:10.1007/978-94-6091-355-6_3

KEY TERMS AND DEFINITIONS

IAU: The International Association of Universities (IAU) founded in 1950, is a worldwide membership-led non-governmental organization. It comprises more than 650 higher education institutions and organizations in some 130 countries. IAU is an official partner of UNESCO. Its permanent Secretariat, the International Universities Bureau, provides Policy Statements on issues of global importance for higher education. Each Statement is the product of extensive drafting and international consultations. Recently the Statements have been formally adopted by the General Conference of the Association, though in the past some emanated from international Round Tables or meetings of the Administrative Board.

Inazo Nitobe: Inazo Nitobe (1862-1933) was agricultural economist, writer, educator, diplomat, and politician. In 1877 he entered the newly founded Sapporo Agricultural College in the northern island of Hokkaido and graduated in 1881. In 1884, he went to study in the United States, first at Allegheny College in western Pennsylvania, and then at Johns Hopkins University in Baltimore. After earning his doctorate in agricultural economics in Germany, he returned to Japan in 1891 to assume an assistant professorship at the Sapporo Agricultural College. He served as a professor of law at Kyoto Imperial University and Tokyo Imperial University, Headmaster of the First Higher School (the preparatory division for the Tokyo Imperial University), and the first president of Tokyo Women's Christian University. He was an Under-Secretary General of the League of Nations from 1919 to 1926. He was a prolific writer and exerted a powerful influence on Japanese intellectuals and students. He wrote many books in English and is most famous in the West for his work Bushido: The Soul of Japan.

GPA: Grade point average, which is a measure of a student's academic achievement at institutions such a college or university; calculated by dividing the total number of grade points received by the total number attempted.

Chapter 3
Plan Ceibal as Where Technology Accelerates Pedagogy

Miguel Brechner
Centro Ceibal University, Uruguay

ABSTRACT

This chapter describes how the government of Uruguay believes that all children have the right to have technology at their fingertips and that all children have the right to connectivity and access to the internet, that it is as important to have electricity and running water as to have access to the internet, and that it would have a high impact on the country's technological deployment and, obviously, on education and teaching. The parts of the chapter are concerned with the technology and pedagogy relationship: how to improve pedagogy through technology, the importance of teaching English and math online with help of education inspectors and the teachers using modern platform—virtual classrooms, books in digital format, digital technology laboratories—that allows collaborative work, work on projects, logical thinking, an online assessment system. All these integrated tools transfer to the biggest investments which the author calls "global learning network."

INTRODUCTION

Plan Ceibal was created in Uruguay in 2006 and its implementation started in 2007 (Plan Ceibal, 2006, 2007). It was not designed as an ICT programme or a laptop programme, but as an inclusion programme. From the first day, we in government believed that all children have the right to have technology at their fingertips and that all children have the right to connectivity and access to the Internet. In a school, it is as important to have electricity and running water as to have access to the Internet. Besides, all students should have the same devices throughout the national territory. Our focus was always on inclusion. We were aware that by the time this was established, it would have a high impact on the country's technological deployment and, obviously, on education and teaching. In a very short synthesis, our vision is to have a good Internet connection in schools, work with computers and use "the cloud" as the main basis for services.

DOI: 10.4018/978-1-5225-3395-5.ch003

I would like to share where we are, how we are and what we are doing.

INTERNET AT THE STUDENT'S PLACE OF STUDY

Uruguay is a small country (Wikipedia, 2018). If I look at the figures from Argentina and Colombia – fortunately Costa Rica, which is also a small country, Colombia overwhelms me with 8 million devices and I'd rather not hear how many there are in Argentina. In Uruguay, we have 566,000 beneficiaries (Plan Ceibal, 2015). This means all public-school students between first and ninth grade, or between the first class of elementary or primary school and the third class of secondary school, depending on how this is defined in different countries. We cover all students and all teachers, with 80% of them having Internet availability greater than 95%, and with 98% of students accessing the Internet at their place of study. We developed this basic infrastructure strongly since 2007. The work in the field of technological deployment (Croteau & Bergeron, 2001) allowed us to have optical fibre in all urban educational centres, with high-quality video conferencing equipment in those places, and access to the Internet in neighbourhoods requiring priority care or deprived neighbourhoods, in housing complexes, in hospitals, in public squares; i.e., wherever there is a Ceibal student, there is access to the Internet (see Figure 1).

The Internet is fundamental for the development of this plan, we believe in "the cloud" and we are working increasingly on "a centralised cloud infrastructure".

Figure 1. Deployment of WiFi and video network
Source: Plan Ceibal (2016)

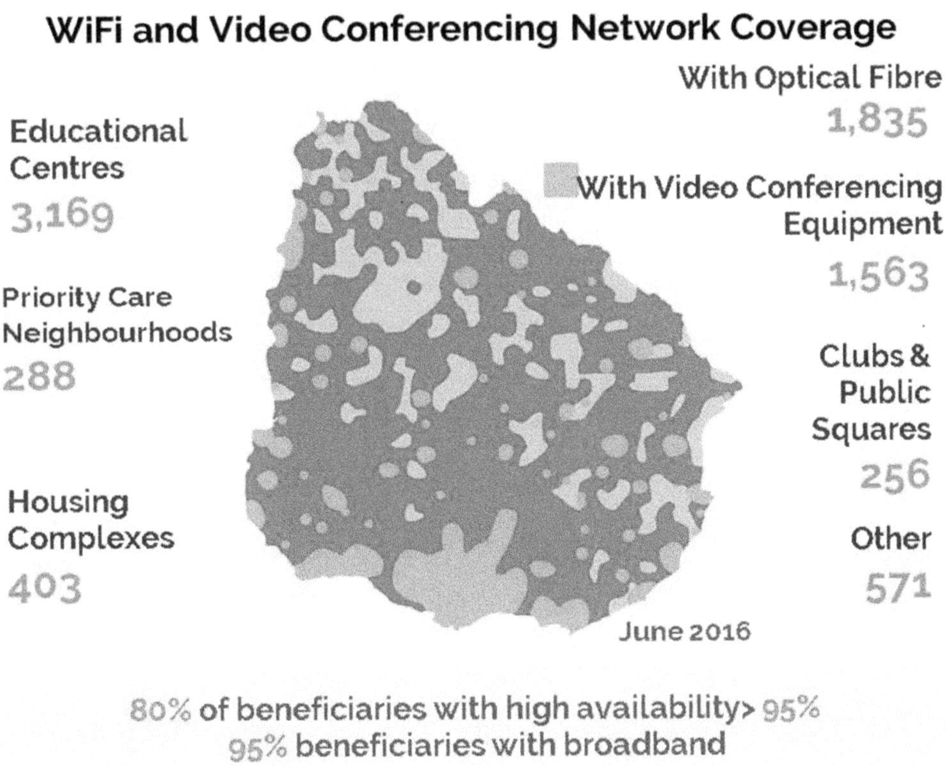

We deliver tablets in the first, second, and third grade of elementary schools; laptops which work both as a laptop and tablet in the third grade; and common laptops in the first grade of secondary school or seventh grade.

COMPUTER ACCESS PER STUDENT

In terms of computer access per person, it started to grow in 2007, and today we have no digital divide between the richest and the poorest (see Figure 2). If I would have to synthesise something in a graphic about Uruguay and Plan Ceibal, I would choose this one.

Next, we are going to see computer access by age range and by quintile; what the access to technology was like in 2007 (see Figure 3). Clearly, between the ages of 6 and 13 there was a difference from 9 to 90%; between 14 and 24 years, from 9 to 86%. If you look at what it is like today (see Figure 4), there is clearly no digital divide, but equity. This is the best synthesis of our definition. Where we do have a great difference, is among those over 65, because where it says that 44% have access to technology in the poorer sections of the population, this refers to computers held by the children of Plan Ceibal in homes where there are grandparents aged over 65. Just as a side comment, in 2016 we started a programme to deliver tablets to retirees over 65.

Figure 2. PC access, according to per capita income quintiles. Total for the country, in percentage of households
Source: ECH-INE, 2016

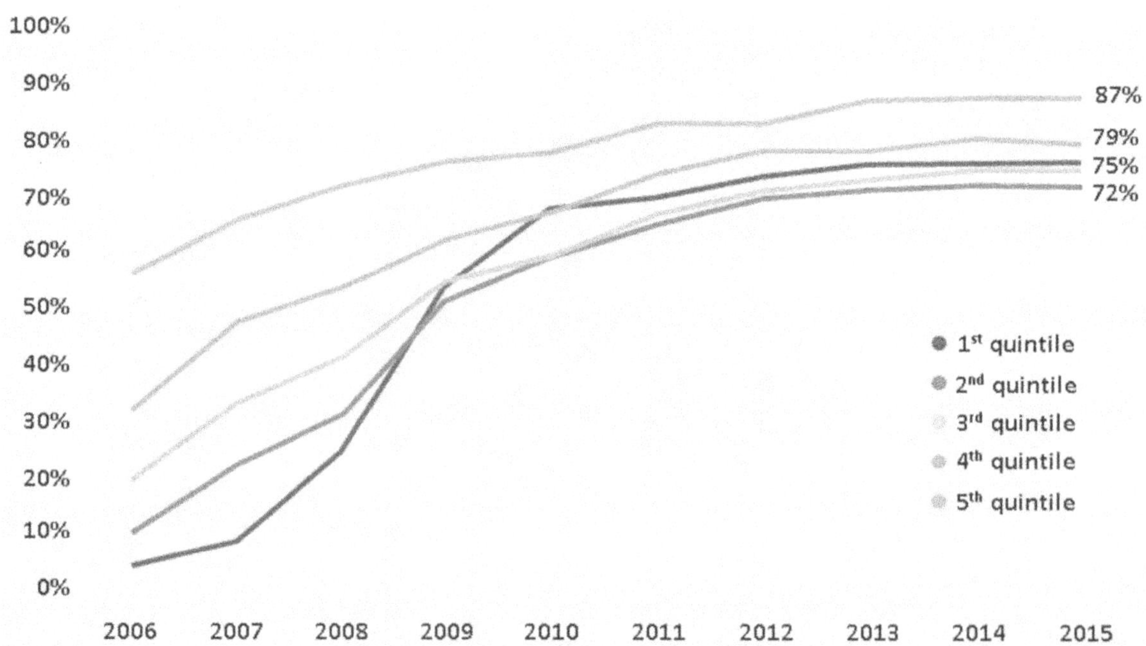

Figure 3. PC access by age group, according to per capita income quintiles. Total for the country, in percentage of the population, 2007
Source: ECH-INE 2007

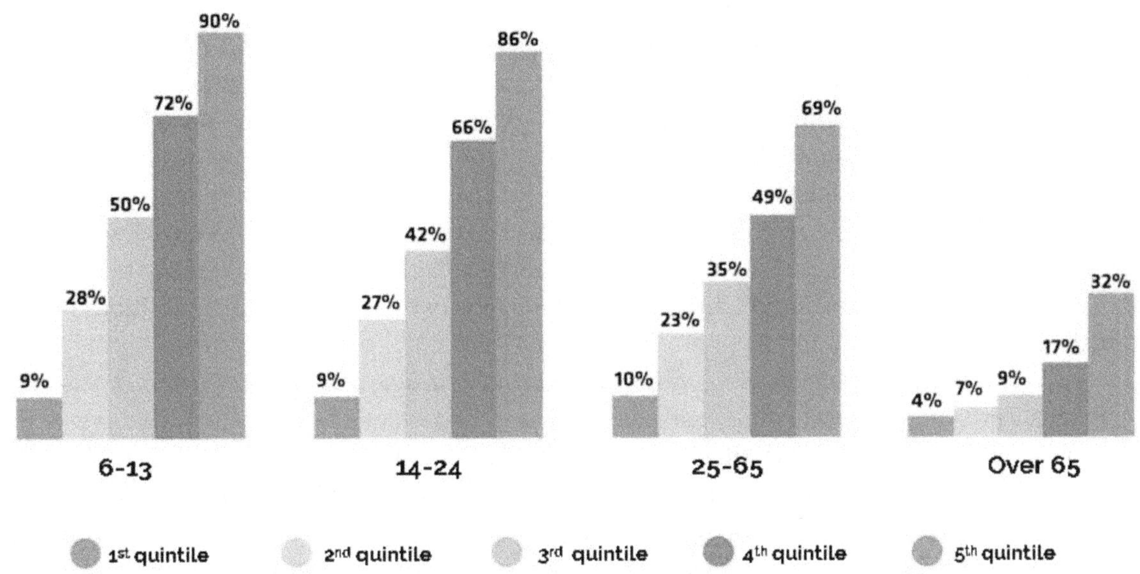

Figure 4. PC access by age group, according to per capita income quintiles. Total for the country, in percentage of the population, 2017
Source: ECH-INE 2007

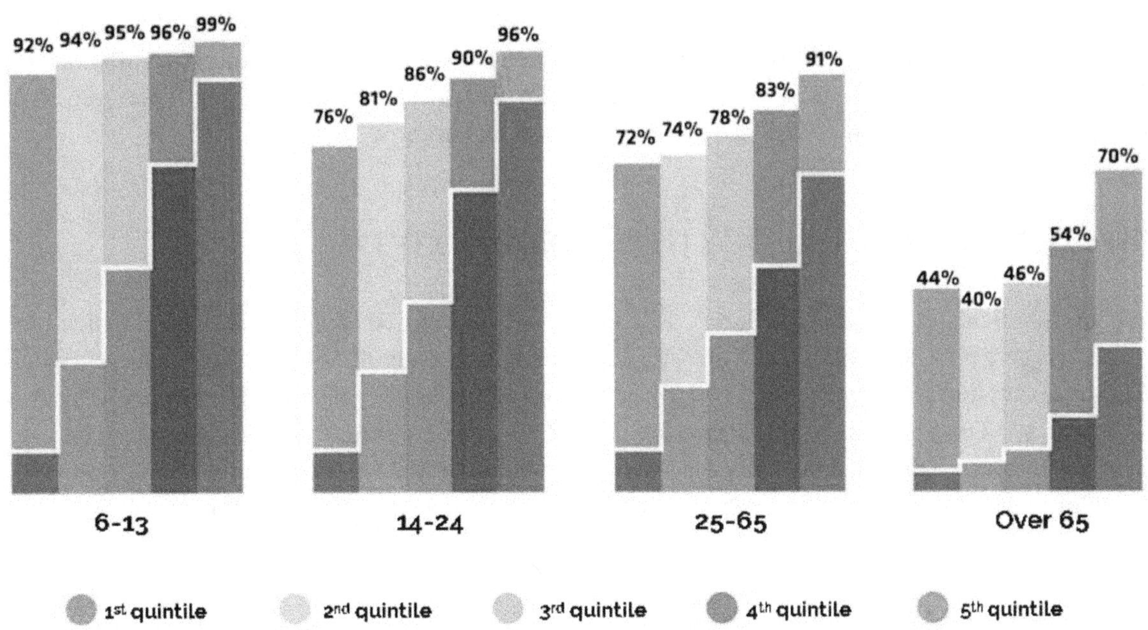

TECHNOLOGY AND PEDAGOGY

The essential issue, as I understand it, lies in the discussion of technology and pedagogy. I've spent all my live in the world of ICT, but, if truth be said, I would like to hear less the term "ICT" and much more the term "pedagogy". The main question is always: Why has technology had so much influence in our lives and so little influence on the educational world? When we started Plan Ceibal, there was no iPhone and there was no iPad. When we think about how technology has influenced our lives and how little it has influenced education, we should ask ourselves why. I have no doubt that one of the main reasons is that what matters to technology sellers, is selling technology. They don't care what technology is used for.

Over the last 20 years, there have been different programmes, here and there. With a few exceptions, notably that of Costa Rica, where they always knew clearly what they wanted to do with technology, in other parts of the world digital blackboards were implemented, with an expenditure of billions of dollars that made no sense at all.

Technology is not an instrument to solve the problem, but is an instrument that will help others to solve the problem. This issue with technology and pedagogy is so important, that technology wants everybody to adapt to it, when it should be the other way around. It is technology that should adapt to the world.

At this meeting, there has been much talk about training. How many of us use Facebook? How many Twitter? How many WhatsApp? How many training courses did WhatsApp provide? Just imagine a world where everybody would have to receive training to use Twitter or WhatsApp. That's something absolutely crazy. Why should people be trained to use a platform? Why is it necessary to receive training to use a computer? We have lived too many years influenced by the world of engineering. I am proud to be an engineer. But now we are starting to live in a world where technology is part of an art.

My mother is 90 years old and she uses Skype. Can we afford to have people who aged 90 without Skype? They use the computer to talk to their grandchildren and great-grandchildren. Technology simply has to adapt to teachers, not teachers to technology. We must improve the ability of teachers to use and understand technology, and not just train them to use a computer or some specific software. If the software is not easy to use, it shouldn't be used. Just like that. We have to look for software that is easy to use. And this is my vision of how we have to focus on technology.

HOW TO IMPROVE PEDAGOGY USING TECHNOLOGY

There are two fields in technology. One field that is clearly a solution to the problem and which cannot be solved if there is no technology; and another field that is an accelerator of pedagogy, which is what we should be discussing today: *how to improve pedagogy using technology*. But not how we are introducing ICT in schools. There should be no discussion about the fact that schools need to have devices and access to the Internet. There can be some discussion on who should do it. I propose to have a specialized agency for technology and pedagogy innovation in charge of the solution of this problem.

Teaching English

I would like to mention some examples, and we should think whether we are talking about technology to solve a problem or whether we are talking about technology to improve pedagogy. And as we delve deeper into this issue, we will realise the action to be taken and who should take it. Here are some examples. In Uruguay, we don't have sufficient English teachers. So, we asked ourselves: how are we going to solve the problem of teaching English? Well, we decided to put optical fibre into schools, and to employ a remote teacher who is in Uruguay but in a different city, in Argentina, in the Philippines, in England and in Colombia, where we have people teaching via video conference. But not by means of a web software. It is a virtual class. The video conference is of such high quality that teachers interact with their students from their remote location. The local teacher, who doesn't speak English, because most Uruguayan teachers don't know English, works with that teacher; one class session is face-to-face and two class sessions are conducted with the use of computers. In 2012, we tested this system with 1,000 students; in 2013, we had 25,000 students; in 2014, 50,000; in 2015, 77,000; and in 2016, 80,000 (see Figure 5).

Figure 5. Teaching of English
Source: Plan Ceibal, 2016.

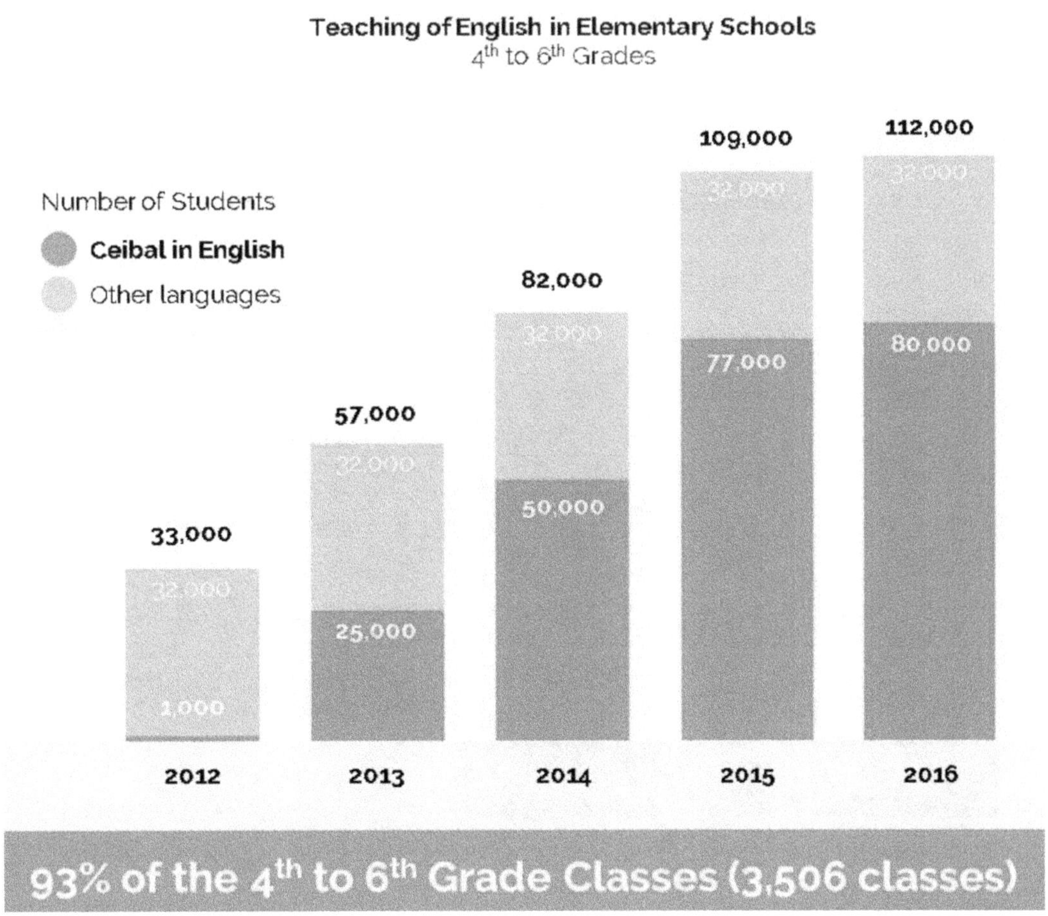

In addition, the face-to-face programme has about 32,000 teachers. For the first time ever, 93% of fourth, fifth and sixth grade of students in Uruguayan elementary public education are learning English. We wouldn't be able to do this without the video conferencing technology and the English teaching methodology we agreed upon with the remote teachers. In this case, technology clearly solved a problem for which there was no solution. What are the results? We tested the students last year, and both those who have face-to-face classes and those who have remote classes, learn English. Let's look at the graphs: 57% of kids in the sixth grade are close to A2 in English standards, 22% are in A1+, and 15% in A1- (see Figure 6).

I say it once again, technology helped us solve a problem. We had to be very innovative, but now we have universalised the teaching of English in this way. Of course, we had to build the optical fibre and video conferencing infrastructure for this. There is no new pedagogy to teach English. We teach English using the old method.

Let's see another approach: we purchased an adaptive math platform with 100,000 activities, covering from the third to the ninth grade. What is happening in this case? Students go to school, the teacher gives them assignments to do, and the teacher doesn't have to prepare math exercises anymore because students log in to the platform. As they move on the platform, things become more difficult. As exercises get more difficult, the platform helps students and if necessary, lowers the level of difficulty to the level of student understanding. At the end of the day, the teacher receives from each student what they did, where they stopped and what their difficulty was. This is obviously a very modern platform; we hired a German company that adapted it to Uruguay working with Uruguayan teachers.

Figure 6. Levels reached in the adaptive test. In percentage of students who took the test, 2014
Source: Plan Ceibal, 2014

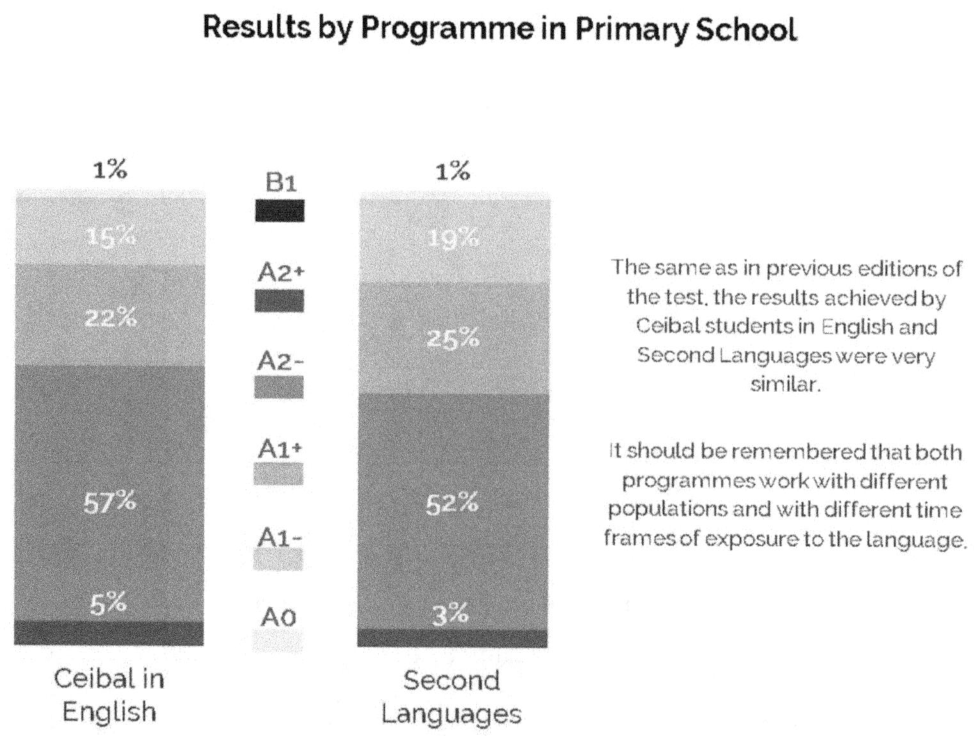

Teaching Math

In 2016, more than 150,000 students worked on the platform and submitted almost 40 million exercises. Let's take a look at the ownership graphs (see Figure 7). They are mind blowing. We organise a math proficiency competition twice a year. The prizes to motivate the students: in 2014, it was to go to the World Cup; in 2013, to see the games of the qualifying round. Current prizes are for the educational centres, for the children as a group. In 2016, the number of people using the platform has increased by almost 65% compared to those using it in 2015. The adoption of technology is slow, but is faster when it is easy to use. This takes the burden of preparing exercises off the teachers' shoulders and also gives them the ability to see the level of each of their students concerning problem solving. In other words, customising education without technology is absolutely impossible. The only tool that allows teachers to have a vision about the difficulties faced by each one of their students, and even more so when classes are numerous, is technology.

This is technology, not new pedagogy. It is math exercises in a dynamic that allows the student to do them at home, at the club, allowing them to ask questions and seek help, but it is nothing new. We worked with education inspectors and the teachers, but without gigantic training being necessary.

Figure 7. Growth of the use of the Adaptive Mathematics Platform (PAM)
Source: Plan Ceibal, 2016.

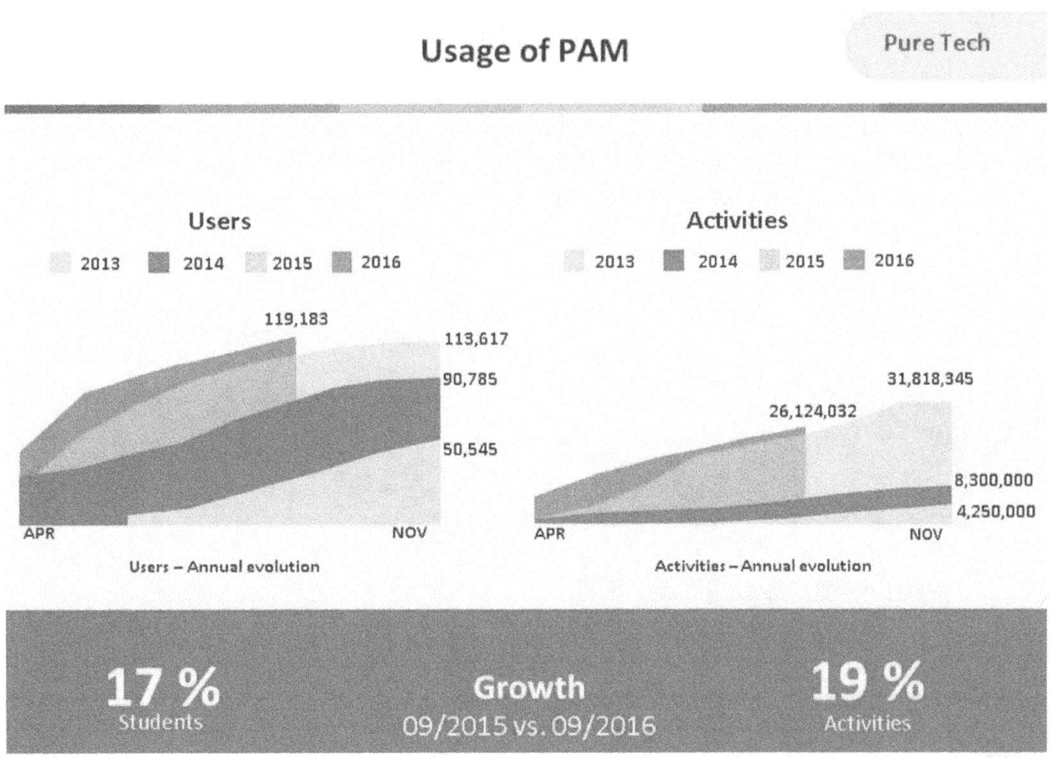

Virtual Classroom

Another clear example is the management of the virtual classroom. We have a Facebook-like tool where students work, where discussions take place, where homework is left for them to be done... and none of us received training on how to upload photos to Facebook. It takes a little while to learn how to do it, but you learn. And in our case, it is the same. This platform manages classroom content, homework, educational materials, and again, the growth achieved in one year compared to the previous one is mind blowing. Why? Because as technology starts to be quite transparent and simple to use, teachers, who are the backbone of any change, take ownership. Not to mention the kids pushing forward (see Figure 8).

Books in Digital Format

Another very clear example of technology not changing pedagogy is the use of books. What do I mean by this? Books in digital format. Ceibal bought the rights to the texts, to almost all texts and most of the reading books, and all members of Plan Ceibal have access to the texts for free. The level of downloads has increased sharply. Why did it increase? Not because the books improved. Because the software to read the books improved. Previously, it was a bother to read a book, so that nobody read. Today, the

Figure 8. Growth in the use of CREA 2: Schoology
Source: Plan Ceibal, 2016.

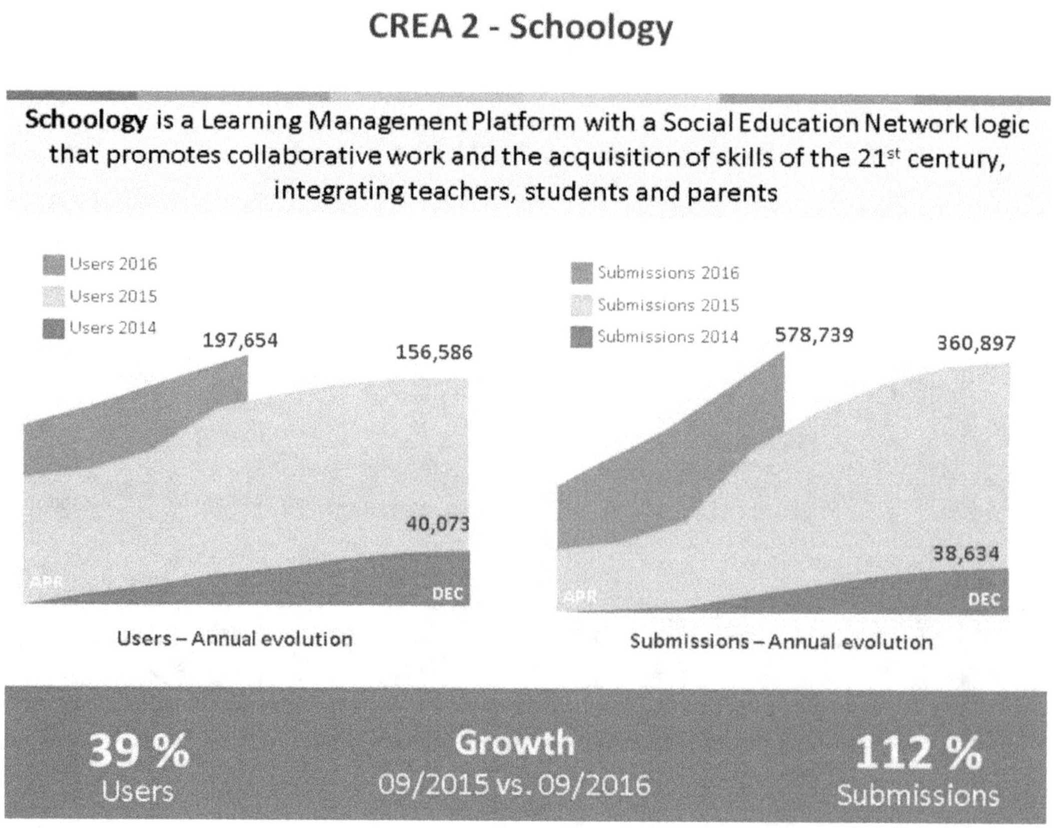

software has improved and people can read it. We made several calls for tenders, we did not find any software, we built software based on open software, which improved the reading performance. Good software is crucial. Easy-to-use software is crucial. Now, what is our strategy with the software? We have a clear mandate: to reduce the divide and to support the education system in everything related to technology to respond to its needs. Thus, our mandate is not to be a software producer. Our mandate is to make software available as soon as possible. When good software exists abroad or at local level, we make a call for tenders, and if the software features all that is necessary and the price is appropriate, we purchase it. When no software exists, then we create it ourselves or have it created by somebody. Half of the products we have today are made in Uruguay by Uruguayan companies, and the other half was purchased by us in different parts of the world.

Digital Technology Laboratories

There is a tool we call "LabTeD", the digital technology laboratories, that allows collaborative work, work on projects, logical thinking. What is it? It is robotics; programming; video games; physics, chemistry and science sensors, audio-visual management and 3D printing. What started just as a project, today it has a very large scope. We have delivered 4,600 robotics kits, which means that in most middle schools or high schools, there is one robotics kit every four students. It started out being voluntary, and today there are many thousands of students working on this. The development has been such, that for the first time ever, we are in midst of a pilot project in the whole department of Maldonado to change the traditional computer science or office automation syllabus from what was in the past, to work of this kind. This actually is a change in pedagogy. This actually is a change in the way to work with students. We conducted a series of evaluations and there is great motivation among teachers. It doesn't relate to the computer science teacher alone; it involves all of them, the computer science, physics and chemistry teachers, discovering the virtues of collaborative work or networking. The level of enthusiasm about this is great. We organised programming courses with thousands and thousands of students who enrolled to learn. This is a change in technology that accelerates the change in pedagogy.

An Online Assessment System

Another example is an online assessment system. We have an online assessment system where teachers can bring together immediately all the results of their class, where they can see the issues their students have, where they can see what caused the error. The level of participation is very high, and although it started on a voluntary basis, today it is used by practically the entire public and private education system. Plan Ceibal platforms are also available for private education. The online assessment is used by practically more than 80 percent of the teachers in public and private education.

FUTURE RESEACH DIRECTION: GLOBAL LEARNING NETWORK

And finally, one of the biggest investments we are making is called "Global Learning Network", an experiment we are doing in seven countries, with 100 educational centres per country. These countries are the United States, Canada, Uruguay, the Netherlands, Finland, Australia and New Zealand. The idea is to work together. To find out how technology can accelerate pedagogy. Michael Fullan, the renowned educator and reformer of several learning projects in different countries, is leading this project. We are facing a great dilemma. We all want to educate a student with critical, collaborative thought, having good communication, having a didactic use of ICT, being a good, creative, imaginative citizen, and with character. But, where are the rubrics to measure this? How are we going to measure whether somebody is collaborative? We can continue making speeches, our politicians live by making speeches. But somebody has to do the actual work. This project is working on that and it is something very complicated. To have students work in teams, on projects about various topics and to share these topics with other countries in the world and see where the technology did help and where it did not. But technology shouldn't be the core. It should be extremely peripheral. I have no doubt at all that this is a gigantic challenge.

Finally, from the point of view of the provision of services, technology has to help in a number of fields, including digital planning, access to a wealth of historical information on learning. Currently there is a very large collection of data to work on, and we must be able to detect student issues well in advance. An electronic attendance list, which is something trivial, takes time; but in Uruguay, this is something already being done.

There are two more points to be added. One is costs. We spend about USD 100 per student per year on the programme. This includes the laptop and exchanging the laptop, but if you look at the percentages (see Figure 9), USD 37 of the USD 100 correspond to the laptop's depreciation; USD 14 to the network; USD 8 to video conferences; USD 16 to laptop support and logistics; i.e. technical support costs money, support areas; servicing the devices costs money.

Figure 9. Implementation and sustainability
Source: Plan Ceibal, 2015.

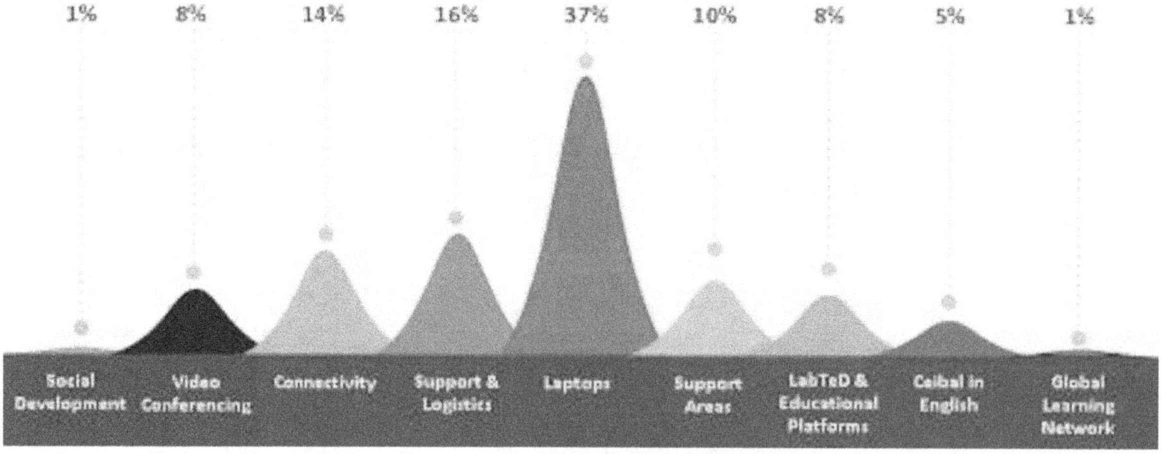

CONCLUSION

Finally, what are the challenges? We are talking about what we did. Today, the biggest challenges account for much of what nobody is doing. We are very concerned because reading comprehension is a big problem in Uruguay, and I believe in all of Latin America, and no good adaptive platform for reading comprehension exists. We believe that instead of discussing some general software issues, we should be thinking at regional level why, if we undoubtedly are the region with the most laptops in schools, there is no reading platform where the student starts reading a book and the book adapts to the student's understanding ability. This is not just a dream, it is something possible to achieve. It requires a very complex software development which will take money and time, but it is a necessity which at least we in Uruguay have and are working very hard on it. What for? To find areas where technology will help to improve that understanding. Because today, when a child goes to class and is not able to read as well as the child sitting next to them, we are facing a difficulty that technology does not solve. It will solve it when it adapts to the child. Technology provides adaptive platforms, not matter what grade a child is in, allowing students to understand what they need to understand and to produce what they need to produce, something that is very clear in the case of mathematics.

There are many challenges, especially concerning educational platforms. Above anything, improving "the quality of the cloud". Although today we have optical fibre available, it is not enough. We need to improve on interaction. This implies having laptops and tablets with the best WiFi chips because otherwise video won't work as it should. Students use video, video and more video, and this consumes bandwidth and computer video chip capacity.

Although we have already been working for eight years on this, we hope to be here still a few more years and contribute to greater equity and better learning in our country.

REFERENCES

Ceibal, P. (2006). *What is Plan Ceibal?* Retrieved 18.12.2017 from https://www.ceibal.edu.uy/en/institucional

Ceibal, P. (2014). *Plan Ceibal over the years – 2014*. Retrieved 18.12.2017 from https://www.ceibal.edu.uy/en/institucional

Ceibal, P. (2015). *Plan Ceibal over the years – 2015*. Retrieved 18.12.2017 from https://www.ceibal.edu.uy/en/institucional

Ceibal, P. (2016). *Plan Ceibal over the years – 2016*. Retrieved 18.12.2017 from https://www.ceibal.edu.uy/en/institucional

ECH-INE. (2007). *Uruguay Investment and Promotion Agency, 2007*. Retrieved from http://www.uruguayxxi.gub.uy/

ECH-INE. (2016). *Uruguay Investment and Promotion Agency, 2016*. Retrieved 18.12.2017 from http://www.uruguayxxi.gub.uy/

Wikipedia. (2018). *Uruguay*. Retrieved 18.12.2017 from https://en.wikipedia.org/wiki/Uruguay

ADDITIONAL READING

Croteau, A.-M., & Bergeron, F. (2001). An information technology trilogy: Business strategy, technological deployment and organizational performance. The Journal of Strategic Information Systems. 10. 77-99. Retrieved from https://www.researchgate.net/publication/222693883_An_information_technology_trilogy_Business_strategy_technological_deployment_and_organizational_performance

KEY TERMS AND DEFINITIONS

Global Learning Network: New project of the Plan Ceibal.

On-Line Platform: The basic hardware (computer, servers, network) and software (operating systems, web-applications) for collective project work over the internet.

Pedagogy Relationship: Special kind of personal relationship between adult and child or adult and student that is different from other personal relationships.

Plan Ceibal: The program created in Uruguay in 2006 as an inclusion program.

Technological Deployment: A part of technology trilogy (together with business strategy and organizational performance) based on the strategic impact of the information system department, the technological architecture, the information system performance evaluation and technological scanning, contributes directly to organizational performance for the strategic activities' analyzer, while it contributes indirectly to organizational performance.

Chapter 4
Engineers for Industry:
Challenges, Solutions, and Future Ideas

Robin Clark
University of Warwick, UK

Jane Andrews
Aston University, UK

ABSTRACT

The need for a reliable supply of engineering talent is accepted globally, but in many parts of the world the many challenges mean that this is not easily achieved. Even if the graduate supply is a reality, often there are concerns about the quality of the engineers entering the workforce. This chapter will explore this landscape, and after identifying the many challenges, explore solutions and potential ideas for the future of engineering education and the university/industry collaboration.

INTRODUCTION

The current environment in which engineering educators are operating is both complex and demanding. The particular context explored within this paper is the UK, although a more global perspective will be taken wherever possible. The needs of industry are a primary driver when it comes to developments in engineering education, but increasingly the policies of government and the forces of the tertiary education marketplace are adding to this complicated picture (Perkins, 2013).

This paper will discuss some of the ways in which UK higher education institutions are addressing these challenges. Although the primary focus of the work will be the present, the paper will conclude with some more strategic thinking, contemplating what can be done to create a more robust foundation on which to build future innovation. Throughout, the key theme will be the employability of the engineering graduates being created and nurtured within the engineering education process.

DOI: 10.4018/978-1-5225-3395-5.ch004

BACKGROUND

Tertiary education, not just in the UK but across the globe, is in a state of continual change – whether that be rationalisation to balance budgets as has happened over much of Europe in recent years, the changing demands of students or the need to innovate to ensure the development of industry ready graduates. This paints a complex picture and engineering educators are firmly situated at the centre of this change.

If we consider the UK, all of the previously mentioned forces are acting on the higher education sector along with pressure to ensure that the student experience is of the highest quality. In the UK this is captured by the annual National Student Survey that all final year students in every university are asked to respond to (HEFCE, 2017a). The results of this survey are crucial as they contribute a significant amount to the University League Table rankings that are published annually and in this last year the award of recognition by way of the Teaching Excellence Framework (TEF) (HEFCE, 2017b).

In order to create a 'Sustainable Future for Higher Education' (UK Government, 2010), legislation led to undergraduate student fees being raised from £3000 per year of study to £9000. With the arrival of the TEF, these fees will now increase further and this was demonstrated in 2017 when fees were raised to £9250. The Government White Paper that introduced the TEF in May 2016, 'Success as a Knowledge Economy' (UK Government, 2016), made many proposals all with the aim that "universities should produce well equipped students ready to contribute to society and business". Clearly employability is at the forefront of current UK Government thinking when it comes to the value and impact of Higher Education. This was taken further in the autumn of 2017 when the UK Government published its future plans for the TEF (UK Government, 2017). Although still called the TEF, the full title going forward is 'Teaching Excellence and Student Outcomes Framework' thus placing the future destination of university graduates firmly at the centre of the evaluation exercise.

The TEF has been perceived in some circles as yet another measure and a way to interfere in university learning and teaching. Taking a more optimistic and constructive view, it really should be seen as an opportunity. With the framework subjecting learning and teaching to a similar level of scrutiny as is already done for university research, the value of teaching professionals and the contribution they make is sure to rise.

STEM (SCIENCE, TECHNOLOGY, ENGINEERING, MATHS) GRADUATES

So, in thinking about STEM graduates, exactly what does industry want?

In one word – everything!

They want a sound technical background but also a wide range of interpersonal, personal and business skills to supplement that STEM understanding. Also, they are not always consistent. Sometimes one area is valued more than another and this will vary from company to company.

Despite the 'demands' industry often places on universities and the criticism that is sometimes levelled at them concerning the quality of the graduates being produced, industry engagement in the educational process can be very variable. It is almost as though there is a reluctance to get too involved, as in it not being their responsibility.

The UK Institution of Mechanical Engineers produced a report in 2011 (IMechE, 2011), of which the key messages are still as relevant today as when the report was written, in particular the recommendations.

The recommendations are split into those for government, those for employers and those for the profession. Focussing on those for employers, they are all about investing in skills, engaging in dialogue with education providers and promoting engineering as a fulfilling career. The key message being 'work together' as the results will be of greater value to all involved. It sounds simple, but the collaboration between industry and universities is not occurring as widely as it should and, when it does occur, the effort needed for success is considerable.

Within many UK universities with a clear focus on employability, this challenge has been tackled from both the student viewpoint and from the industry position. One particular project conducted as part of the UK National HE STEM Programme produced two toolkits for use by students (Andrews et al., 2012) and universities / industry (Clark & Andrews, 2012a) to promote greater awareness and engagement. The student toolkit captures the key characteristics employers are asking that students possess on graduation, in addition to their technical knowledge. Each skill is identified, explained and ways to develop it presented. Examples include team working, problem solving and global citizenship. In that way students and course leaders are given easy to follow guidance on what needs to be embedded within the engineering and applied science study programmes. The universities / industry toolkit explores the features of strong engagement with respect to the learning and teaching space and identifies actions that each can consider in order to foster even stronger relationships.

In many UK universities there is an aim to have the majority of STEM students experience an industry placement of some type combined with greater industry input to course design and development. That way, the graduates produced will be ready for industry. As has been stated earlier, this is not trivial and it does require each side to acknowledge the need to work together and this is where the second piece of work I wanted to share with you comes in.

Following a study of the literature and a series of interviews with industry representatives, the universities / industry toolkit resulted in the development of an Engagement Model (Figure 1). Although when looking at the model the different elements around developing networks, strengthening relationships and managing the employer / academic interchange more effectively appear to be straight forward, actually realising a positive outcome requires a major effort if the effects are to be sustainable.

INNOVATIVE ENGINEERING EDUCATION

For effective and innovative engineering education to be realised, the starting point requires the need to take a holistic view. Although the main focus of this paper is higher education, it is important to see engineering education as a series of steps towards the ultimate goal of becoming an engineering leader in industry. This means the need to value the inspiring of young children when they are 4 years old as much as the engagement with upskilling of mature engineering professionals.

At primary level (4-11 years old) both boys and girls tend to be equally engaged in engineering activity when they are introduced to it. The problem is that many children don't experience this opportunity. Primary Engineer (Primary Engineer, 2017) is a key provider in the UK and work exploring the impact of the experiences they offer suggests that there is a solid foundation on which to build (Clark and Andrews, 2010). The clear need is to spread the activity more widely and ensure a smoother transition from primary to secondary education.

Figure 1. Universities/industry engagement model

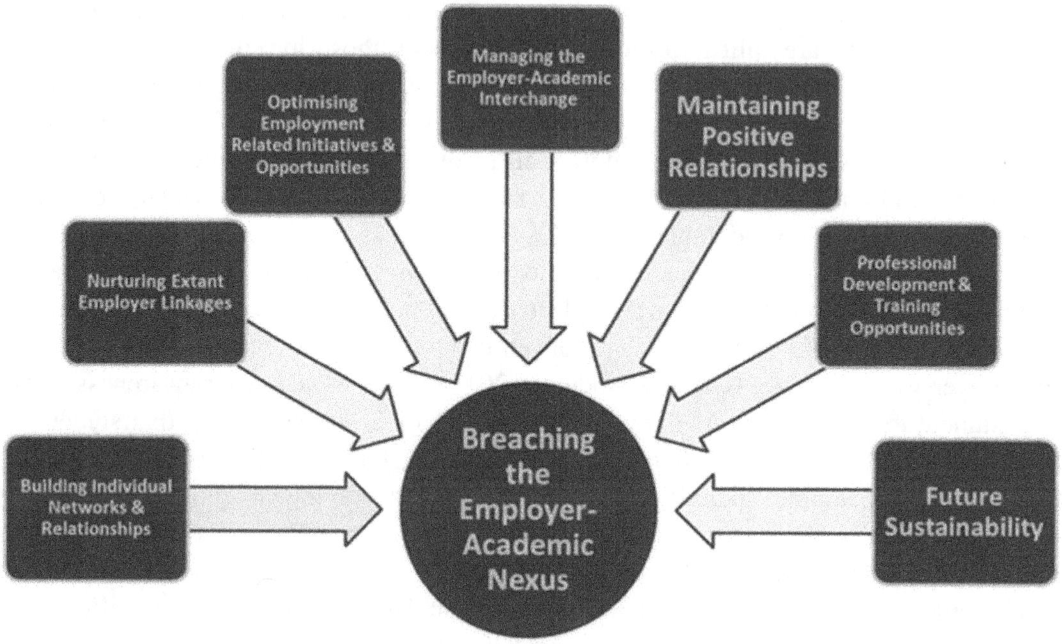

In the UK it is these transition points and the consequent changes in learning and teaching approach that prove most problematical. The Institution of Mechanical Engineers 'When STEM?' report (IMechE, 2010) suggested very clearly that the primary to secondary transition needs particular attention as from 12 years on, interest in STEM subjects, particularly amongst girls, drops quite markedly. The typically dry delivery, lack of practical opportunities and shortage of qualified teachers all compound this problem. The challenge of engaging girls in engineering is explored further by Clark and Andrews (2016).

Moving into university, issues around understanding exactly what engineering is, grasping the idea of independent learning and the socialisation need are all dominant. 'Belonging' is highlighted as a key consideration, something that comes through strongly in the 'What Works' Programme Report, the results of a major study on transition into university conducted in the UK (Thomas, 2012).

One particular approach which has proven helpful in addressing the student retention and progression challenge in higher education is that of peer mentoring. This is where typically 2nd year students are paired with 1st years and they support the students as they transition into university, initially socially and later academically. The study conducted, the largest undertaken looking solely at peer mentoring, involved 6 universities and adopted a multiple case study approach, employing surveys, interviews and observations.

The result was the Transition + Peer Mentoring Model shown in Figure 2 that identifies the key features of an effective peer mentoring approach within a university. The features of the model allow institutions to tailor their process to meet their needs and environment but encourage them to consider each of the features proactively such as the need to train and support the mentors, to take care in matching and ensure that each mentor has no more than 3 to 5 mentees. The model has been used by several institutions both within the UK and globally (Clark et al, 2013).

Figure 2. Transition + peer mentoring model

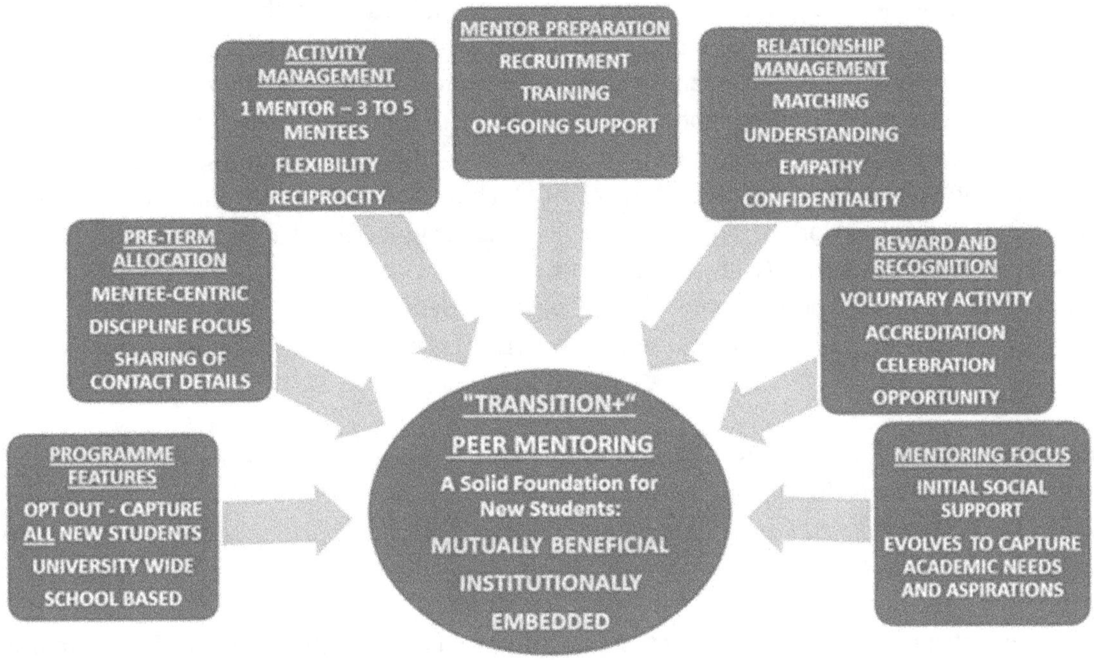

Approaches such as peer mentoring are important considerations when it comes to considering the quality of the student university experience. This can then impact the National Student Survey results referred to earlier. With good results an institution can demonstrate to potential students that it is a great place to come and study, but more importantly, the data can feed into the universities continuous improvement plan. In recent years, the scores for the 'assessment and feedback' part of the survey have been low nationally. Consequently, many universities have clear objectives to address this particular area of what they do.

In order to ensure the 'Teaching on my course' rating of the survey is high, innovation in learning and teaching is widely encouraged in most institutions. The Royal Academy of Engineering report from 2007 entitled 'Educating Engineers for the 21st Century' (RAEng, 2007) suggested that project and problem based learning should be important elements of a relevant engineering degree programme. The PrBL and PBL literature is extensive but it can be argued that there is a need to take a slightly broader view of this proposal by suggesting to colleagues that variety and activity in learning is important, alongside the consideration of authentic or real world learning experiences. Staying with one 'novel' approach can often be as problematical as retaining a wholly didactic approach to learning and teaching.

One particular active learning approach adopted within the Mechanical Engineering and Design programmes at Aston University is the use of the CDIO framework. CDIO stands for Conceive Design Implement Operate and is very much based on the engineering design process. Students work in teams and create an artefact to satisfy a brief – cars, wind turbines and valves are three such examples. In addition to promoting innovative and self-motivated student learning opportunities, the CDIO framework provides guidance for staff on what they should be considering.

In addition to employers recognising the impact of this approach on graduates, the evaluation of the impact has suggested that students are also able to see and realise the benefit of a more active and authentic approach to engineering education (Clark & Andrews, 2012b; Clark & Thomson, 2015).

Another area where innovation in engineering education is readily apparent is the development of more flexible work based learning opportunities that cater to working professionals who desire or are required to take their education to the next level. Bespoke company programmes are becoming increasingly popular, often based on a blended approach of block release and on-line materials. One particular programme in the UK is a work based MSc in Professional Engineering that is part of the UK Engineering Council Gateways initiative (Glew & Elsworth, 2010; Clark et al, 2014).

The programme has not only operated for UK based students, company cohorts from overseas have also enrolled. A typical comment from a student on one of these overseas programmes is given in the quote below:

An important fact is that learning is both an active and a reflective process. What really stands out from my learning experience is the lifelong learning. Up until I left university, I had always thought that university was going to be the end of my educational years. I have now learnt that going to university was not just to learn how to be an engineer, but also to learn how to learn.- Student, BP Angola

A final area that needs consideration is that of diversity and in particular in the UK, the gap in attainment between white students and those from a black minority ethnic background. The work conducted in this area suggests that there is still much to do to realise some real improvements. Where courses involved BME students in co-creation activity and where BME attainment was specifically discussed, some promising results have been demonstrated. Sadly the subject is one that is often not discussed as it makes teachers and management uncomfortable. The important conclusion is that if you have student diversity, embrace it and respond to it if you want all of your students to be successful (Andrews & Clark, 2015).

MAKING ENGINEERING EDUCATION WORK

With so much happening in higher education, with the unique challenges of the STEM subjects and with the pressure on staff to do their discipline research, it is important to find a way to engage staff in the teaching improvement process, but in a way that is accessible to them.

The starting point has been Biggs Constructive Alignment (Biggs, 2011) that suggests the need to ensure a coherency between the learning outcomes a teacher wants to achieve, the teaching method employed and the assessment approach. This has now been extended to suggest that the coherency needs to take into account what happens to a student before they enter university and how their university experience will connect with the career path they wish to follow on leaving university. This is captured in a model developed by the paper authors called the RVS model (Clark & Andrews, 2014).

In communicating with staff, the simple equation R+V+S = Student Success has been very effective. To add some explanation:

R = relationships – between staff and students, students and student, staff and staff etc
V = variety, in other words doing different things in the class and not just lecturing
S = synergy, which captures the extended alignment idea.

The premise is that for engineering educators, by focussing on the R, V and S they will become effective teachers and create engaging and innovative learning experiences for students. At the same time, they are embracing some important ideas from the pedagogical literature without having to read, distil and interpret for themselves, something that is of considerable value in the current environment.

In order to support this further, Aston University has established a research centre, ASEC – Aston STEM Education Centre, to promote scholarship in learning and teaching, a way to share and network, to provide evidence for decision making and for individuals, a way to gain professional recognition and promotion (ASEC, 2017). The paper authors were the founders of ASEC which was an extension of a smaller Engineering Education Research Group.

In the wider UK, from a Special Interest Group first set up in 2009, there is now a thriving independent UK and Ireland Engineering Education Research Network that has close to 100 members and it will hold its 5th Annual Symposium in London in November 2017. It has been a long journey to recognition for the participants, but the Network website is now firmly established and hosted by the Royal Academy of Engineering (EERN, 2017).

Looking further afield, in Europe there is SEFI (European Engineering Education Society), globally we have REEN (Research in Engineering Education Network) and several countries have their own national Engineering Education Societies. In Europe, EU projects have been a valuable way of strengthening the networks and bringing institutions together. With the establishment of Centres and Networks, there is now, more than ever, an opportunity to share, collaborate and learn. All of this suggests that a degree of recognition for the work on innovative engineering education is being realised. The exchange of ideas across institutional and national boundaries can only make the community stronger and help us understand our respective challenges and thinking better.

LOOKING TO THE FUTURE

Looking forward, the drive to produce ever more capable graduates, relevant courses and stronger industry links has become the path to follow. Engineering Education has reached a crucial point in its evolution and the ways of the past are increasingly no longer relevant.

In moving towards a new vision, a 'Learning Future', a significant culture change is needed. The culture change is one that will affect all of the stakeholders in the engineering education process as we strive to put the student truly at the centre of the learning experience. We can do this by creating a defined but flexible framework within which students can determine their own learning programme. Similar to work based learning, blended learning and other more innovative pedagogies, a student will not only define the subjects they want to study, but also the way they want to study. This will require learners to be truly independent such that flipped pedagogies and flexible learning and assessment can be introduced and be successful. At present, student engagement concerns can make the implementation of some novel approaches to learning and teaching problematical.

Initially there will be much work to be done, not only with staff and students, but with industry, professional bodies, schools and parents to explain the value of this new approach and to engage them in its development.

In working towards a realisation of this vision, it is important to identify priority projects that will lay the foundation for a successful and sustainable implementation of this new framework. Examples might include, but not limited to, learning analytics, greater industry engagement, the redefinition of assessment and looking at how we can best measure the impact of learning. The ultimate goal, support by scholarship and research undertaken in Centres such as ASEC, is to provide opportunities for all – learners, industry and the wider society.

FUTURE RESEARCH DIRECTIONS

Having explored the landscape it becomes clear that the main direction for future research must be focused on a greater understanding as to what works and what doesn't work when considering innovation in engineering education. If we are willing to accept that the innovation is needed to make the subjects more attractive to potential students and more relevant to what engineering graduates need to know, understand and demonstrate to be effective in industry, then we must ensure that we are studying and measuring the impact of this innovation. In order to do this we must embed the idea of Engineering Education Research in our Engineering Schools and Departments across the globe.

To be effective, the work undertaken in this space needs to be valued and recognised. With visibility and the promotion of sharing, a multitude of small scale interventions and action research projects can become a global understanding that has the potential to benefit students, universities and industry to the fullest extent. To take steps in this direction will require commitment from senior university leadership and funding from national and international bodies to support the work. Until now this has been lacking for many, with work taking place as a result of a grass roots commitment to bring about change.

CONCLUSION

Working in Engineering Education and Engineering Education Research requires passion, a passion that says it is ok to think out of the box if you want to benefit your students and the industry they will be working in. A passion that will encourage a person to persist. The Einstein quote - "If at first the idea is not absurd, then there is no hope for it"- should drive us towards ever more novel approaches and away from conventional engineering education teaching.

Wherever we are in the world, the higher education environment is challenging. In this paper it has been demonstrated that much work and creative thinking is taking place and that because of this, Engineering Education and Engineering Education Research has a bright future. The May 2017 event 'New Approaches to Engineering Education' at the IET in London is one such event that has sought to challenge the status quo (IET, 2017). Whatever steps we take, we must consider industry, we must be student focused and we must be flexible and innovative. If we do this, then everyone will win.

ACKNOWLEDGMENT

The authors would like to thank those colleagues, too numerous to mention, that have stimulated the thinking and work presented in this chapter.

REFERENCES

Andrews, J., & Clark, R. (2015). *BME SDG Final Report*. York, UK: Higher Education Academy.

Andrews, J., Clark, R., & Higson, H. (2012). *Enhancing Employability: Making the Most of University*. Report for National HE STEM Programme.

ASEC. (2017). *Aston STEM Education Centre*. Retrieved 11/8/17 from http://www.aston.ac.uk/eas/research/groups/asec/

Biggs, J. B. (2011). *Teaching for quality learning at university: What the student does*. McGraw-Hill Education.

Clark, R., & Andrews, J. (2010). Researching Primary Engineering Education: An Exploratory Study from a UK Perspective. *European Journal of Engineering Education*, *35*(3), 585–595. doi:10.1080/03043797.2010.497551

Clark, R., & Andrews, J. (2012a). *Innovation in the Academic / Vocational Interchange: Developing and Achieving Good Practice in Employer Engagement*. Report for National HE STEM Programme.

Clark, R., & Andrews, J. (2012b). Engineering the Future: CDIO as a tool for combating retention difficulties. In M. Rasul (Ed.), *Developments in Engineering Education Standards: Advanced Curriculum Innovations* (pp. 143–155). IGI Global. doi:10.4018/978-1-4666-0951-8.ch008

Clark, R., & Andrews, J. (2014). Relationships, Variety & Synergy [RVS]: The Vital Ingredients for Scholarship in Engineering Education? A Case-Study. *European Journal of Engineering Education*, *39*(6), 585–600. doi:10.1080/03043797.2014.895707

Clark, R., & Andrews, J. (2016). A Community Based Participatory Research Study into Why Some Girls Don't 'Do' Engineering. *International Journal of Engineering Education*, *32*(6).

Clark, R., Andrews, J. & Gorman, P. (2013). Tackling Transition: The Value of Peer Mentoring. *Journal of Widening Participation and Lifelong Learning*, *14*, 57-75.

Clark, R., Glew, B., & Andrews, J. (2014). Developing the Engineering Talent Pipeline using Work Based Learning. *Engineering Leaders Conference*.

Clark, R., & Thomson, G. (2015). Exploring the Form and Impact of a CDIO Implementation – A Case Study. *European CDIO Meeting*.

EERN. (2017). *UK and Ireland Engineering Education Research Network*. Retrieved from https://hefocus.raeng.org.uk/research-network/

Glew, B., & Elsworth, T. (2010). *Development of a work-based learning MSc course which incorporates the development and demonstration of professional engineering competence standards*. Engineering Education Conference 2010, Birmingham, UK.

HEFCE. (2017a) *NSS Results 2017*. Retrieved from http://www.hefce.ac.uk/lt/nss/results/2017/

HEFCE. (2017b). *Teaching Excellence Framework*. Retrieved from http://www.hefce.ac.uk/lt/tef/

IET. (2017). *New Approaches to Engineering Higher Education.* Retrieved from http://www.theiet.org/policy/panels/education/22-05-17.cfm

IMechE. (2010). When STEM? Institution of Mechanical Engineers.

IMechE. (2011). *Meeting the Challenge: Demand and Supply of Engineers in the UK.* Institution of Mechanical Engineers.

Perkins, J. (2013). *Review of Engineering Skills.* Retrieved from https://www.gov.uk/government/uploads/system/uploads/attachment_data/file/254885/bis-13-1269-professor-john-perkins-review-of-engineering-skills.pdf

Primary Engineer. (2017). Retrieved from http://home.primaryengineer.com/

RAEng. (2007). *Educating Engineers for the 21st Century.* London: Royal Academy of Engineering.

Thomas, L. (2012). *What Works – Final Report.* Retrieved from http://www.phf.org.uk/publications/works-student-retention-success-final-report/

UK Government. (2010). *Securing a Sustainable Future for Higher Education.* Retrieved from https://www.gov.uk/government/publications/the-browne-report-higher-education-funding-and-student-finance

UK Government. (2016). *Success as a Knowledge Economy: teaching excellence, social mobility and student choice.* Retrieved from https://www.gov.uk/government/publications/higher-education-success-as-a-knowledge-economy-white-paper

UK Government. (2017). *Teaching Excellence and Student Outcomes Framework Specification.* Retrieved from https://www.gov.uk/government/publications/teaching-excellence-and-student-outcomes-framework-specification

ADDITIONAL READING

IET. (2017) *New Approaches to Engineering Higher Education* http://www.theiet.org/policy/panels/education/22-05-17.cfm Accessed 11/8/17

Johri, A., & Olds, B. (Eds.). (2014). *Cambridge Handbook of Engineering Education Research.* Cambridge: Cambridge University Press; doi:10.1017/CBO9781139013451

The Proceedings of the IET Conference offer an insight into the latest UK thinking in terms of innovation in Engineering Education. The collection of papers explores a range of topics of relevance to the ongoing debate. UK and Ireland Engineering Education Research Network (2018) *Proceedings of the 5th Annual Symposium – 'Time for Change'* https://hefocus.raeng.org.uk/network-events/ Accessed 28/2/18

The Proceedings of the UK & Ireland EERN Symposium held in London in November 2017 offer a clear and up to date account of the pressing research issues being explored by the UK and Ireland Research Community. The collection again offers insight into a range of topics currently of interest.

Trevelyan, J. P. (2014). *The Making of an Expert Engineer*. London: CRC Press/Balkema - Taylor & Francis. doi:10.1201/b17434

Wankat, P. C., & Oreovicz, F. S. (2015). *Teaching Engineering* (2nd ed.). West Lafayette, IN: Purdue University Press.

Williams, B., Figueiredo, J. D., & Trevelyan, J. P. (Eds.). (2013). *Engineering Practice in a Global Context: Understanding the Technical and the Social*. Leiden, Netherlands: CRC/ Balkema. doi:10.1201/b15792

KEY TERMS AND DEFINITIONS

Employability: The employability of engineering graduates once they complete their course of study is a key consideration across the globe and one that needs to be explored within the engineering education process.

Engineering Education Research: Increasingly there is an awareness that, in order to promote greater understanding of the value and impact of innovation in engineering education, a more scholarly / research-based approach is needed. To that end, institutional. national and international groups have formed to promote and conduct engineering education research (EER).

Industry Collaboration: The collaboration between industry and universities is essential if we are to ensure that graduating students can contribute fully when they start their careers. Unfortunately, the collaboration not occurring as widely as it should and, when it does occur, the effort needed for success is considerable. The university / industry toolkit explores the features of strong engagement with respect to the learning and teaching space and identifies actions that each can consider in order to foster even stronger relationships.

Innovative Engineering Education: The generic used term for new approaches to engaging students in the study of engineering subjects. It covers all aspects of the education process from design and implementation to assessment and reflection.

SEFI (European Engineering Education Society): The European Engineering Education Society offers a community of practitioners and scholars who have a mutual interest in the development and greater understanding of engineering education.

STEM: The term used to describe the combined subjects of science, technology, engineering, maths. This term is used most often in schools but increasingly in the tertiary education space.

Section 2

Quality and Standards

Chapter 5
The Impact of Rankings on Russian Universities' Student Choice

Liliya Ravilevna Yagudina
Kazan National Research Technical University, Russia

ABSTRACT

On the basis of its own hypotheses and analysis of foreign studies, the author determines how the informational and motivational functions of rankings have impact on Russian universities' student choice. The analysis of the rankings position of universities and some of their key performance indicators (the foreign students number, the quality of applicants) showed there is not any direct correlation between them. According to the author, in order to maximize the effectiveness of rankings, it is necessary to improve the ranking methodology, to develop universities' decision-making processes based on the rankings results, and to create customer culture on the rankings results using.

INTRODUCTION

To improve the quality and relevance of training and teaching, which is the primary mission of the European Higher Education Area (EHEA), requires new collective efforts to be exercised by the EHEA countries against the backdrop of the current transformation of higher education (Yerevan Communique, 2015).

Expansion and growing diversity of stakeholders in the education system, as a result of academic and career mobility of students, graduates and teachers, makes the tasks faced by the universities more challenging in terms of identifying the consumers' requirements, assuring education quality, reporting and providing open and transparent information about the quality level.

Development of an independent assessment of quality as an institutional component of national education systems is a part of the most important issues involved in renewal of the EHEA.

DOI: 10.4018/978-1-5225-3395-5.ch005

The author assumes that an independent education quality assessment would be an effective component of national system for education quality assurance, if it were to perform guaranteeing, informational, motivational, educational, and consultative functions. This chapter focuses on researching implementation of informational and motivational functions of such type of education quality assessment as ranking. Relating to higher education rankings, the content of these functions looks as follows.

The informational function consists of informing consumers about the relative, comparative level of the universities' reputations, satisfying the needs of each individual person and society as a whole so that to provide the information they are interested in. Under the conditions, where universities are becoming rightful players in the global market, rankings are designed to furnish information for the processes of potential consumers' educational choice. Some researchers doubt the value of implementing this particular function of rankings to develop transparency and openness of the education system, in view of those undesirable consequences, which they produce, due to their inability to provide reliable and up-to-date information (Rauhvargers, 2011).

The motivational function of ranking consists, firstly, of the ability to motivate universities to enhance the reputational characteristics and to reinforce the universities accountability to society. Capability of rankings to influence the development of the education quality remains controversial; they are often criticized for simplistic approach to the issues of universities' activities, for resorting to this approach instead of thorough and scrupulous work on quality control; rankings are also criticized for pressure on universities which causes the latter to lose their individuality and national diversity. Indeed, it has to be admitted that rankings are more of a reputational management tool rather than a quality management one. Secondly, the motivational function of ranking lies in its capability of impacting the consumer preferences of the players in the market for educational services. This very impact is what this paper is devoted to.

BACKGROUND

Studies targeted at comparative analysis of the impact of rankings on higher education show that, there are many things in common in the way people in different countries react to rankings, make management decisions and act on them.

Governments exploit the motivational function of rankings to promote universities in the world education market, making them more competitive and seeking to achieve recognition of the national educational system abroad, ensuring them with appropriate funding. E. Hazelkorn underlines exclusively motivational function of rankings which play the role of social accountability, compelling universities to stick to and adopt the best practices, and based on the researcher's empirical study in Germany, Japan and Austria, there is an opinion that rankings serve as both overt or covert factors causing changes in the higher education system (Hazelkorn, 2009).

With respect to universities, the informational function of rankings is expressed, on the one hand, in providing information base for the universities themselves in identifying the best practices, and on the other hand, a university position in ranking enables the stakeholders to ascertain this university potential as a possible partner.

According to the university presidents' opinion poll in 2006, nearly 50% of the universities exploited their ranking position for advertising purposes, while 63% of the respondents said that rankings were especially helpful in recruiting students (Hazelkorn, 2011). The websites of the universities which occupy winning positions in the rankings of various levels demonstrate that such results are widely used to represent the university.

In studies dealing with impacts of rankings on access, choice and students' opportunities it was noted that a positive influence making the total number of applicants at the universities grow in Canada (Mueller & Rockerbie, 2005), Germany (Helbig & Ulbricht, 2010), the U.K. (Matzdorf & Greenwood, 2015), the U.S. (Espinosa, Grandall, & Tukibayeva, 2014) and other countries. Chevalier & Jia (2016) concluded that one standard deviation in ranking results causes an increase in the number of applicants by an average 4.3%.

Though the research undertaken by Higher Education Research Institute of California evidences that the choice of a university is more impacted not by the university position in rankings but by parents' advice, the reputation of the university among the peers, etc., they found that the importance of rankings was generally on the rise. The number of students, who regarded the ranking results as a very important factor in shaping their decision on the choice of the university, had risen between 1995 and 2006 by 56.2%; at the same time, there was not such a significant change observed in the growth of importance for other important factors (Higher Education Research Institute University of California [HERI], 2007).

Subsequent studies undertaken by Matzdorf & Greenwood (2015) also demonstrate dependence of students' choice on the position of the university in the league tables: 42% (Definitely agree) and 41% (Mostly agree) of the students polled agreed with the statement "My choice was strongly influenced by League Tables".

The majority of the studies are focused on establishing the factors which influence reference to these rankings, and their authors arrive at the conclusion that rankings do influence decision making of highly talented students and that such decisions are caused by gender, nationality (ethnicity), country of origin, social and economic status of the family, applicants' aptitude and the chosen path of studies.

It can be considered as unambiguously established by empirical research that there is a linkage between ranking of a university and quality of applicants for enrollment: applicants with high aptitude tended to choose universities under the influence of rankings (Clarke, 2007; Ehrenberg, 2002; Griffith & Rask, 2005; Yonezawa, Akiba & Hirouchi, 2009; HERI, 2007). Rankings are also primarily needed to students from high income families (Clarke, 2007; Federkeil, 2002; McDonough, Antonio, Walpole, & Perez, 1998).

More and more foreign students tend to choose universities under the influence of ranking results. This correlation was experimentally hit upon by a number of researchers; for example, foreign applicants in the U.K. were more sensitive to the information about the quality of education in comparison with local students: a single standard deviation in ranking results causes an increase in the number of applicants by an average 7.4% (Chevalier & Jia, 2016).

Therefore, an overview of the results yielded by previous research enables the author to isolate the following key tendencies in the impact of rankings on consumers:

- During the last decade, on the whole, one has been able to observe growing importance attached to rankings when consumers of educational services decide as to where it would be best to get a university degree;
- Implementation of motivational function performed by rankings depends on the family income, the level of education results at the previous level of education, an applicant's place of residence at the time that he or she makes the choice.

THE RESEARCH STUDY

Methodology

The main section of the work is devoted to investigation of the tendencies defined above in the impact of rankings with respect to Russian universities.

Data was analyzed mainly from the following sources:

- Results from the QS World University Rankings;
- Informational and analytical materials based on the results of Monitoring of efficiency of higher education institutions (HEIs) conducted by The Ministry of Education and Science of Russia;
- Ranking of the Higher School of Economics (HSE) and Novosti Russian Information Agency (RIA Novosti);
- National Ranking of Universities (Interfax Agency);
- Results from the survey carried out by the author (532 responses). First year students studying for a bachelor degree at universities in Moscow, St. Petersburg, and Kazan took part in the survey. Without claiming for universality of the survey, given such an insignificant number of respondents, nevertheless, the author believes that this sampling is sufficient to identify the key trends in the matter under consideration. The survey consisted of closed (structured) questions.

Study Findings

The first task of the survey consists in determining the students' awareness of existing education rankings and analyzing motivation of students' choice of education caused by rankings.

The question aimed at finding out former university applicants' awareness of existing different rankings of education would be justifiable. Analysis of responses allow to state that the Ranking of the HSE and RIA Novosti as well as National Ranking of Universities (Interfax Agency) are the most popular ones among Russian university applicants; Figure 1 demonstrates that 29.15% and 22.92% of the respondents respectively are familiar with them.

55% of those polled indicated that the university position in rankings impacted the choice of the institution for receiving education, and besides this, the majority of them, in their choice, relied on the data provided by the Ranking of the HSE and RIA Novosti and by National Ranking of Universities (26.76% of the students in each case). The answers are summarized in Figure 2.

The respondents were less informed about world university rankings. Perhaps, it can be explained both by the lack of interest to education abroad and by orientation of rankings on different consumer audiences, defining the originality of ranking criteria list.

Figure 1. Students' awareness of the existing education rankings

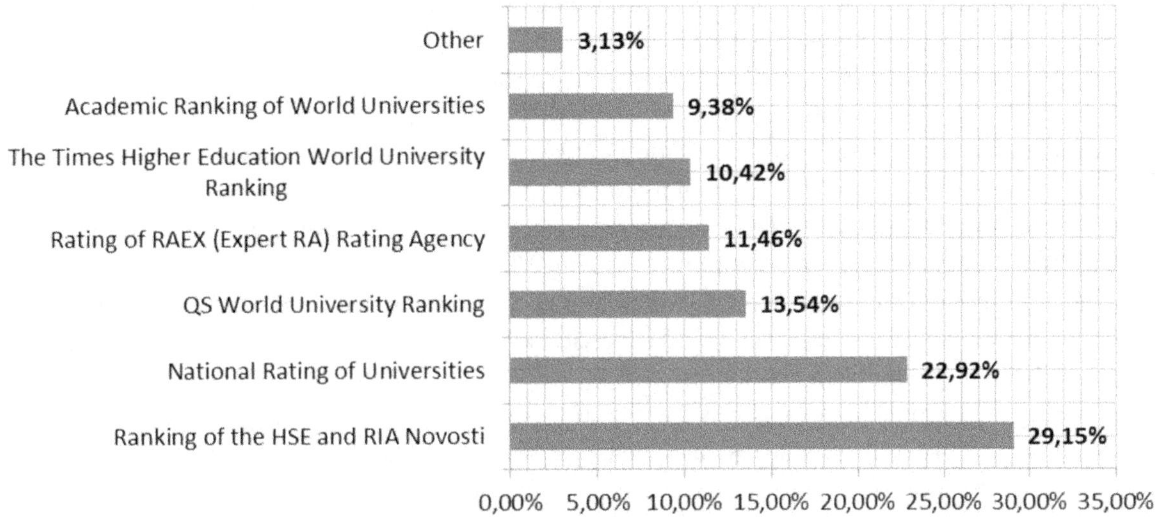

Figure 2. The impact of higher education rankings on Russian university applicant's choice

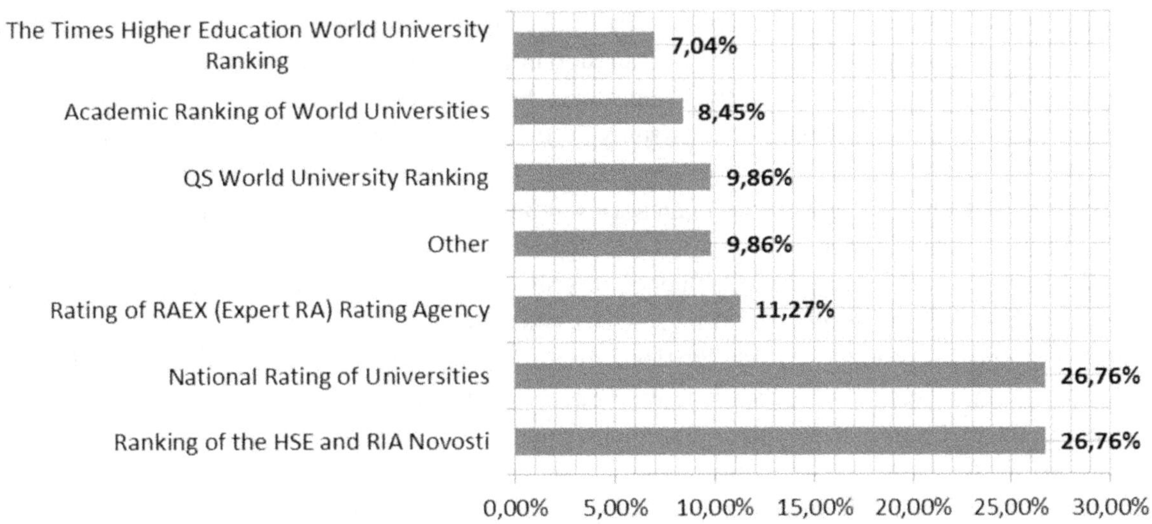

The second task of the survey is devoted to determining the motivational potential of rankings with respect to foreign students and the students with a high average of points earned in the Unified State Test.

Proceeding from the assumption that when choosing the university, an applicant has available ranking results for the previous year, commonly announced in autumn, the author determined the impact of Russian universities positions in rankings on the choice of foreign students by comparing universities position in QS World University Rankings for 2011, 2012, 2013, 2014 and the universities indicators in terms of E3 "International activities" in the Monitoring of efficiency of HEIs for 2012, 2013, 2014, 2015 (monitoring data bank contains information about performance results of organizations, beginning from 2012). The E3 "International activities" indicator is the specific gravity of foreign students' proportion, studying for a bachelor degree, master degree, pursuing post-graduate studies in the total number of students (adjusted contingent).

An array of universities under analysis comprised 11 Russian educational organizations, which have been featured in the QS rankings since 2011. The results of the analysis give to the author every reason to speak about the lack of direct correlation between ranking position of a Russian university and specific gravity of foreign students, though a number of universities do demonstrate steady growth of this indicator (Moscow State Institute of International Relations, Tomsk Polytechnic University, Tomsk State University, Higher School of Economics, Kazan (Volga Region) Federal University, Ural Federal University). It is obvious that use of such an indicator as the number of foreign students enrolled for the first year to the university in the survey could have yielded more accurate results, however this indicator was not available in public open data bases.

Underlying the following analysis was the hypothesis, resting on the results of the above mentioned surveys, conducted in foreign countries and consisting of the fact that change in ranking position of the university was the motivating factor in the choice of university made by applicants with high aptitude, in the case those with a high average of points earned at the UST. Due to the fact that in this particular research the author dealt with Russian students choosing an institution for training for a bachelor degree, the results of the National Ranking of Universities were used in these analysis. Based on E. Hazelkorn's statement regarding the changes in quality of new enrollees caused by the change of the university position in rankings by 10 or more points (Hazelkorn, 2014), the author made the analysis grounded on the data of the Russian universities included in the top hundred universities in the National Rankings of 2014/15 academic year and those that rose up by 10 and more positions in the rankings between 2013/14 and 2014/15 academic years. Short period under study is stipulated by the condition that National Rankings before 2013/2014 academic year included only classic and national research universities that had no drastic change in positions. Data on other categories of universities did not appear in National Rankings until 2013/2014 academic year.

In this respect, we determined the influence of university ranking on the average points earned at the UST (publicly funded enrollments) in 2014, 2015 based on the data provided by the Monitoring of the quality of enrolled students by universities carried out by the HSE and Novosti Russian Information Agency. Only 13 universities proved to meet such criteria and the link between the position of the university in the rankings and the increase in the average number of points earned in the Unified State Test was observed only in 61.54% of the cases.

In order to make the obtained results more accurate, the author expanded the selection of educational institutions by including the universities quoted in the National Rankings for 2014/15 academic year in the top hundred positions and those that rose by 5-9 positions in the rankings between 2013/14 and 2014/15 academic years. The number of the universities under analysis increased to 23 and the link between the position of the university in rankings and the increase in the average number of points earned in the Unified State Test constituted 60.87% of all the cases.

SOLUTIONS AND RECOMENDATIONS

Thus, the results of this survey allow the author to make a conclusion that such ranking methodology under the conditions of existing culture of perceiving rankings, the informational function of the rankings is implemented in the most complete manner due to its visibility and ease of interpretation which enable the consumer to obtain the results of universities assessments as an output finished product. The most effective implementation of this function is possible in case there is certainty about specific target audience for each particular kind of rankings or, vice versa, in case of development of such information systems that recognize diversity of social cultural environment and social procurement and that perform the function of constructing independent rankings. Implementation of the latter, enabling the development of the basis for an individual educational path before entering the educational process, would also allow the motivational potential of the rankings to be enhanced with respect to applicants' choice of an institution for the studies.

The results of the research are not sufficient for an unambiguous confirmation that the motivational function of rankings is implemented in line with worldwide trends as a whole, though even in the current conditions it makes sense to talk about the emerging tendency that a university ranking position can influence the choice of Russian university applicants with high educational performance results.

In order to maximize the efficiency of rankings through effective implementation of their informational and motivational functions, it could be suggested that, firstly, ranking methodologies should be improved to make them more receptive to the consumer's needs, secondly, the mechanisms for integration of rankings into the universities' system of reputational management should be developed, and thirdly, communication policy should be formulated to call the attention of stakeholders to rankings as well as the culture of utilizing the ranking results by their end users.

FUTURE RESEARCH DIRECTIONS

It is clear that for a more accurate determination of trends it is necessary to continue the analysis of subsequent years. The author will continue to follow this research using the coming results of Rankings and Monitoring mentioned above. Of course, it is desirable to supplement the study with a survey of foreign students and students with high scores received at the Unified State Exam.

CONCLUSION

The survey and the analysis of rankings positions of Russian universities and some of their key performance indicators (foreign students' number, the quality of applicants) led to the conclusion that there is no any direct correlation between them.

In view of their inherent limitations, the universities rankings cannot be regarded as a fully-fledged type of independent assessment of education quality and their applications must be as follows: formulation of priority activities for educational systems and identification of the best practices, assurance of educational institutions accountability to the society and their communications, increased motivation for universities to achieve their key performance indicators.

In order to maximize the effectiveness of rankings, it is necessary to improve ranking methodology, to develop universities decision making process based on the rankings results and to develop customers culture on using rankings results.

REFERENCES

Chevalier, A., & Jia, X. (2016). Subject-Specific League Tables and Students' Application Decisions. *Manchester School*, *84*(5), 600–620. doi:10.1111/manc.12124

Clarke, M. (2007). The Impact of Higher Education Rankings on Student Access, Choice and Opportunity. *Higher Education in Europe*, *32*(1), 59–70. doi:10.1080/03797720701618880

Ehrenberg, R. (2002). Reaching for the brass ring: The U.S. News & World Report Rankings and competition. *The Review of Higher Education*, *26*(2), 145–162. doi:10.1353/rhe.2002.0032

Espinosa, L. J., Grandall, M., & Tukibayeva, M. (2014). *Rankings, Institutional Behavior, and College and University Choice Framing the National Dialogue on Obama's Ratings Plan*. Retrieved May 2, 2017, from www.acenet.edu/news-room/Documents/Rankings-Institutional-Behavior-and-College-and-University-Choice.pdf

Federkeil, G. (2002). Some Aspects of Ranking Methodology - The CHE Ranking of German Universities. *Higher Education in Europe*, *27*(4), 389–397. doi:10.1080/0379772022000071878

Griffith, A., & Rask, K. (2005). *The influence of the U.S. News and World Report collegiate rankings on the matriculation decision of high-ability students*. Retrieved May 2, 2017, from http://digitalcommons.ilr.cornell.edu/cgi/viewcontent.cgi?article=1029&context=cheri

Hazelkorn, E. (2009). Attitudes to Rankings: Comparing German, Australian and Japanese Experiences. In *Addressing Critical Issues on Quality Assurance and University Rankings in Higher Education in the Asia Pacific*. Retrieved May 2, 2017, from http://arrow.dit.ie/cgi/viewcontent.cgi?article=1002&context=cserart

Hazelkorn, E. (2011). Globalization and the Reputation Race in *Rankings and the Reshaping of Higher Education: the Battle for World Class Excellence*. Palgrave MacMillan. Retrieved May 2, 2017, from http://arrow.dit.ie/cgi/viewcontent.cgi?article=1010&context=cserbk

Hazelkorn, E. (2014). The Effects of Rankings on Student Choices and Institutional Selection. In *Access and Expansion Post-Massification: Opportunities and Barriers to Further Growth in Higher Education Participation*. London: Routledge. Retrieved May 2, 2017, from http://arrow.dit.ie/cgi/viewcontent.cgi?article=1015&context=cserbk

Helbig, M., & Ulbricht, L. (2010). *Perfekte Passung: Finden die besten Hochschulen die besten Studenten? Gefälligkeitsübersetzung: Perfect fit: do the best universities find the best students?* Retrieved May 2, 2017, from www.wzb.eu/sites/default/files/u128/scan_perfekte_passung_0.pdf

HERI. (2007). *College Rankings & College Choice: How important are college rankings in students' college choice process?* Los Angeles: Higher Education Research Institute University of California. Retrieved May 2, 2017, from https://www.heri.ucla.edu/PDFs/pubs/briefs/brief-081707-CollegeRankings.pdf

Matzdorf, F., & Greenwood, J. (2015). *Student choice, league tables and university facilities.* Retrieved May 2, 2017, http://shura.shu.ac.uk/10423/1/EuroFM2015_StudentChoicesLeagueTables%26Facilities_final.pdf

McDonough, P., Antonio, A., Walpole, M., & Perez, L. (1998). College Rankings: Democratized College Knowledge for Whom? *Research in Higher Education, 39*(5), 513–537. doi:10.1023/A:1018797521946

Mueller, R., & Rockerbie, D. (2005). Determining demand for university education in Ontario by type of student. *Economics of Education Review, 24*(4), 469–483. doi:10.1016/j.econedurev.2004.09.002

Rauhvargers, A. (2011). *Global university rankings and their impact.* Retrieved May 2, 2017, from http://www.eua.be/Libraries/publications-homepage-list/Global_University_Rankings_and_Their_Impact.pdf?sfvrsn=4

Yerevan Communique. (2015). *Paper presented at Ministerial Conference, Yerevan.* Retrieved May 2, 2017, from www.ehea.info/Uploads/SubmitedFiles/5_2015/112705.pdf

Yonezawa, A., Akiba, K., & Hirouchi, D. (2009). Japanese university Leaders' perceptions of Internationalization: The Role of Government in Review and Support. *Journal of Studies in International Education, 13*(2), 125–142. doi:10.1177/1028315308330847

KEY TERMS AND DEFINITIONS

Monitoring of Efficiency of Higher Education Institutions: (Conducted by The Ministry of Education and Science of Russia) The annual monitoring of HEIs implies the evaluation of HEIs based on a list of the key performance indicators, such as the Unified State Test score of enrolled students, graduates' employment, scientific results, international activity, financial sustainability, and academic staff salaries.

National Ranking of Universities (Interfax Agency): A special project of the information agency "interfax," created with the support of The Ministry of Education and Science of Russia to develop and test new mechanisms of an independent assessment of Russian universities.

Ranking of the HSE and RIA Novosti: Ranking of Russian higher education institutions based on the average Unified State Test score of enrolled students conducted by Higher School of Economics and Novosti Russian Information Agency.

Unified State Test (Edinyi Gosudarstvennyi Eksamen): High school final and University entrance exam taken upon completion of 11th grade in Russia.

Universities' Ranking: A listing which compares universities by various combinations of various factors.

Chapter 6
Comparison of Academic and Professional Recognition Systems of Engineering Degrees in Bologna Countries:
Case Studies From Cyprus and Russian Federation

Lyudmila Zinchenko
Bauman Moscow State Technical University, Russia

Marios Evangelos Kassinopoulos
Cyprus University of Technology, Cyprus

ABSTRACT

Academic and professional recognition of engineering degrees is an important problem in higher education and human resources mobility. The chapter presents a review of academic and professional recognition systems features in Cyprus and Russia. Both Russia (non-EU-member country) and Cyprus (EU-member country) are Bologna countries, use similar education curricula, and will potentially follow the qualification framework in the European Higher Education Area. However, national qualification frameworks are different. The chapter discusses the academic and professional recognition systems features in Cyprus. Then the Russian system of engineering degrees is explained and the academic and professional recognition approach is clarified. Case studies for both countries are outlined. A comparison of the academic and professional recognition systems features in Cyprus and Russia is given.

DOI: 10.4018/978-1-5225-3395-5.ch006

INTRODUCTION

The subject of academic and professional recognition of engineering degrees is an important issue not only in the European Higher Education Area (Rauhvargers, 2008; Bologna, 2014) but also in the whole engineering academic community (Marginson, 2009; Allen, & Van der Velden, 2011; Chung, 2011; Teichler, 2013). In some countries like US (Froyd, Wankat, & Smith, 2012) this problem is solved and controlled at a satisfactory level. Unfortunately in the Bologna Area the problem of recognition is still under an interesting but long discussion and a general policy applied to all Bologna countries is not established yet while the legal framework for the recognition was elaborated by the Council of Europe and UNESCO. This obstacle directly affects negatively the international student mobility and employability of graduates in the Bologna community (Kassinopoulos, 2004).

This paper deals with the academic and professional recognition of engineering degrees in two European-Bologna countries Cyprus and Russia with different national qualification frameworks. Cyprus is a very small country E.U. member, with a high number of engineering degree holders, most of which have received their degrees abroad in foreign universities. Russia is the largest country in the world, non E.U. member with many high ranking universities offering high quality engineering courses. The great majority of engineering degree holders are graduates of Russian or Soviet Union Universities. Cyprus shows extremely high shares of outgoing mobile students (over 50%) as well as high levels of incoming mobile students (over 30%). It should be noted that Greece is the main country destination in the student mobility. In opposite, Russian higher educational system demonstrates low shares of incoming mobile students (in average, 5%) and outgoing mobile students (in average, 1%). Table 1 illustrates our case studies countries indicators. Therefore, it seems to be interesting to compare recognition systems in the two countries.

Table 1. The case study countries indicators

Indicator	Cyprus	Russia
Area	9 251 km^2	17 098 242 km^2
EU member	Yes	No
Bologna country	Yes	Yes
Human population	1,100,000	146,500,000
Human population density	120, 8/km2	8, 4/km2 (max 4626/km2, min 0.07/km2)
Student population	31,000	4,766,000
Official languages	Greek Turkish	Russian
Number of Universities in the TOP 300 Universities list	N/a	2
BSc duration	4 years	4 years

In this manuscript first it will be explained in detail the academic and professional recognition system in both countries, and then it will follow a comparison and discussion of the strengths and weaknesses of the two systems. The discussion will include as an example the methodology applied for the design of an Electrical Engineering course in a new University in Cyprus, which was based on the local requirements for Academic and Professional recognition. Then the methodology of Electrical Engineering curriculum design in Bauman Moscow State Technical University will be discussed. This will be followed by a comparative analysis and discussion of the two systems and finally the overall conclusions will be formulated.

BACKGROUND

One of the important issues presently under discussion among the Bologna stakeholders is the recognition of qualifications which is directly related to the European Qualification Framework for lifelong learning (EQF LLL) (The European Qualifications Framework 2014) and the Qualifications Framework in the European Higher Education Area (QF-EHEA) (Bologna, 2014). The EQF LLL and QF-EHEA serve as instruments in comparing the national qualification systems. Knowledge, skills and competences are the main criteria to indicate the learning outcomes in each qualification. The eight EQF levels cover all possible qualifications starting from lower studies to the most advanced PhD level. The QF-EHEA assists for the classification of academic qualifications and describes the three cycle systems (Kassinopoulos, 2007). The first cycle qualification QF-EHEA requires 180-240 ECTS credits and corresponds to the 6th level EQF LLL, the second cycle QF-EHEA corresponds to 90-120 ECTS credits and links with the 7th level EQF LLL. The third cycle corresponds to the 8th level EQF LLL. The Salzburg II Recommendations have been published in 2010 but no specific number or range of ECTS credits is defined for the 3rd cycle (EUA Council for Doctoral Education, 2014).

As a part of the European Commission-funded Lifelong Learning programme the European Area Recognition Manual is available online (The EAR Manual, 2014). More information about each EHEA country recognition system can be found via the ENIC Network (ENIC-NARIC, 2014).

It is noted that knowledge can be acquired with formal, non-formal and informal studies and as it has been seen qualifications can be classified at various levels. The recognition of non- formal and informal studies and the classification of these studies is a problem that is not solved yet in all Bologna countries (Sadler, 2013).

It is reminded that first cycle degrees BSc according to Bologna Process are awarded to successful graduates after 3 or 4 years studies associated to 180 or 240 ECTS credits respectively. In some countries like France and Greece this formula is not respected for engineering degrees and these are awarded after five years studies which are associated to level 7 EQF LLL (Master level-second cycle). Some countries like Portugal use also an integrated master curriculum. In this case an engineering master degree is granted after a 5-years study cycle.

In the paper we will discuss only the recognition system of engineering studies BSc qualifications at level 6 EQF LLL (first cycle degrees QF-EHEA) acquired after formal studies in two Bologna countries, Cyprus and Russia.

CYPRUS RECOGNITION SYSTEM

The first university in Cyprus was established in 1992, and before this date only certain colleges were offering higher studies at the technician engineering diploma level. As a result most of the Cypriot students were studying abroad and after their graduation they were returning home with engineering degrees from a variety of universities from different countries around the world. Most of them were studying in Greece, UK, U.S. and other European countries. Even after 1992 when more state and private universities were established, Cypriots continued to follow higher studies abroad, others by tradition for better quality degrees, others because they could not find place in the local universities and others for specialized studies not offered in Cyprus. This created a serious problem of recognition of these foreign degrees, especially for graduates wishing to work in the public administration.

In order to solve efficiently this problem, the local government established by law a governmental Council responsible for the academic recognition of local and foreign degrees, applying a strict and detailed recognition methodology which is briefly explained in the next paragraphs. The name of this Cypriot Council for the Academic Recognition of Study Degrees is KYSATS (Cypriot Council for the Academic Recognition of Study Titles 2014). The responsibility for the professional recognition of the local and foreign Engineering degrees was given by law to another professional body named ETEK (The Cyprus Scientific and Technical Chamber) (Technical Chamber of Cyprus, 2017). ETEK is a Public Law Body with an elected Governing Body and is the statutory Technical Advisor of the State and the umbrella organisation for all Cypriot Engineers. The Chamber is governed by a thirty-member council elected directly by ETEK members for a three years term of office.

Features of academic and professional recognition are discussed in more detail below

CYPRUS ACADEMIC RECOGNITION SYSTEM

The council of KYSATS is appointed by the council of Ministers and it consists of 7 members. One professor of the main local university (chairperson), a representative from the office of Attorney General, a representative of the Minister of Education and four other professors, one from a local university and three others from different universities and different countries. The council term is three years and it is renewable. The recognition by KYSATS takes one of the following two forms, a) Recognition of Equivalence and b) Recognition of Equivalence and Correspondence (Kassinopoulos, & Dodridge, 2004). Both forms are explained below.

Recognition of Equivalence

For this recognition it is examined if the structure and organisation of the foreign institution is Equivalent to a reference model of university. It is granted to a degree after examination of the degree provider institution and the academic program offered. More specifically the council examines the length of studies, the entry requirements, the evaluation methodology of students and the learning and teaching procedures. As reference for comparison, is taken the University of Cyprus or other local state Institution. It is required also that the whole program of study should have been accomplished in an accredited institution and in the case of transfer students a substantial part of the academic program has been completed at the institution which awarded the Degree.

The content of the academic program under evaluation is not examined in detail in the Recognition of Equivalence. It is examined only the level of knowledge and whether the learning outcomes reflect the level of the qualification of the degree. The skills and competences are examined in detail in the second step of the examination which is the recognition of Correspondence.

Recognition of Equivalence and Correspondence

This recognition is provided to a degree which has already satisfied the recognition of equivalence and is directly related to the examination and comparison of the content of the study program of the degree under examination. More specifically it is examined whether the study program under examination corresponds with the study program of a local state university, which is taken as reference. The recognition is granted to the applicant if in addition to the requirements for Equivalence, the program under examination includes at least the two thirds of the course taken as a criterion for comparison, including the core courses. In this examination it is also examined the level of the skills and competences including practical work and experience provided by the academic program of the foreign university as well as the correspondence of the Learning Outcomes (LO) with the local reference institutions program LO. In the case that the program under evaluation does not fully satisfy the recognition requirements the applicant may be required to attend and pass exams on an additional number of courses which do not exceed 10 semester subjects. Otherwise the application is rejected.

For the cases in which a corresponding program for comparison is not offered in the local state universities, the reference program is taken from a state university in Greece or from a recognized institution from another EU country.

It is noted that once the engineering degree of a specific university is approved, all engineering degrees originated from that university are immediately recognized. So KYSATS holds a list for all universities the degrees of which have been recognised so far. This is very helpful for prospective students wishing to study abroad and interested to find a suitable university which offers degrees recognized in Cyprus.

CYPRUS PROFESSIONAL RECOGNITION SYSTEM

The license to work as a professional Engineer in Cyprus is acquired by the membership at ETEK. The basic requirements, for a graduate to become member of ETEK are:

- To hold a university degree or other comparable qualification in any field of Engineering which permits him/her to practice the profession in the country in which it was obtained and
- Is recognized by the Chamber in accordance with the Law or the regulations pursuant to the ETEK Law.

The main criteria and the framework regarding the recognition by ETEK of an Engineering degree are given below.

The university which awarded the engineering degree is a recognized Institution in the local country and the degree is granted with a full academic recognition (Recognition of Equivalence and Correspondence) in Cyprus. The duration of study should be at least four years. In case that the applicant has done a part of his/her studies in a Higher Education Institution non university level, at least two years out of four should have be done in a university.

The engineering degree holder should have followed successfully at least 12 semester-subjects (or 6 year-subjects) of the basic branch of studies for which the application is done. The choice of the subjects to be taken is defining not only the recognition but also the specialization of the prospective engineer. More information and an example will be discussed in the next paragraphs.

There are also other criteria which refer to postgraduate studies and other special cases with equivalent degrees, but this paper is limited to first cycle degrees only.

It is noted also that in the case of local engineering degree program recognition, ETEK besides the approval or rejection of a program, gives advises (not requirements) regarding the introduction of other subjects like Economics and Management and other comments regarding the proposed subjects.

CASE STUDY: DESIGN OF AN ELECTRICAL AND ELECTRONIC ENGINEERING PROGRAM IN THE CYPRUS UNIVERSITY OF TECHNOLOGY

In this paragraph the case of the design of an Electrical/Electronic Engineering program in the new Institution Cyprus University of Technology will be discussed. Graduates can acquire a degree with professional recognition in the fields of Electrical Engineering, or Electronic Engineering or both.

The ETEK requirement as it is explained below in more detail, states that the professional recognition with the specialisation of say Electrical Engineer is provided to a graduate of an Engineering school who has followed successfully 12 semester subjects with the following combination: 4 semester subjects of the branch of Electrical Engineering (column A in Table 2), 4 semester subjects of any branch of Electrical or Electronic engineering (columns A or B in Table 1) and finally 4 semester subjects common to Electrical and Electronic engineering braches (columns A or C in Table 2). It is noted that column C includes subjects which are considered common to Electrical and Electronic engineering. To clarify this requirement ETEK provides a list of subjects of these 3 groups. An indicative list is provided in Table 2. An analogous requirement exists for the recognition of the specialisation of Electronic Engineering.

The study program was first designed in a way to respect the academic recognition requirements and the Bologna process principles. Later a pool of elective subjects was defined in a way to give the possibility to students to choose the combination of subjects required by ETEK for the specialisation(s) they wanted to follow. It is obvious that membership to ETEK which is equivalent to the professional recognition in one of the two specializations Electrical or Electronics Engineering (or both) is granted to the graduate who has taken an appropriate combination of the subjects offered by the program.

RUSSIAN RECOGNITION SYSTEM

The first Russian university was established in 1724. The first technical Russian University (now National Research University Bauman Moscow State Technical University (BMSTU)) was established in 1830. In 1915 the department of Electrical Engineering of BMSTU has been founded. The first private University in Russia was established in 1991. Now 18 Russian Universities are in the QS World University Rankings list.

Table 2. Subject classification list of ETEK

N0	Course	A- Electrical	B- Electronic	C - Both
1	Circuit / Circuit Analysis			X
2	Electronics			X
3	Electromagnetic fields and Waves			X
4	Control Systems			X
5	Numerical Analysis			X
6	Digital Logic		X	
7	Tele/Communications		X	
8	Introduction to Photonics		X	
9	Digital Signal Processing		X	
10	Power Electronics		X	
11	Advanced Electronics		X	
12	Power Systems Analysis	X		
13	Electrical Machines /Drives	X		
14	Electrical Power	X		
15	Generation/Transmission	X		
16	Electrical Services	X		
17	Power Electronics Renewable Energy Systems	X		

As a result of this long tradition in engineering education, most engineering degree holders are graduates of Russian Universities or former Soviet Union Universities. A distinguishing feature of the first cycle qualification is that all BSc curricula in Russia allow the continuation of education at the next cycle. Mainly BSc diploma holders continue their education and acquire MSc and Dr diplomas that correspond to 7th and 8th level EQF LLL. The first cycle qualification BSc in Russia corresponds to 240 ECTS credits (4 years) and second cycle qualification MSc to 120 ECTS credits (2 years). It should be noted that exams between cycle qualifications are mandatory that allows exploiting a soft recognition of diplomas issued by different Russian Universities. During last decade due to more active cooperation between Russian and foreign Universities an increase in mobility for study purposes has been achieved and some Russian citizens graduated Universities abroad and returned to Russia. This created a problem of recognition of their foreign degrees. Now the Federal Centre Glavexpertcenter (Glavexpertcenter 2014) that was established in 1997 is the sole entity responsible for recognition of diplomas and degrees that were obtained from Universities abroad. Glavexpertcenter performs recognition of equivalence and correspondence only.

RUSSIAN ACADEMIC RECOGNITION SYSTEM

Academic recognition in Russia is discussed in more detail below. Foreign engineer degree holders can be classified as follows. The first group of engineering degree holders hold dual diplomas from a Russian University and from a foreign University that is a partner University of the correspondent Russian University. This is possible thanks to specially designed curricula that allow receiving dual diplomas. An academic and professional recognition for this group is similar to recognition of Russian University diplomas and degrees.

The second group of engineer degree holders hold diploma from high ranking foreign Universities that have been included in one of the academic rating systems: Academic Ranking of World Universities, QS World University Ranking, The Times Higher Education World University Ranking among 300 Top Universities (List TOP 300). From 2012, all engineering degrees granted by these Universities are immediately recognized according to the Federal Law. The duration of studies for these top universities is not examined.

There are bilateral agreements between Russian Federation and some countries about mutual recognition of diplomas and degrees. These foreign diplomas are recognised automatically according to these agreements.

The rest of engineer degree holders have to apply to Glavexpertcenter to recognize their diplomas or degrees. During the recognition process experts compare the curriculum of a foreign university with a corresponding curriculum according to Federal Standards for Higher Education issued by Ministry of Education and Science of Russian Federation. If an application is unsuccessful the applicant may be requested to pass additional exams. Each application is examined independently and it is not generated a list of universities the degrees of which have been recognised.

RUSSIAN PROFESSIONAL RECOGNITION SYSTEM

Initial professional recognition in Russia requires only a Russian and /or foreign diploma with academic recognition. Only some engineering specializations require an additional certificate, for example, for high voltage construction etc. However, these certificates are not issued by Russian Universities. Both Russian graduates and foreign graduates have to apply to special training centres to receive these additional certificates.

It is mentioned that additional tests can be required for some positions, e.g. in the public administration. However, these tests are in law subjects and not in engineering.

Further steps in professional engineer recognition are possible. However, they vary significantly from an employer and may not be recognized by another employer.

It should be noted that sometimes BSc degrees are not recognized by an employer as an engineering diploma. It hinders a professional recognition for the first cycle qualification and forces BSc diploma holders to continue their education to reach MSc level.

Finally, the professional recognition system in Russia is under progress and some more actions are required to harmonize this area.

CASE STUDY: DESIGN OF AN ELECTRICAL ENGINEERING PROGRAM IN BAUMAN MOSCOW STATE TECHNICAL UNIVERSITY

The curricula design for Electrical Engineering degree by the Department of Design and Technology of Electronic Devices, BMSTU is discussed in this part of the paper. Usually curricula content, competences, skills, knowledge and learning outcomes are dictated by the Federal Standards for Higher Education issued by Ministry of Education and Science of Russian Federation. However, 38 Russian Universities including Lomonosov State University, Saint Peterburg State University, National Research Universities and Federal Universities can design their own curricula. Nevertheless, they have to correspond to the Federal Standards for Higher Education.

According to the Federal Standards the structure of the curricula contains three main parts (Shakhnov, Vlasov, & Zinchenko, 2012). The first part includes human and social courses that are either mandatory such as Foreign Language, History, Economics etc. or optional courses such as Russian Language, Political Science etc. This part is designed to enhance the global competence (Rajala, 2012). The second part contains courses that are general for engineering education (Rajala, 2012). It includes Mathematics, Physics, Chemistry, Informatics etc. The third part covers professional education in the field of Electrical Engineering. The courses such as Circuit Theory, Circuit Design, Technology, Material Science etc. provide required professional skills, competence and knowledge. An important part of engineering education in the mentioned above curricula is Practical Training (14 ECTS credits) that is planned during 2 months after the third year. It should be noted that this Practical Training course introduces students to real engineering tasks and provides a smooth transfer from University to professional activity. It is in line with the 3D global engineer approach (Chang, Atkinson, & Hirleman, 2009; Rajala, 2012) and provides training in professional competence. The designed curricula are linked with ECTS credits that simplify their recognition at the international level as 240 ECTS credits BSc in EE curricula.

A COMPARATIVE ANALYSIS OF ACADEMIC AND PROFESSIONAL RECOGNITION IN CYPRUS AND RUSSIA

The recognition systems in Cyprus and Russia explained previously represent very close similarities regarding the academic recognition and partly a different approach regarding the professional recognition. A comparison and further discussion of the two systems are given below.

ACADEMIC RECOGNITION COMPARISON

The specific requirements for the academic recognition of local and foreign degrees are directly related to the quality of the Institutions offering the degrees and to the correspondence of the content of the program of study compared with a reference model. In Cyprus, the main requirements are the evaluation of the quality of the Institution offering the degree under examination and the comparison of the academic program with the program of a main local university taken as reference. In Russia the academic recognition is directly granted to institutions included in the List TOP 300 ranked universities and for the others it is given after evaluation by the Federal Centre Glavexpertcenter. The evaluation is based on the reference program of Federal Standards for Higher Education. This evaluation has been proved in general very efficient and no major academic or practical problems have been met so far.

In Cyprus engineering programs at BSc level with 3 years studies are not recognized as a degree level. This is an important problem in practice because in order to work in the Cyprus administration or in public education a 4-year recognized degree is required.

On the contrary, in Russia a 3-years degree is recognized if this diploma is granted by a University from the List TOP 300 mentioned above. However, some European Universities, that are not in the List TOP 300, propose engineering degrees with 3 years curricula as well. In this case a diploma holder has to apply to GlavexpertCenter. In general the mentioned problem can be overcome by mutual exchange of students between Russian and European Universities and dual diploma curricula design.

Table 3 summarises similarities and differences between BSc recognition in Cyprus and Russia.

PROFESSIONAL RECOGNITION COMPARISON

The professional recognition of an engineering degree in Cyprus is provided by an official Professional Body (ETEK) and is granted to engineering degree holders if the following three main requirements are satisfied:

- Professional Recognition in the country offering the engineering degree (for foreign degrees).
- Academic full Recognition in Cyprus.
- Suitable combination of subjects in the academic program of the degree related to the specialisation required.

In Russia there is no an official Professional body or Organisation to grant the Professional recognition. The main requirement for the professional recognition is the academic recognition provided officially and for some special cases additional certificates. Even though, certain employers have the possibility to ask for further qualifications, like MBA, in order to provide professional licence to an engineering degree holder.

Table 3. Summary of similarities and differences

Bologna Country	BSc-3 year programs, 180 ECTS	BSc-4 year program, 240 ECTS
Cyprus	No Recognition	Recognition after evaluation
Russia	TOP 300 universities: Recognition Non TOP 300 universities: Recognition after Evaluation	TOP 300 universities: Recognition Non TOP 300 universities: Recognition after Evaluation

Comparing the two systems we note that in the Cyprus system there is a detailed examination of the content of the programs of the degrees under evaluation and this ensures a high level standard of the engineering profession. However it can be noted that in the list of the combination of subjects required for recognition, no laboratory sessions or final projects are included. As it was explained by the chamber evaluation committee, ETEK assumes that the presence of these very important parameters for engineering programs is assured by the fact that the program has already acquired academic recognition.

The professional recognition in Russia is more flexible and mainly is defined by employers. However the fact that the academic recognition is based on a reference model of the Federal Standards for Higher Education, ensures that the level of the engineering profession is high. This can be deduced also by the fact that on the contrary to the Cyprus approach, laboratory sessions, course projects and final project are mandatory components in curricula design in Russia. The only weakness that is noted is the absence of an official professional body like ETEK in Cyprus, IEE and others in the UK and other professional bodies abroad which could control better the standards and quality of the engineering profession in the country.

CONCLUSION

The Bologna process has made studies in Europe more attractive for prospective students (Fino Solovyev, & Zinchenko, 2013) and it created new opportunities for international mobility of students, academic staff and professionals. Among the obstacles of mobility, that the Bologna policy makers make an effort to remove, is the problem of academic and professional recognition. This paper examined and compared the recognition system of engineering degrees in two Bologna countries with different characteristics Russia and Cyprus and reached at certain conclusions which are given below.

Both recognition systems follow the same two general principles which are the examination of the quality of the institution providing the engineering degree and the standard of the content of the academic program offered. This evaluation is applied in a different way in the two countries which satisfies the local requirements and the country's academic tradition.

Another point which is discussed is the duration of studies of engineering degrees. It is noted that in Russia, engineering BSc degrees of 3 years duration can be recognized something which is not possible in Cyprus.

As the final point we may state that for 4-year BSc engineering degree programs of studies done in Russia, academic and professional recognition by KYSATS and ETEK respectively is required. Similarly as Cyprus Universities are not part of the list TOP 300 universities, engineering degrees taken at Cypriot universities should be evaluated by the Russian Federal Centre Glavexpertcenter for recognition.

ACKNOWLEDGMENT

Lyudmila Zinchenko acknowledges the support by RFBR (grant RFBR #16-06-00404).

REFERENCES

Allen, J., & Van der Velden, R. (Eds.). (2011). *The flexible professional in the knowledge society*. Dordrecht, The Netherlands: Springer; doi:10.1007/978-94-007-1353-6

Bologna Process - European Higher Education Area. (2017). Retrieved April 12, 2017 from http://www.ehea.info

Chang, Y., Atkinson, D., & Hirleman, E. D. (2009). International research and engineering education: impacts and best practices. *Online J. Global Eng. Educ.*, *4*(2), Article 1.

Chung, C. (2011). Changing engineering curriculum in the globalizing world. *New Horizons in Education*, *59*(3).

Cypriot Council for the Academic Recognition of Study Titles. (2017). Retrieved April 14, 2017 from http://www.kysats.ac.cy

ENIC-NARIC. (2017). Retrieved April 5, 2017 from http://www.enic-naric.net

EUA Council for Doctoral Education. (2017). Retrieved April 13, 2017 from http://www.eua.be/cde

Fino, H., Solovyev, V., & Zinchenko, L. (2013). Challenges for students mobility between European and Russian Universities. *Proc. ICL*. 10.1109/ICL.2013.6644629

Froyd, J. E., Wankat, P. C., & Smith, K. A. (2012). Five major shifts in 100 years of engineering education. *Proc. IEEE*, 100. 10.1109/JPROC.2012.2190167

Glavexpertcenter. (2017). Retrieved April 5, 2017 from http://nic.gov.ru

Kassinopoulos, M. (2004). Student and staff mobility in EHEA during the first years of Bologna Process. ICEE 04.

Kassinopoulos, M. (2007). The three cycle structure in European Higher Education Institutions. *Proc. ICEE 07*.

Kassinopoulos, M., & Dodridge, M. (2004). Academic and professional recognition of UK degrees in Cyprus. *Proc. ICEE 04*.

Marginson, S. (2009). Global perspectives and strategies of Asia-Pacific research universities. *Evaluation in Higher Education*, *3*(2), 1–43.

Rajala, S. (2012). Beyond 2020: preparing engineers for the Future. *Proc. IEEE*, 100. 10.1109/JPROC.2012.2190169

Rauhvargers, A. (2008). Implementation of the Lisbon Recognition Convention in the countries participating in the Bologna Process. In A. Rauhvargers & S. Bergan (Eds.), New Challenges in Recognition, Council of Europe Higher Education Series No. 10 (pp. 13–27). Academic Press.

Sadler, D. R. (2013). The futility of attempting to codify academic achievement standards. *Higher Education*, *7*. doi:0.100710734-013-9649-1

Shakhnov, V., Vlasov, A., & Zinchenko, L. (2012). O metodicheskom obespechenii ingenernogo obrazovaniya v sovremennuh usloviyah [Features of Engineering Education]. Vysshee obrazovanie - Higher Education in Russia, 3, 104-108.

Technical Chamber of Cyprus. (2014). Retrieved April 18, 2017 from http://www.etek.org.cy

Teichler, U. (2013). Opportunities and problems of comparative higher education research: The daily life of research. *Higher Education, 11*. doi:10.100710734-013-9682-0

The EAR Manual. (2017). Retrieved April 7, 2017 from http://www.eurorecognition.eu/emanual

The European Qualifications Framework. (2017). Retrieved April 9, 2017 from http://ec.europa.eu/ploteus

ADDITIONAL READING

Aleksandrov, A. A., Neusipin, K. A., Proletarsky, A. V., & Fang, K. (2012). Innovation development trends of modern management systems of educational organizations, In Proc. *2012 International Conference on Information Management, Innovation Management and Industrial Engineering* (pp.187-189).

Clark, R., & Andrews, J. (2010). Researching Primary Engineering Education: An Exploratory Study from a UK Perspective. *European Journal of Engineering Education, 35*(3), 585–595. doi:10.1080/03043797.2010.497551

Gavrilina, E. A., Zakharov, M. A., Karpenko, A. P., Smirnova, E. V., & Sokolov, A. P. (2017). Software System META-3 for Quantitative Evaluation of Student's Meta-competencies on the Basis of Analysis of his or her Behavior in Social Networking Services. *Procedia Computer Science, 103*, 432–438. doi:10.1016/j.procs.2017.01.012

Gray, P. J., Patil, A., & Codner, G. (2009). The Background of Quality Assurance in Higher Education and Engineering Education. In P. Gray (Ed.), *A. Patil A., P* (pp. 3–25). Engineering Education Quality Assurance. doi:10.1007/978-1-4419-0555-0_1

Kassinopoulos, M., & Zinchenko, L. (2014). A Comparison of Quality Assurance Systems in Bologna Countries for Engineering Education. A Cyprus and Russia Case Study. In M. H. Bakr & A. Elsharabasy (Eds.), *iCEER2014-McMaster conference digest* (pp. 194–197). McMaster.

Shakhnov, V. A., Zinchenko, L. A., Rezchikova, E. V., & Glushko, A. A. (2015). An opportunity in engineering education: Russian BYOD tendencies: BMSTU case study. In Proc. *2015 International Conference on Interactive Collaborative Learning (ICL)* (pp. 299-304).

Shakhnov, V. A., Zinchenko, L. A., Rezchikova, E. V., & Verstov, V. A. (2017). Distinctions of a learning content for education in the field of nanotechnology. *International Journal of Nanotechnology, 14*(7-8), 690–697. doi:10.1504/IJNT.2017.083443

Willmot, P., & Smirnova, E. V. (2014). East-West cooperation for the enhancement of teaching and learning engineering. In Proc. *Joint International Conference on Engineering Education and the International Conference on Information Technology, ICEE/ICIT -2014*,(pp. 500-507).

KEY TERMS AND DEFINITIONS

Academic Recognition: A recognition of periods of study or qualifications issued by an educational institution.

Bologna Country: A signatory of the Bologna Accord.

Bologna Process: A voluntary higher education reform process with the aim of making higher education systems compliant and enhancing their international visibility.

Curriculum: A curriculum is a planned sequence of learning experiences.

Curriculum Design: Design includes consideration of aims, intended learning outcomes, syllabus, learning and teaching methods, and assessment.

Engineering Degree: Academic degree in engineering.

Professional Recognition: A recognition of an individual's professional status.

Chapter 7
Changes in the Engineering Competence Requirements in Educational Standards

Aleksandr Vasilyevich Berestov
Moscow Engineering Physics Institute, Russia

Gennady Konstantinovich Baryshev
Moscow Engineering Physics Institute, Russia

Aleksandr Pavlovich Biryukov
Moscow Engineering Physics Institute, Russia

Ilya Igorevich Rodko
Moscow Engineering Physics Institute, Russia

ABSTRACT

This chapter presents prognostic analysis results concerning the changes in the engineering competence requirements. It is noted that professional competences of future experts in this field are undergoing certain changes related to the need for operating complex systems and working in a team in uncertain contexts in order to support and ensure good management throughout the entire high-tech systems life-cycle. It has been established that certain technological areas of the National Technology Initiative (NTI), which is being implemented now, are not provided with the educational training programs by the adopted Federal National Educational Standards (FNES). This chapter also focuses on the role of Worldwide CDIO Initiative international engineering standards of education in the development of new engineering competence assessment tools to enhance the national system of educational standards and includes National Research Nuclear University MEPhI's own educational standards in higher education as an example.

DOI: 10.4018/978-1-5225-3395-5.ch007

INTRODUCTION

The research on the changes concerning professional engineering competence requirements is directly connected to the vision of the high-tech industry future, which will demand relevant competences for development, analysis of development strategies, foresights, road maps exploring promising areas of technological development, research focus and innovation implementation. For its implementation, this vision, shaped by the expert community, should be provided with the relevant methodological standards on the one hand and with enhanced staff training and the creation of a relevant system to form qualifications and competences on the other hand (IBIS, 2014).

First of all, it should be mentioned that the very term 'engineering' is understood quite broadly starting from its general meaning as a perfect synonym to the creation of promising and competitive engineering systems to the narrow and specific interpretation as 'business-consulting', i. e. monitoring projects which create engineering systems in order to raise their saleability and value. This makes the issues related to defining engineering and industrial design professional competence requirements multidimensional and multifold (Osmakov & Pastukhov, 2015).

BACKGROUND

The Financial Terms Dictionary (Financial Terms Dictionary, 2017) gives the following definition of engineering: engineering is a field of activities concerning the issues of the creation of industrial sites, infrastructure etc., primarily in the form of rendering of commercial engineering and consulting services.

In terms of pre-production, according to Construction Guidelines 80-12.2000: Methodological Recommendations on Working Out of the Investor's (Customer's) Conditions (Requirements) during the Preparation for Contract Tendering, engineering means engineering and consulting services related to pre-production and ensuring of normal production and sales (Construction Guidelines, 2000).

Several normative documents suggest the following definitions of engineering and related terms:

- **According to GOST R (Russian National Standard) 54147-2010:** Strategic and Innovation Management. Terms and Definitions - 3.1.14 Engineering: Research, design-and-engineering, computational and analytical activities, preparation of feasibility studies, working out organizational recommendations (GOST R 54147-2010, 2010);
- **According to GOST R (Russian National Standard) 54147-2010:** Strategic and Innovation Management. Terms and Definitions - 3.1.15 Innovation Engineering: the totality of works and services aimed at creating an innovation project including the conception, implementation, promotion and diffusion of innovations;
- **According to GOST R (Russian National Standard) ISO 15704-2008:** Industrial Automated Systems. Standard Architecture and Enterprise Methodology Requirements - 3.7. Enterprise Engineering: discipline used for any works related to the creation, change or reorganization of any enterprise (GOST R ISO 15704-2008, 2008);

- **According to GOST R (Russian National Standard) ISO 19439-2008:** Enterprise Integration. Enterprise Modelling Basis - 3.21 Enterprise Engineering: discipline used to complete any tasks related to the creation, modification or reorganization of any enterprise (GOST R ISO 19439-2008, 2008);
- **According to GOST R (Russian National Standard) 54136-2010:** Industrial Automation Systems and Integration. Guidelines for Standards Implementation, Structure and Glossary: - 3.31 Enterprise Engineering: discipline used to complete any tasks related to the creation, modification or reorganization of any enterprise (GOST R 541-2010, 2010).

Thus, "engineering" can be interpreted as a tool for efficiently selling a product in the market which can include, in some cases, the promotion of the product in the market as well as its development. At the same time, it is during the development stage that the implementation of innovative technology should be elaborated for all the stages of its life-cycle, which is a prerequisite for the creation of a competitive product. In this regard, engineering can be considered as a major, but not single, sufficient condition for the final cost efficiency of its sales in the market. That is why the high level of professional training of design engineers should be combined with a number of skills inherent to experts of engineering companies.

The systematic approach to the development of future skills and high-tech expert training should rely on the improvement of skills related to planning, coordination of mutual and meta-competences, interindustry and interproject mobility and ability to put into practice the life-cycle concept in real production. It is planned to encourage experts to translate their own professional competences into other industries and markets along with the development and acquisition of new competences.

The Russian Ministry of Economic Development and the Ministry of Industry and Trade introduced foresight-research to analyse the strategic vision of high-tech industries development, including the key skills in engineering and industrial design (Strategy Partners Group & Minpromtorg RF, 2015).

In these types of research, the Ministry of Industry and Trade and Rosstat (Federal State Statistics Service) define engineering as rendering of engineering and consulting services related to pre-production, ensuring production and sales of the products, servicing of construction and operation of industry, infrastructure and other sites, based on a contract with the customer.

The core of the engineering activity is formed by the fulfillment of engineering tasks related to the design of products, production system, pre-production and maintenance of production, construction and sites operation. Engineering covers the complete life-cycle of an industrial product - from the conception, test model development, commissioning, including its technical equipment, to marketing, supply to the market, after-sales service and subsequent utilization.

The specifics of engineering are in a project-based approach.

Current global trends put forward new demands for the skills of engineering experts. Multidisciplinary thinking, knowledge of new technologies, command of foreign languages and development of strategic thinking are mentioned among new engineering competences requirements.

ANALYSIS OF ENGINEERING SKILLS REQUIREMENTS IN THE IMPLEMENTATION OF THE NATIONAL TECHNOLOGY INITIATIVE

The National Technology Initiative (NTI) is a programme aimed at shaping fundamentally new markets and creating conditions for the global technology leadership of Russia by 2035. The NTI includes integrated solutions to define the key technologies, necessary changes in norms and rules, efficient financial and personnel development measures, involvement and encouragement mechanisms for experts with the necessary skills. The choice of technologies is based on major global trends, with network technologies being the priority. Such technologies are people-focused, as people are the end consumers (ASI, 2017).

The basis for the current systematic work is the NTI matrix which comprises new markets, new technologies, institutes and infrastructure/resources. The NTI areas include a group of the markets of the future:

- EnergyNet (distributed power generation from personal power to smart grid and smart city) (ASI, 2016);
- FoodNet (system of personal production and food and water delivery);
- SafeNet (new personal security systems);
- HealthNet (personal medicine);
- AeroNet (distributed systems of unmanned aerial vehicles);
- MariNet (distributed systems of unmanned maritime transport);
- AutoNet (distributed network of unmanned management of road vehicles);
- FinNet (decentralized financial systems and currencies);
- NeuroNet (distributed artificial elements of consciousness and mentality).

The NTI roadmaps through 2018 shape the ecosystem of future markets. It is important to mention that not all NTI roadmaps have been approved at this moment. As for the implementation of approved roadmaps (NeuroNet and others), this has recently started. Working groups have not entirely formulated all the skills requirements which are necessary to put the roadmaps projects into practice. Table 1 shows a version of engineering skills requirements from the point of view of the NTI working groups participants.

We have analysed the ways to improve the national system of federal state higher educational standards in engineering. For this end, engineering higher education training programmes, including bachelor, master and specialist degrees (in groups of specialization 09.00.00, 10.00.00, 11.00.00, 12.00.00, 13.00.00, 14.00.00., 15.00.00, 16.00.00, 17.00.00, 18.00.00, 19.00.00, 20.00.00, 21.00.00, 22.00.00, 23.00.00, 24.00.00), were compared to the list of higher education specializations and training programmes relevant to priority areas of the modernization and technological development of the Russian economy approved by the Russian Governmental Decree #7-p of 6 January 2015 (Russian Government Decree, 2015), as well as with the list of "breakthrough technologies" defined in the National Technology Initiative (NTI).

The results of the analysis show that not all of the NTI technological areas are provided with educational programmes within the framework of engineering Federal National Educational Standards (FNES). For example, as for the following technological areas:

Table 1. Example of future necessary engineering skills from the point of view of the NTI working groups participants

Market NTI	Engineering Skills	Year 2020, Number of People	Year 2035, Number of People
AutoNet	Development of mathematical models and OS for UAS	1 800	25 500
	Transport management systems development (logistics, Big Data etc.)	1 600	29 000
	Development and improvement of applications, controllers and UTV sensor system	1 200	14 000
	Communication and information transmission system development	400	8 000
	UTV design and construction development and modernization	500	9 000
AeroNet	UAS electronic components development and design and their integration with mechanical components	3 400	11 400
	UAS engines prototyping, design and development	2 400	8 400
	UAS OS and applications development	2 500	8 500
EnergyNet	Power grid information platform and applications development	2 200	17 000
	Mathematical models development	500	1 900
	Energy carrier design development	200	3 800
	Power physical components and sensors design	2 000	15 000
NeuroNet	Big data analysis	1 300	6 200
	Cognitive technologies	2 600	12 300
	Built-in biocybernetics systems engineering	2 600	12 300
	Humanoids and UAVs design	6 600	30 800

- Sensory;
- Mechabiotronics;
- Bionics;
- Genomics and synthetic biology;
- Neurotechnologies,

they cannot practically be compared to the ensured Federal National Educational Standards in the higher education in engineering. Consequently, it is in these areas that the federal national educational standards in the higher education of engineering can be improved.

APPROACHES TO THE DEVELOPMENT OF NEW ENGINEERING SKILLS IN COMPLIANCE WITH INTERNATIONAL ENGINEERING EDUCATIONAL STANDARDS

One of the requirements for creating a competitive product is a deep understanding of the methodology and principles of the product life-cycle concept, which covers a class of homogenous products, some of specimen (samples) of which are at different stages of its life-cycle.

In this regard, the application of the approaches and provisions of Worldwide CDIO Initiative (Conceive - Design - Implement - Operate) is crucial for engineering education programmes. The approach suggested in CDIO is aimed at strengthening the practical skills of future engineers as well as introducing the system of practice-focus and project-based teaching (Crawley, 2001).

The CDIO Initiative suggests organizing the learning process in engineering programmes so that graduates may demonstrate:

- Deep theoretical and practical knowledge of the technical foundations of their engineering profession;
- Ability to create and operate new products, processes and systems in demand;
- Understanding of the importance and strategic role of scientific and technological development of society.

In general this, approach indeed provides for an integrated mastering of engineering.

The application of various evaluation methods of engineering skills provides for a wide range of learning techniques and enhances the reliability and adequacy of evaluation data. An example of requirements application to the basic higher education training programmes, made in compliance with international standards aimed at necessary engineering competences implementation, is the creation of an organisation's own educational standards surpassing the level of FNES 3+ generation.

A number of National Research Nuclear University MEPhI educational standards are based on Article 2, paragraph 7 and Article 11, paragraph 10 of Federal law "On Education in the Russian Federation" No. 273 - FZ of 29 December 2012. They are also compliant with the international engineering education requirements of Worldwide CDIO Initiative and the best practices of Russian and foreign universities, Bologna Declaration fundamental provisions, public and occupational requirements, including the requirements of the international accreditation of educational programmes (FEANI and others), GOST (Russian national standard) ISO 9001-2011 and professional industry standards requirements as well as employers' requirements.

The specialized core higher education training programmes are modernized in MEPhI so that they comply with international Worldwide CDIO Initiative standards, which ensures the flexibility and individualization of education with the use of new educational technologies. This modernization is aimed at improving the skilled professional engineering training on the basis of practice-focused and project-based teaching methods which rely on interdisciplinarity and multidisciplinarity principles as well as on fundamental physics and mathematics education. Group project work during interdisciplinary development and projects for a real customer are recognized as some of the most efficient forms of practice-focused education.

The focal point is the problem of synthesis of educational and innovation process in the contemporary research university. An important part of transformation of a contemporary research university – for example, National Research Nuclear University MEPhI (Moscow, Russia) with the aid of the 5 top 100 competitiveness program is increasing the quality of engineering education (mostly with the methodological support of international standards, such as CDIO), and provide synthesis of educational and innovation process. The last one is also crucial for successful implementation of CDIO.

As a result of the pilot project of implementation CDIO standards in MEPhI for nuclear engineering education, new principles of education were developed and approbated (Baryshev, Berestov & Konashenkova, 2016). Some of them concerned application of new digital instruments for student's obtaining practical skills in analogue electronics – important to design contemporary information measuring systems for nuclear technologies.

An approach to teaching the disciplines related to analogue electronics and the elaboration of analogue devices is changing now. Today not only the knowledge of the theory and several basic principles is significant, but also the knack of applying them and getting a high-quality and competitive product (Bozhko, Baryshev, Maksimkin, Voronin & Kondratyeva, 2016). The essential part of staff training is practice, when a student perceives the gained theoretical knowledge firsthand through the training tasks, gets and polishes up the working skills. In our case these skills are in development, engineering and construction of electronic devices.

It is necessary to provide students with corresponding infrastructure, tools and methodology for developing their practical skills, in order to solve the aforementioned tasks. The laboratory facilities for analogue electronics have been constructed in the same way as the collection of laboratory practice has been created with special methodological guidelines. The Laboratory Practical Manual as a basic instrument for students to get-to-know with new instruments required special attention due to the peculiarities of the standards for engineering education being implemented.

The outcome of application of the developed Analogue Electronics laboratory module resulted in increased effectiveness of educational process. Students that take nuclear educational engineering programs have a possibility to obtain practical skills in analogue electronics in a better and faster way. They have contemporary hardware, universal virtual devices and proper methodological support to grow into qualified specialists that can successfully solve the task of development of contemporary information measuring systems for nuclear industry with the application of obtained skills and competences.

And this is the focal point where educational and innovation process in a contemporary research university come to a synthesis – because practical tasks of development of information systems for quality and safety control of operation of nuclear reactors and facilities are typical tasks for students graduate works for several educational programs in MEPhI. The universality of the laboratory module developed allows students not only to perform training tasks to obtain and develop practical skills in the area of analogue electronics – but also to solve real practical hi-tech industrial engineering tasks with the aid of the same methodological, hardware and software digital instruments.

There are advances in other Russian leading universities concerning development of new engineering skills in compliance with international engineering educational standards, such as the CDIO standards. For example, in Tomsk Polytechnic University there is a project of modernization of bachelor programs in engineering in accordance with international standards of engineering education (Boev, 2014). The project is aimed at elaboration of guidelines for development of undergraduate learning outcomes and curricula based on best practices and international standards of engineering education for SKOLKOVO priority areas (IT, Space, Energy Efficiency, Nuclear Technology and Biomed) defined by the Federal Law n.244-FL, September 28, 2011. The project has been started with the International Workshop in Boston (USA). The main idea of the Workshop was to identify and compare possible approaches to undergraduate engineering education development considering modern trends in science and technology and current industry challenges. Studying and analysis of requirements of international standards to engineering graduates' attributes resulted in recommending the ABET criteria to be used as a framework of learning outcomes for curricula under development. CDIO Syllabus and Standards are chosen to be used as guidelines for curriculum design and implementation by project partners.

Another case is from NUST MISiS: the Modernizing English Language Teaching project (Frumina, 2013). The project aim is to contribute to the development of NUST MISiS as an international research university by introducing a new model of English language learning for Bachelor students. The success factors of the project implementation are:

- Strategic partnership with Cambridge University;
- Attention and support from the university administration (sufficient resourcing and long term sustainability);
- New model of program management;
- Support on the part of the students (the role of regular surveys and meetings);
- The scope of the project and the type of innovations to attract young motivated teachers seeking professional development.

In Moscow Aviation Institute CDIO is implemented for new educational programs in the faculties of aviation technologies, robotics and intellectual systems and others (Mirzoyan, 2013).

CONCLUSION

Thus, the prognostic analysis results allow us to make the following conclusion.

The development of engineering skills is directly connected to the need for support and management throughout the entire high-tech product life-cycle. Professional skills of future engineering experts are undergoing certain changes related to the necessity of operating complex systems and working in a team in uncertain contexts in order to support and ensure the management throughout the entire life-cycle of high-tech systems.

The development of requirements for the engineering and industrial design of professional skills to implement the NTI road maps projects is now at the initial stage. Meanwhile, working groups of promising markets have a certain concept of the changes in professional skills of such experts taking into account the key importance of these skills for putting into practice the promising projects.

Some NTI technological areas are not provided with educational programmes by the adopted engineering FNES.

A promising way to successfully complete the set tasks is to primarily improve Russian educational standards in the leading universities, taking into account CDIO Initiative principles and approaches.

REFERENCES

ASI. (2017). *Natsionalnaya tehnologicheskaya initsiativa* [National Technology Initiative]. Retrieved from: http://asi.ru/nti/

ASI. (2016, January 27). *National Technology Initiative. EnergyNet Market Development Road Map.* Prepared for discussion at the Government Expert Council meeting. Moscow, Russia: Agency for Strategic Initiatives, RVC.

Baryshev, G. K., Berestov, A. V., & Konashenkova, N. A. (2016). The Application of Online Team Project Training in Nuclear Engineering Education. In A. Chugunov, R. Bolgov, Y. Kabanov, G. Kampis, & M. Wimmer (Eds.), *Digital Transformation and Global Society. DTGS 2016. Communications in Computer and Information Science* (Vol. 674, pp. 370–379). Cham: Springer; doi:10.1007/978-3-319-49700-6_35

Boev, O. (2014). *BEng Project: Outputs and Further Development.* Retrieved from MyBookLibrary Web site: http://www.mybooklibrary.com/beng-project-outputs-and-further-development.html

Bozhko, Y. V., Maksimkin, A. I., Baryshev, G. K., Voronin, A. I., & Kondratyeva, A. S. (2016). Digital Transformation as the Key to Synthesis of Educational and Innovation Process in the Research University. In A. Chugunov, R. Bolgov, Y. Kabanov, G. Kampis, & M. Wimmer (Eds.), *Digital Transformation and Global Society. DTGS 2016. Communications in Computer and Information Science* (Vol. 674, pp. 386–391). Cham: Springer; doi:10.1007/978-3-319-49700-6_37

Construction Guidelines. (2000). *MDS 80-12.2000: Metodicheskie rekomendacii po razrabotke uslovij (trebovanij) investora (zakazchika) pri podgotovke podryadnyh torgov* [Construction Guidelines 80-12.2000: Methodological Recommendations on Working Out of the Investor's (Customer's) Conditions (Requirements) during the Preparation for Contract Tendering. Gosstroy (Russian State Committee for Construction)]. Moscow, Russia: SUE Centre of Construction Design Products.

Crawley, E. F. (2001). *The CDIO syllabus: a statement of goals for undergraduate engineering education.* Cambridge, MA: The Department of Aeronautics and Astronautics, Massachusetts Institute of Technology.

Finansovyj slovar'. (2017). [Financial Terms Dictionary]. Retrieved from: http://dic.academic.ru/dic.nsf/fin_enc/13778

Frumina, E. (2013). *Modernizing English Language Teaching: a case of a pilot project at NUST MISiS. NUST MISiS. The report by Elena Frumina, Advisor to Rector NUST MISiS, made at the CDIO Academy training program.* Moscow, Russia: SkolTech.

GOST R 541-2010. (2010). *GOST R (Russian national standard) 541-2010: Industrial Automation Systems and Integration. Guidelines for Standards Implementation, Structure and Glossary.* Moscow, Russia: Standartinform. (in Russian)

GOST R 54147-2010. (2010). *GOST R (Russian national standard) 54147-2010: Strategic and Innovation Management. Terms and Definitions.* Moscow, Russia: Standartinform. (in Russian)

GOST R ISO 15704-2008. (2008). *Promyshlennye avtomatizirovannye sistemy. Trebovaniya k standartnym arhitekturam i metodologiyam predpriyatiya* [GOST R (Russian national standard) ISO 15704-2008: Industrial Automated Systems. Standard Architecture and Enterprise Methodology Requirements]. Moscow, Russia: Standartinform. (in Russian)

GOST R ISO 19439-2008. (2008). *Integraciya predpriyatiya. Osnova modelirovaniya predpriyatiya* [GOST R (Russian national standard) ISO 19439-2008: Enterprise Integration. The basis of Enterprise Modelling]. (in Russian)

Strategy Partners Group & Minpromtorg RF (Ministry of Industry and Trade of the Russian Federation). (2015). *Forsajt-issledovanie v oblasti perspektivnyh professional'nyh kompetencij specialistov v oblasti inzhiniringa i promyshlennogo dizajna* [Foresight-research in promising engineering and industrial design professional skills]. Authors. (in Russian)

Tech. (2014). *World Industry Report: Global Engineering Services Market.* London, UK: Technavio.

Mirzoyan, L.A. (2013). *Primeneniye CDIO v Mosckovskom Aviatsionnom Institute* [Implementation of CDIO in Moscow Aviation Institute]. The report by Lolita A. Mirzoyan, Head of Department of Educational Programs NUT MAI, made at the CDIO Academy training program. Moscow, Russia: SkolTech.

Osmakov, V. S., & Pastukhov, V. A. (Eds.). (2015). Inzhiniring I promyshlennyi dizain [Engineering and Industrial Design]. Moscow, Russia: Onebook.ru. (in Russian)

Russian Government Decree. (2015). *Rasporyazhenie Pravitel'stva RF ot 06.01.2015 N°7-r «Ob utverzhdenii perechnya special'nostej i napravlenij podgotovki vysshego obrazovaniya, sootvetstvuyu-shchih prioritetnym napravleniyam modernizacii i tekhnologicheskogo razvitiya rossijskoj ehkonomiki»* [Russian Government Decree of 6 January 2015 #7-p "On Approval of the Higher Education Specializations and Programmes List Which Corresponds to the Priority Areas of Modernization and Technological Development of the Russian Economy"]. Author. (in Russian)

Chapter 8
CDIO Standards Implementation and Further Development in Russia

Alexander I. Chuchalin
Tomsk Polytechnic University, Russia

ABSTRACT

Russian experience in the implementation of CDIO (conceive, design, implement, operate) standards for modernization of BEng programs focused on graduate training for complex engineering activity are considered. The CPD program "Applying CDIO Standards in Engineering Education" for managers and faculty staff at Russian HEIs is described. Further development of the CDIO concept for MSc and PhD engineering programs design are discussed taking into account the priorities of innovative and research engineering activities. The FCDI (forecast, conceive, design, implement) standards focused on MSc program graduate training for innovative engineering activity and FFCD (foresight, forecast, conceive, design) standards focused on PhD program graduate training for research engineering activity are presented.

INTRODUCTION

More than a hundred universities around the Globe implement *CDIO Standards* to engineering education. The standards are based on the CDIO (Conceive, Design, Implement, Operate) approach to training basic engineering education program graduates to prepare them for *complex* engineering activities at all stages of the life cycle of technical products, processes and systems (Crawley et al, 2014).

The CDIO approach is widely used, as it is consistent with the requirements of International Engineering Alliance (IEA) Standards (*IAE Graduate Attributes and Professional Competences*) to the engineering HEI's graduate learning outcomes (LOs) and competences of *Professional Engineers*. The CDIO Standards allows the design and implementation of BEng programs in accordance with the criteria for accreditation of engineering programs in the countries - signatories of the *Washington Accord*, including the accreditation criteria of the Association for Engineering Education of Russia (Chuchalin, 2012; Brodeur, 2012).

DOI: 10.4018/978-1-5225-3395-5.ch008

CDIO STANDARDS IMPLEMENTATION

The CDIO Standards have become popular in many Russian engineering HEIs.

In 2010 the Tomsk Polytechnic University (TPU) began to implement CDIO Standards to modernization of BEng engineering programs. In 2011 TPU formally joined the CDIO Initiative as the first Russian university member. In 2014 the first cohort of Bachelors, trained at TPU under the CDIO programs, graduated. The CDIO programs at TPU were accredited by the Association of Engineering Education of Russia (EUR-ACE Labeled) and by Accreditation Board for Engineering and Technology (USA), the most prestigious engineering organization in the world. Thus, the high quality of the TPU CDIO programs was officially confirmed. Following the example of TPU a dozen Russian HEIs have implemented the CDIO Standards and formally joined the CDIO Initiative.

In 2013 TPU and Skolkovo Institute of Science and Technology (Skoltech) designed and piloted a joint CPD program "Applying CDIO Standards in Engineering Education" in order to ensure proper preparation of managers and faculty staff at Russian HEIs to use the CDIO approach. The CPD program was based on the experience and best practice of engineering programs' modernization at universities participating in the CDIO Initiative. In total, 27 experts from 6 Russian HEIs: TPU, Skoltech, Ural Federal University, Astrakhan State University, Moscow Institute of Physics and Technology, Tomsk State University of Control Systems and Radio-electronics and 5 foreign HEIs: KTH Royal Institute of Technology, Chalmers University of Technology (Sweden), Delft University of Technology (The Netherlands), Technical University of Denmark, Massachusetts Institute of Technology (USA) were involved in the design of teaching and learning (T&L) materials, as well as in the program implementation (Chuchalin, Tayurskaya, & Malmqvist, 2015).

The CPD program consist of 5 modules designed in compliance with the CDIO model: Conceive, Design, Implement, Operate. At the initial stage, the trainees define the particular engineering program (course) to be improved by applying the acquired knowledge and skills. While studying the sections of the 1st module, the trainees pass through the 'Conceive' stage of the particular engineering program (course). They develop program (course) objectives and learning outcomes necessary for future professional activity and make them aligned with the key stakeholders (employers). Individual projects of the 2nd and 3rd modules are focused on the 'Design' and 'Operate' stages of the particular engineering program and its elements. During the 4th module at the 'Implement' stage, the trainees develop assessment methods and criteria of students learning outcomes achievement, as well as evaluate the particular engineering program for its compliance with CDIO Standards. The 5th module is devoted to faculty staff development for CDIO implementation and enhancing their competency for teaching.

The first version of the CPD program "Applying CDIO Standards in Engineering Education" was scheduled for delivering during a 16-week academic semester and involved 3 face-to-face interactive lectures and practical classes, 2 Internet on-line webinars and 4 individual assignments given as homework. The pilot test of the CPD program started in Spring semester of 2013–2014 academic year. The face-to-face sessions were organized in Russian and foreign universities with experience of the CDIO Standards implementation. Broadcasting of on-line Internet webinars were arranged by TPU instructors. All organizational, T&L materials based on learning management system (LMS) *Moodle* were available on the TPU website.

The CPD program was characterized by a great amount of independent and individual work of trainees in collaboration with consultants from TPU. The CPD program was accomplished by the final project aimed at design of particular engineering programs using modules/courses modernized by the trainees. As a result, over 60 engineering programs in 24 Russian HEIs (including 7 "national research universities" and 7 "federal universities") were re-designed or upgraded (Chuchalin et al, 2012; Chubik, & Zamyatina, 2013; Petrovskaya, 2013; Chuchalin, 2013; Lunev, Zaripova, & Petrova, 2013; Lunev, Fedotova, & Rybakov, 2014; Rechistov, & Plotkin, 2014; Rebrin, Sholina, & Berestova, 2014; Muratova, & Tayurskaya, 2014; Geraschenko, & Efremov, 2014; Marchenko, Osipova, & Amautov, 2016).

Feedback of participants focused on the CPD program quality evaluation and also data for its improvement and upgrading was provided during the delivery of the program. The idea of developing an on-line version of the CPD program was proposed by participants to prepare more Russian HEIs faculty members and administrators for CDIO programs implementation. In 2016 an on-line version of the CPD program "Applying CDIO Standards in Engineering Education" within a massive open on-line course (MOOC) format based on the *EdX* platform was designed. For the on-line version, all T&L materials were redesigned to meet the requirements of the MOOC format.

Special scenarios for on-line presentations of the CPD program were elaborated based on the following principles: T&L materials were decomposed into separate and logically related themes, modules and short fragments to be learned in chorological order during the weeks of a semester; segmented T&L materials were prepared in video, audio, and text formats based on the recommendations for *EdX/MOOC* format; each fragment of T&L materials was provided with the tests for self-assessment; T&L materials were supported by individual assignments, methodological guidelines and recommendations for their accomplishment and assessment; T&L materials were accompanied by references to the information related with the corresponding section, theme, module or fragment; T&L materials were complemented with explicit self-study instructions and options for communicating with other the participants and the instructor.

Creating the on-line version of the CPD program "Applying CDIO Standards in Engineering Education" in *MOOC* format has become a step forward in continuing the professional development of Russian HEIs faculty members and administrators. The on-line format allows the CPD program availability on a continuous basis and its expansion to a wide audience geographically dispersed over many time zones.

To reach this goal on the *EdX* platform, developers used as much as possible, the on-line version accumulated training resources and experience of the leading universities – participants of CDIO Initiative. Currently, the on-line version of the CPD program is being tested and is expected to be launched in 2017. For the future, it is foreseen that a blended format of delivery could also be provided to further support the application of the CDIO Standards in Russian engineering HEIs.

CDIO STANDARDS FURTHER DEVELOPMENT

As mentioned above, the graduates of BEng programs are mainly trained for *complex* engineering activities: Conceive, Design, Implement and Operate technical products, systems and processes, solving a wide variety of technical and other issues at all stages of the engineering product life cycle. Solving complex engineering problems requires *basic* knowledge of mathematics, natural sciences, engineering and other sciences, as well as specific technical, economic, administrative and other knowledge, including interdisciplinary knowledge in the area of specialization.

Analysis of the Russian Federal State Educational Standards (FSES) and the Association for Engineering Education of Russia (AEER) accreditation criteria shows that most core competencies of engineering BEng program graduates (>60%) are related to engineering activities at the stages of *Implementation & Operation* of technical products, processes and systems. About 25% of intended LOs are focused on *Design*, and a little more than 10% of competencies enable Bachelors to participate in activities at the *Conceive* stage.

Masters in engineering are prepared mainly for *innovative* engineering activity aimed at the development and design of *new* technical products, systems and technologies for human needs to get social and (or) economic impact, and therefore demanded and competitive. Innovative engineering is an interdisciplinary activity. It requires *deep* fundamental and applied knowledge, based on the analysis and synthesis of technical products, systems and processes with the use of mathematical models of the high level with more emphasis on interdisciplinary knowledge. A half of the engineering MSc program graduate LOs (50%) corresponding to the FSES requirements and the AEER accreditation criteria, are related to *Design*, about 25% are focused on *Implement & Operate*, and about 25% of competencies enable Masters to participate in activities at the *Conceive* stage.

The PhD engineering program graduates are prepared mainly for *research* activity in the field of *engineering sciences*. It aims to *generate new knowledge* and to transform fundamental knowledge into applied knowledge for its subsequent use in engineering, as well as *scientific support* of the development of new technical products, systems and technologies based on research results. According to the FSES requirements the majority (>60%) of expected PhD holder competencies is connected with the preparation for research activities at the stage of *Conceive*, about 25% of LOs focus on activities at the stage of *Design*, and a little more than 10% of competencies refer to *Implement & Operate* stages.

A group of experts from leading Russian universities (including participants of the CPD program "Applying CDIO Standards in Engineering Education") was asked to assess the applicability of *CDIO Syllabus v2* to engineering MSc and PhD programs. The results of the survey conducted in 2016 showed that there were problems of adaptation of the CDIO approach to graduate and postgraduate engineering programs (Chuchalin, Daneikina, & Fortin, 2016).

The development of conceptual and methodological models, similar to CDIO Syllabus and CDIO Standards, but aimed at the MSc and PhD engineering programs design considering the peculiarities of *innovative* and *research* engineering activities, was recommended. Following the recommendation required models were developed at TPU under the leadership of the author of the paper.

In the formation of a list of intended LOs for engineering MSc program, it was proposed to use the abbreviation FCDI (Forecast, Conceive, Design, Implement) instead of the abbreviation CDIO (Conceive, Design, Implement, Operate). The absence of "*Operate*" in a new abbreviation indicates that this kind of engineering activity (operation and maintenance of products, processes and systems) is not a priority for MSc program graduates. The presence of "*Forecast*" emphasizes the importance of forecasting potential needs of society in the new products, processes and systems. A two-level list of intended LOs (FCDI Syllabus) for graduate engineering programs was developed (Table 1).

In the formation of a list of intended LOs for engineering PhD programs it was proposed to use the abbreviation FFCD (Foresight, Forecast, Conceive, Design). The absence of "*Implement*" in the abbreviation indicates that participation in the production of products, processes and systems is not a priority for PhD program graduates. The presence of "*Foresight*" emphasizes the importance of technological foresight to anticipate potential needs of society and to create a scientific basis for conceiving and design-

Table 1. The list of intended LOs for graduate engineering programs

FCDI Syllabus
1. INTERDISCIPLINARY SCIENTIFIC AND TECHNICAL KNOWLEDGE
1.1. In-depth knowledge of mathematics and natural sciences
1.2. In-depth knowledge of engineering and methods of innovative activity
2. PROFESSIONAL COMPETENCES AND PERSONAL QUALITIES
2.1. Analytical study and solution of innovative problems
2.2. Experimentation, research and acquisition of deep knowledge
2.3. Systematic innovation thinking
2.4. Attitude, critical analysis and creativity
2.5. Ethics, equity and other types of liability
3. PERSONAL COMPETENCES: TEAMWORK AND COMMUNICATIONS
3.1. Team leadership
3.2. Communication
3.3. Communication in foreign languages
4. FORECASTING, CONCEIVING, DESIGNING, AND IMPLEMENTING SYSTEMS IN THE ENTERPRISE, SOCIETAL AND ENVIRONMENTAL CONTEXT – THE INNOVATION PROCESS
4.1. Societal and environmental context
4.2. Enterprise and business context
4.3. Forecast and innovation management
4.4. Conceive
4.5. Design
4.6. Implementation
4.7. Leadership in innovative technical enterprise
4.8. Innovative technological entrepreneurship
5. PEDAGOGICAL ACTIVITY
5.1. Development and implementation of educational resources

ing new products, processes and systems in the research activity. A two-level list of the intended LOs (FFCD Syllabus) for postgraduate engineering programs was developed as well (Table 2).

Based on FCDI Syllabus and FFCD Syllabus it is possible to generate detailed lists of intended LOs and to design the optimal structure, content, and technology of implementation and evaluation of the quality of MSc and PhD engineering programs.

In order to achieve the intended LOs (competencies of graduates), given in FCDI Syllabus and FFCD Syllabus, it was proposed to develop standards that define the requirements to graduate (FCDI Standards) and postgraduate (FFCD Standards) engineering programs by analogy with the CDIO Standards (Table 3).

Based on the CDIO Standards, FCDI Standards and FFCD Standards it is possible to develop, design, implement and evaluate undergraduate, graduate and postgraduate programs aimed at preparing graduates for *complex, innovative* and *research* engineering activities, respectively. Following proposed standards, a new generation of BEng, MSc and PhD programs are being designed at Tomsk Polytechnic University.

Table 2. The list of the intended LOs for postgraduate engineering programs

FFCD Syllabus
1. NEW SCIENTIFIC AND TECHNICAL KNOWLEDGE
1.1. New knowledge in the field of basic and applied sciences
1.2. New knowledge in the field of engineering and research methods
2. PROFESSIONAL COMPETENCES AND PERSONAL QUALITIES
2.1. Analytical study and solution of scientific problems
2.2. Experimentation, research and generation of new knowledge
2.3. Systematic scientific thinking
2.4. Attitude, critical analysis of the scientific data and results of own research
2.5. Ethics, equity and other types of liability
3. PERSONAL COMPETENCES: TEAMWORK AND COMMUNICATIONS
3.1. Research team leadership
3.2. Communication
3.3. Communication in foreign languages
4. FORESIGHTING, FORECASTING, CONCEIVING, AND DESIGNING IN THE ENTERPRISE, SOCIETAL AND ENVIRONMENTAL CONTEXT - THE RESEARCH PROCESS
4.1. Societal and environmental context
4.2. Enterprise and business context
4.3. Foresight and innovation management
4.4. Forecast
4.5. Conceive
4.6. Design
4.7. Leadership in the research enterprise
4.8. Research entrepreneurship
5. PEDAGOGICAL ACTIVITY
5.1. Design and delivery of Higher Education programs

CONCLUSION

The CDIO Standards were implemented in Russian HEIs extremely successfully when modernizing engineering BEng programs and training graduates for *complex* engineering activity. Special CPD program "Applying CDIO Standards in Engineering Education" have been developed and implemented to train Russian HEIs faculty members and administrators. On-line version of the CPD program in *MOOC* format is being tested to be implemented starting with 2017. As a result of the analysis of generalized national and international requirements to competences of engineering MSc and PhD program graduates, as well as expert examination of the characteristics and priorities of *innovation* and *research* engineering activities the FCDI Standards and FFCD Standards have been developed at Tomsk Polytechnic University by analogy with the CDIO Standards. The Standards are being piloted to design new generation of 3-cycle programs in electrical and nuclear engineering.

Table 3. The standards for undergraduate, graduate and postgraduate engineering programs

CDIO Standards	FCDI Standards	FFCD Standards
Standard 1 CDIO – Context of Engineering Education. Adoption of the principle that product, process, and system lifecycle development and deployment—Conceiving, Designing, Implementing and Operating are the context for undergraduate engineering education.	*Standard 1 FCDI – Context of Engineering Education.* Adoption of the principle that innovative product, process, and system lifecycle design and development – Forecasting, Conceiving, Designing and Implementing are the context for graduate engineering education.	*Standard 1 FFCD – Context of Engineering Education.* Adoption of the principle that creation of scientific basis for the development and design of innovative product, process, and system lifecycle – Foreseeing, Forecasting, Conceiving and Designing are the context for postgraduate engineering education.
Standard 2 CDIO – CDIO learning outcomes. Specific, detailed learning outcomes for personal and interpersonal skills, and product, process, and system building skills, as well as disciplinary knowledge, consistent with program goals and validated by program stakeholders.	*Standard 2 FCDI – FCDI learning outcomes.* Specific, detailed learning outcomes for personal and interpersonal skills, and innovative product, process, and system designing and developing skills, as well as interdisciplinary knowledge and teaching skills, consistent with program goals and validated by program stakeholders.	*Standard 2 FFCD – FFCD learning outcomes.* Specific, detailed learning outcomes for personal and interpersonal skills, and abilities to create scientific basis for innovative product, process, and system design and development, as well as transdisciplinary knowledge and pedagogical skills, consistent with program goals and validated by program stakeholders.
Standard 3 CDIO – Integrated Curriculum. A curriculum designed with mutually supporting disciplinary courses, with an explicit plan to integrate personal and interpersonal skills, and product, process, and system building skills.	*Standard 3 FCDI – Integrated curriculum.* A curriculum designed with mutually supporting interdisciplinary courses, as well as innovation and teaching activities with an explicit plan to integrate personal and interpersonal skills, and innovative product, process, and system design and development skills based on forecasting the needs of stakeholders.	*Standard 3 FFCD – Integrated curriculum.* A curriculum designed with mutually supporting transdisciplinary courses, as well as research and pedagogic activities with an explicit plan to integrate personal and interpersonal skills, and abilities to create scientific basis for innovative product, process, and system design and development using the methods of technological foresight.
Standard 4 CDIO – Introduction to engineering. An introductory course that provides the framework for engineering practice in product, process and system building, and introduces essential personal and interpersonal skills.	*Standard 4 FCDI – Introduction to innovative engineering.* An introductory workshop that provides the framework for engineering practice in innovative product, process and system design and development based on forecasting the needs of stakeholders, as well as introduces essential personal and interpersonal skills.	*Standard 4 FFCD – Introduction to research engineering.* An introductory workshop that provides the framework for engineering practice in creation of scientific basis for innovative product, process, and system design and development using the methods of technological foresight, as well as introduces essential personal and interpersonal skills.
Standard 5 CDIO – Design-Implement Experiences. A curriculum that includes two or more design-implement experiences, including one at a basic level and one at an advanced level.	*Standard 5 FCDI – Innovation- Design Experiences.* A curriculum that includes design projects entailing experience in engineering innovations based on forecasting the needs of stakeholders, as well as experience in teaching.	*Standard 5 FFCD – Research- Design Experiences.* A curriculum that includes research projects entailing experience in creation of scientific basis for engineering innovation design based on technological foresight, as well as pedagogic experience in higher education.
Standard 6 CDIO – Engineering Workspaces. Engineering workspaces and laboratories that support and encourage hands-on learning of product, process, and system building, disciplinary knowledge, and social learning.	*Standard 6 FCDI – Innovative engineering workspaces.* Engineering workspaces and laboratories that support and encourage innovative product, process, and system design and development, interdisciplinary knowledge, and social learning.	*Standard 6 FFCD – Research engineering workspaces.* Engineering workspaces and laboratories that support and encourage creation of the scientific basis for innovative products, processes and systems design and development, transdisciplinary knowledge, and social learning.
Standard 7 CDIO – Integrated Learning Experiences. Integrated learning experiences that lead to the acquisition of disciplinary knowledge, as well as personal and interpersonal skills, and product, process, and system building skills.	*Standard 7 FCDI – Integrated Learning Experiences.* Integrated learning experiences that lead to the acquisition of interdisciplinary knowledge, as well as personal and interpersonal skills, and innovative product, process, and system design and development skills based on forecasting the needs of stakeholders.	*Standard 7 FFCD – Integrated Learning Experiences.* Integrated learning experiences that lead to the acquisition of transdisciplinary knowledge, as well as personal and interpersonal skills, and abilities to create scientific basis for innovative product, process, and system design and development using the methods of technological foresight.
Standard 8 CDIO – Active learning. Teaching and learning based on active experiential learning methods.	*Standard 8 FCDI – Active learning.* Teaching and learning based on active learning and innovative methods.	*Standard 8 FFCD – Active learning.* Teaching and learning based on active learning and research methods.
Standard 9 CDIO – Enhancement of Faculty Competence. Actions that enhance faculty competence in personal and interpersonal skills, and product, process, and system building skills.	*Standard 9 FCDI – Enhancement of faculty FCDI – competence.* Actions that enhance faculty competence in personal and interpersonal skills, and innovative product, process, and system design and development skills.	*Standard 9 FFCD – Enhancement of faculty FFCD – competence.* Actions that enhance faculty competence in personal and interpersonal skills, and abilities to create scientific basis for innovative product, process, and system design and development.
Standard 10 CDIO – Enhancement of Faculty Teaching Competence. Actions that enhance faculty competence in providing integrated learning experiences, in using active experiential learning methods, and in assessing student learning.	*Standard 10 FCDI – Enhancement of Faculty Teaching Competence.* Actions that enhance faculty competence in providing integrated learning experiences, in using active and innovative learning methods, and in assessing student learning.	*Standard 10 FFCDI – Enhancement of Faculty Teaching Competence.* Actions that enhance faculty competence in providing integrated learning experiences, in using active learning and research methods, and in assessing student learning.
Standard 11 CDIO – Learning assessment. Assessment of student learning in personal and interpersonal skills, and product, process, and system building skills, as well as in disciplinary knowledge.	*Standard 11 FCDI – Learning assessment.* Assessment of student learning in personal and interpersonal skills, and innovative product, process, and system design and development skills, as well as in interdisciplinary knowledge.	*Standard 11 FFCD – Learning assessment.* Assessment of student learning in personal and interpersonal skills, and abilities to create scientific basis for innovative product, process, and system design and development, as well as in transdisciplinary knowledge.
Standard 12 CDIO – Program evaluation. A system that evaluates programs against twelve *CDIO* standards, and provides feedback to students, faculty, and other stakeholders for the purposes of continuous improvement.	*Standard 12 FCDI – Program evaluation.* A system that evaluates programs against twelve *FCDI* standards, and provides feedback to students, faculty, and other stakeholders for the purposes of continuous improvement.	*Standard 12 FFCD – Program evaluation.* A system that evaluates programs against twelve *FFCD* standards, and provides feedback to students, faculty, and other stakeholders for the purposes of continuous improvement.

REFERENCES

Brodeur, D. (2012). Aligning Accreditation Criteria and Processes with the CDIO Approach. In *Proceedings of the 8th International CDIO Conference*. Queensland University of Technology.

Chubik, P., & Zamyatina, O. (2013). Training Elite Specialists in Engineering and Technologies. *Proceedings of the 9th International CDIO Conference*.

Chuchalin, A. (2012). RAEE Accreditation Criteria and CDIO Syllabus: Comparative Analysis. In *Proceedings of the 8th International CDIO Conference*. Queensland University of Technology.

Chuchalin, A. (2013). TPU-SKOLKOVO Project: Modernization of BEng Programs in Russia. *Proceedings of the 9th International CDIO Conference*.

Chuchalin, A., Daneikina, N., & Fortin, C. (2016). Application of CDIO Approach to Engineering BEng, MSc and PhD Programs Design and Implementation. *Proceedings of the 12th International CDIO Conference*.

Chuchalin, A., Petrovskaya, T., Kulyukina, E., & Tayurskaya, M. (2012). Benchmarking of TPU Academic Standard and CDIO Standards in Engineering Education. In *Proceedings of the 8th International CDIO Conference*. Queensland University of Technology.

Chuchalin, A., Tayurskaya, M., & Malmqvist, J. (2015). Professional Development of Russian HEIs' Management and Faculty in CDIO Standards Application. *European Journal of Engineering Education*, *40*(6), 426–437. doi:10.1080/03043797.2015.1085837

Crawley, E., Malmqvist, J., Ostlund, S., Brodeur, D., & Edström, K. (2014). *Rethinking Engineering Education, the CDIO Approach* (2nd ed.). New York, NY: Springer. doi:10.1007/978-3-319-05561-9

Geraschenko, A., & Efremov, A. (2014). The Targeted Training of the Students Enrolled in Aeronautical Programs. *Proceedings of the 10th International CDIO Conference*.

Lunev, A., Fedotova, A., & Rybakov, A. (2014). Complex Strategy of CDIO Initiative Implementation in a Regional Russian University. *Proceedings of the 10th International CDIO Conference*.

Lunev, A., Zaripova, V., & Petrova, I. (2013). Implementation of CDIO Initiative Approach at a Russian Regional University. *Proceedings of the 9th International CDIO Conference*.

Marchenko, N., Osipova, S., & Amautov, A. (2016). New Role of Employer in the Educational Process of Metallurgy Programme. *Proceedings of the 12th International CDIO Conference*.

Muratova, E., & Tayurskaya, M. (2014). Stakeholders' Evaluation of Learning Outcomes in Educational Programs Improvement. *Proceedings of the 10th International CDIO Conference*.

Petrovskaya, T. (2013). Using CDIO Concept to Develop Engineering Education in Tomsk Polytechnic University. *Proceedings of the 9th International CDIO Conference.*

Rebrin, O., Sholina, I., & Berestova, S. (2014). Interdisciplinary Project for Bachelor Engineering Program. *Proceedings of the 10th International CDIO Conference.*

Rechistov, G., & Plotkin, A. (2014). Computer Engineering Educational Projects of MIPT-Intel laboratory in the Context of CDIO. *Proceedings of the 10th International CDIO Conference.*

KEY TERMS AND DEFINITIONS

Accreditation Criteria: Criteria for evaluation and accreditation of Engineering Programs.

CDIO Standards for BEng Programs: Standards for Bachelor of Engineering degree programs design and implementation focused on graduate training for complex engineering activity (conceive, design, implement, operate).

CPD Program for Faculty: Program for continuing professional development of higher education institution faculty members.

FCDI Standards for MSc Engineering Programs: Standards for Master of Science degree engineering programs design and implementation focused on graduate training for innovative engineering activity (forecast, conceive, design, implement).

FFCD Standards for PhD Engineering Programs: Standards for philosophy doctor degree engineering programs design and implementation focused on graduate training for research engineering activity (foresight, forecast, conceive, design).

Innovative Engineering Activity: Activity aimed at forecast, conceive, design, and implement innovative engineering products, processes and systems.

MOOC Format: Format for massive open online course design and implementation.

Research Engineering Activity: Activity aimed at creating scientific basis for innovative engineering activity.

Chapter 9
The Implementation of Modern Information Technologies in Educational Fields

Anatoly A. Aleksandrov
Bauman Moscow State Technical University, Russia

Andrey V. Proletarsky
Bauman Moscow State Technical University, Russia

Konstantin A. Neusypin
Bauman Moscow State Technical University, Russia

Kai Shen
Beijing Institute of Technology, China

ABSTRACT

As known to all, we are living in an information society today. Learners can acquire knowledge, skill, and ability by participating in teaching, training and learning systems. As the basis of teaching, training, and learning system, the management system is significantly important in the management of online materials and intercommunion between students and teachers. In order to construct more effective systems of management, virtual organizations should be considered with latticed structure, which can easily adapt to external environment changing and may even inspire more innovations in the fields of education.

INTRODUCTION

As for modern education, a mass of information technologies have been put into practice (Aleksandrov, Proletarsky, Ke, & Neusypin, 2012; Kovalenko, 2012). In Japan, for those who accept engineering professional education at Tokyo, Osaka and Kioto universities, vast of high information technologies have been applied. In view of the applying of high educational technologies in USA, numerous of universities can be listed, such as University of Illinois at Urbana-Champaign, University of Pennsylvania, Purdue University, University of Texas, Stanford University, University of California, Massachusetts Institute

DOI: 10.4018/978-1-5225-3395-5.ch009

of Technology, Carnegie Mellon University, Georgia Institute of Technology and etc. With the rapid and booming development in China, new information educational technologies have also been actively suggested to apply in schools, colleges and universities during training, teaching and learning process.

INFORMATION TECHNOLOGY AND INFORMATION SOCIETY

Information technology has become the main pillar of economic development. In terms of the GDP development, information industry comes first. What's more, telecommunication has also become one of main methods to increase the efficiency of production and to strengthen competitive power at home, even in the world.

The development of infrastructure serves to the development of information technology. Above all, we cannot emphasize the importance of infrastructure improvement in the fields of education and science too much. For example, in China redistribution of state revenue and national resources is going on, which makes a benefit to the progress of science and education.

During the struggle for international top ranking, it is also highly required to develop information technology and to build information society. Moreover, the development level of information infrastructure and information technology is one of the most important evaluation indicators.

Information society make it possible for every company, department and unit of society to access to information sources and share their achievements on internet. It should be noted that the exchanging process of information must be guaranteed by laws and regulations in order to keep the technical secret of own country, company and organizations. Therefore, new criteria should be established to assess the development level of society.

Using computers, mobile phones as well as immobile phones, a large number of people all over the world can get access to the internet. Due to technology integration and convergence, such as telecommuting computer, audio-visual technology etc., uniform integrated informational systems have been far and wide applied and popularized in every aspect of society.

Information society gives rise to unitive world economic system, unitive informational living space, global informational infrastructure and global legislation laws (Gorelov, 2007; Knyaginin, Meshkov, & Utolin, 2016). In information society, activities of economy and business are being carried out with the help of information communicative mediums. By utilizing high-technology, virtual economic and financial systems can be established. Moreover, an interrelation exists between the virtual economic and financial systems and the actual-physical economic as well as financial systems, namely, an interactive relation.

Development of Russian Educational Complex

The key problem of socioeconomic development of Russian Federation in contemporary conditions is the low competitiveness of major industries. To some extent, the reason is that the organs of state power and government, organizations and citizens are not fully and not effectively used the opportunities provided by the information society.

One of the most important sectors of the economy is education. Therefore, innovation of contemporary organizational and economic processes in the field of education has been strongly required. Hence, there is a need for organization forms and management methods to develop Russian educational complex

(REC), considering the increasing importance of non-material forms and non-traditional qualitative factors of economic growth. Moreover, the scientific study determines the development of theoretical, methodological, and practical approaches on REC innovation in the information society.

In order to realize the innovative REC development, a management system should be built on the basis of synthesizing goals which comprises of information about external environment, internal state of system and its prediction. Thus, such management system is the symbiosis of synthesizing goals, decision-making, prediction, adaptive control and management, dynamical expert system and so on.

As shown in Figure 1, such advanced intelligent management systems of innovative development of Russian educational complex are complicated functional systems, consisting of some simpler functional subsystems conversely. Based on decision-making block of the intelligent management system, estimation and prediction of future acceptor operations can be made. What's more, during the functional procedure of the intelligent management system, not one integrated complex intelligent system but several simpler intelligent subsystems work together in order to save the system source and reduce the difficulty of system integration.

MANAGEMENT AND ORGANIZATION SYSTEM

It is acknowledged that educational technologies are multi-media presentation, showing object, phenomenon, appearance, etc., as well as other training and teaching information. Furthermore, the basis of teaching, training and learning (TTL) system is the management system which plays an important role in management, spreading and processing of online-materials between students and teachers involved in this TTL system. With the teaching, learning and learning process going on, materials-mentioned are visible and accessible by specified websites. In this TTL system, individual task, project and assignment for exceptional students, as well as educational information based on substantial and communicative materials for all students, are available, downloadable and workable subsequently.

Figure 1. Concept of an intelligent management system of innovative development of Russian educational complex

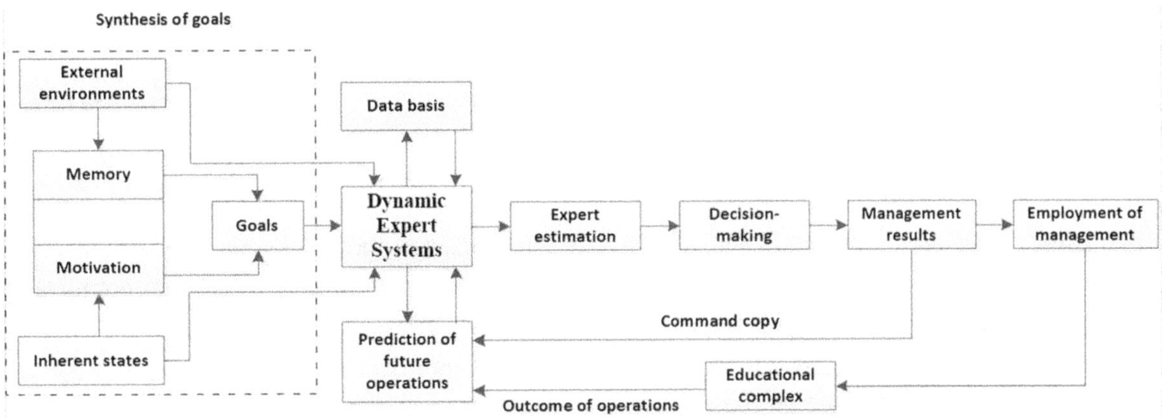

TTL management system can be accomplished by remote management of training program by utilizing the Internet and others networks. Hence, procedure of remote teaching and training project can be realized at a high level. Accordingly, students living in different countries are active to take apart in this high-lever remote interactive TTL system.

The organization systems which deal with the assignment of educational resources and provide a much appreciated service for the country, currently are multi-layered educational institutions, even institutions coordinating with the ministry of education. In general, such institutions belong to national professional bureaucracy. Classical professional bureaucracy secures the coordination and harmonious of high-learning activities for training high-skilled and high-qualified specialists. Moreover, those organization systems are characterized by low-level of bureaucracy and less subsidiary service personnel.

The mentioned features above can be certified by utilizing special lines or trends, such as linear trend, Demark trend and Theil-Wage trend. Linear trends can be written as

$$\hat{z}_{0i} = k_{0i} \cdot t_i + d_{0i} \quad (1)$$

where

$\hat{z}_0$ = predicted values,
k_0, d_0 = slope and constant in linear trends,
t_i = time.

Classical Demark trend has a form as:

$$\hat{z}_{1i}(a_1, b_1) = k_{1i} \cdot t_i + d_{1i} \quad (2)$$

where

$\hat{z}_0$ = predicted values,
k_0, d_0 = parameters of trends,
a_1 , b_1 = bearing points,
t_i = time.

Parameters of treads k_0, d_0 and bearing points a_1 , b_1 can be formally determined as:

Step 1: dividing observational space into several parts according to the tendency of updating states;
Step 2: calculating average values of each part. Average values will function as bearing points a_1 , b_1 in the next step;
Step 3: drawing linear lines from one bearing point a_1 to another bearing point b_1 ;
Step 4: connecting those linear lines and obtain one predicting curve;
Step 5: expressing studied system as predicting curve and predict future states by tracking tendency of established models.

Classical Demark trend is a relatively precise model.

Theil-Wage trend can be expressed as:

$$\begin{aligned} x_t &= a_{1,t} + \varepsilon_t, \\ a_{1,t} &= a_{1,t-1} + a_{2,t}, \\ a_{2,t} &= a_{2,t-1} + v_t, \end{aligned} \tag{3}$$

where

$a_{1,t}$ = values of studied sequence in t moment,

$a_{2,t}$ = increase values from moment $t-1$ to moment t,

ε_t, v_t = time series with zero mathematical expectation and known covariance.

Those trends are some standard mathematic models which can be used not only in the areas of management, but also in the areas of organization of education, economy and society. By applying those standard mathematic models (special trends), the operational efficiency and capability of educational organizations or institutions can be analyzed. In order to meet the need of social progress, it is highly required to propose new paradigm of education, which can realize the transformation from existent traditional educational system to self-developing and self-organizing educational system.

NOVELS OF MANAGEMENT AND ORGANIZATION

Recently, building scientific and teaching complex (STC) at leading university in Russia, such as Bauman Moscow State Technical University (BMSTU), can translate educational institutions of higher learning from deficient-training levels to professional-training levels (Tsibizova, & Neusypin, 2012; Zinchenko, 1988).

In the newly-built STC, the applying of classical management methods brings to the increasing of hierarchical levels and decreasing of the manageability of STC subdivision. Accordingly, due to the imperfection of existent organization and management methods, the increasing of hierarchical levels gives rise to the decreasing of working quality of STC management system.

Comparing with other organization structures, latticed structure can more effective works during the organization and management of complex systems (Russell, 1981; Kim, & Mauborgne, 2005). The characteristics of latticed structure are as follows: (1) in horizontal connections (aspect) maximal progress should be made; (2) in vertical connections (aspect) minimal progress may be made. The efficacy of horizontal connections is much higher than that of vertical connections.

When designing management systems of scientific and teaching center, it should be noted to decrease hierarchical levels of management. Hence, we should make the most of horizontal connections in order to (1) increase the manageability of STC subdivisions (2) decrease hierarchical levels of management. Moreover, vertical connections can be built as less as possible, under the condition of non-decreasing the manageability of STC subdivision. Such modification of STC management system can strengthen STC management structures and increase STC efficacy owning to the increasing of versatility and sensitiveness.

In the times of today's explosion of science and technology, self-organizing and intelligent methods of organization and management should be established for enterprises, companies and educational institutions, in view of human factors. The transformation has taken place from organization based on traditional physical principles to organization based on knowledge and information. New organization forms should own more versatility so as to adapt to the rapidly changing external environment. Furthermore, such new forms may generate new knowledge and may encourage more innovations.

The synthesization of classical organization methods and modern information technologies brings to more advanced organization forms, especially virtual organization. By applying virtual organization, synergizing efficiency of educational complex can be analyzed, and more advanced and intelligent forms of organization can be built during the updated innovative process.

For the sake of effective function of virtual organization, it is highly required to apply central coordination. Central coordination can help participants of virtual organization to proceed according to institutional norms. Moreover, central coordination can ensure the interest concordance of all participants, such as universities, customers and enterprises.

AVIONICS ENGINEERING CENTER

Let's introduce an example of scientific and teaching complex with implementation of information technologies and international cooperation. At Bauman Moscow State Technical University (Ugin & Demihof, 2005), an engineering center "Avionics", functioning as a scientific and teaching complex, was built as a joint project of Bauman Moscow State Technical University, Ramenskoye Design Company (RDC) and Nanjing University of Science and Technology (NJUST) within the framework of the International Russian-Chinese Laboratory "Intelligent Electromechanical Systems" (IRCL).

Avionics Engineering Center has been working on the following main research and educational directions: preparation of bachelors, engineers, masters and highly qualified specialists in the priority areas; improvement of training methods, creating a highly effective system for training technical professions; carrying out joint research projects; participation in scientific competitions, grants, awards, including Grants of Ministry of Education and Science of Russian Federation and P.R. China; joint organization of international conferences and symposia; international cooperation, for example, international laboratory; organization of joint scientific journal, scientific and methodical bases; joint intellectual property rights (patents); joint participation in competitions, grants, awards and research funds.

The equipment of Avionics Engineering Center located in three main areas as follows: the multipurpose multifunctional exerciser based on four-generation aircraft cabin with spread spectrum of functionality: simultaneous work for several students, practicing coalition cooperation, advanced training of flight crews under conditions that are close to combat, practicing dueling and group interaction of several aircrafts, and further complication for training purposes; the Laboratory of Intelligent Systems at Bauman Moscow State Technical University includes several operational working stations for pilots which are adapted to current and future requirements, such as formation of flight mission during multi-conflict flight situation, coalition, interaction and formation of models of external environment; the research is in progress with real navigation and piloting systems and platforms of modeling aerodynamic characteristics.

The research works are carried out in the following directions: the research in the field of aerodynamic characteristics, application of nanotechnology and research of aerodynamic properties, creation of models and working out on modeling complex engineering center; the work on dynamic formation of flight mission on the basis of predictive models with Genetic Algorithms to meet changing operational conditions of aircraft and conditions of our aircrafts against coalition enemies; the development of measurement systems and complex of modern aircraft from the concept of measuring complex synthesis with variable structures which can provide information as accurately as possible under conditions of intensive maneuvering.

Perspective areas of research works are: development of the theory of intelligent control systems; creation of intelligent systems for space applications; development and application of intelligent technologies for control systems of atmospheric flight vehicles; research in the field of control theory; global aggregation and management of multi-site complexes and coalitions.

The primary purpose of Avionics Engineering Center and International Russian-Chinese Laboratory "Intelligent Electromechanical Systems" is to provide scientific and engineering school of world level.

The main tasks of Avionics Engineering Center are: the involvement of prominent scientists to work in IRCL as supervisors to research current and future problems of equipment, technologies and educational programs; the development and coordination of fundamental and applied research works; the increase of level of fundamental education and engineering training of young professionals; the preparation of a new generation of young scientists via scientific subjects; the active involvement of graduate and undergraduate students to participate in international seminars, conferences, symposia; the creation of specialized scientific research stands, simulators; the development of international cooperation in the field of research and educational activities; the organization of international scientific and technical conferences, symposiums, competitions, seminars, exhibitions; the creation of modern laboratory workshop for students and postgraduates; the organization of research works for students and graduate students; the promotion and popularization of engineering knowledge, advanced information technology; the attraction of additional resources for the development and strengthening of material and technical scientific and methodological base; the work on the creation of joint intellectual property right (patents); the joint participation in competitions, grants, awards and research funds.

Therefore, the establishment and successful operation of the engineering center will help to solve one of the most important tasks of present stage – the training of a new generation of young scientists which are capable of creating breakthrough technologies in the defense industry.

SUMMARY

In the fields of modern education, a mass of information technologies have been put into practice. Hence, teaching, training and learning systems can be established on the basis of high information technologies. Therefore, learners all over the world can acquire knowledge, skill, and ability by participating in TTL systems.

The basis of TTL system is the management system of TTL procedure. By applying linear trend, Demark trend and Theil-Wage trend, the efficiency of TTL management systems can be analyzed. In order to obtain more effective management and organization system, latticed structure may be applied. What's more, virtual organization can also be established, which can adapt to external environment and may encourage more innovations.

Based on analysis of domestic and foreign organization methods in educational fields, we propose several recommendations: (1) modifying structures of STC management and organization systems; (2) decreasing vertical connections and increasing horizontal connections as much as possible; (3) applying delegation of authority and decreasing hierarchical levels of management; (4) adopting virtual organization and using central coordination.

Avionics Engineering Center as a joint international project was briefly introduced. Functioning as a scientific and teaching complex of world level, the main tasks of Avionics Engineering Center were illustrated. Then, methods on implementation of perspective scientific-research and educational programs were developed for innovative development of "Avionics" engineering complex. The symbiosis of new proposed methods and programs allows training and getting a set of high-class specialists and better innovative technologies in the defense and aerospace industry.

FUTURE RESEARCH DIRECTIONS

The basis for the further development of the proposed educational concept is the system of advanced specialists training for the industrial complex based on scientific and educational centers (SEC) at Bauman Moscow State Technical University. The advanced training system of specialists was tested at the BMSTU and Ramenskoye Design Company and other enterprises of Russia, China, Vietnam, Myanmar and other countries.

At the BMSTU, specialists are trained on the basis of the traditions of the classical Russian engineering school of polytechnic education that has developed and is constantly developing within its walls, which is based on the deep integration of education, science, innovation, and industrial enterprises.

The requirements of advanced training of specialists are scientifically substantiated by means of analysis of the forecast results obtained on the basis of forecasting models that coordinated with customers and linked with the existing capabilities of the country's educational complex.

When training specialists, great attention is paid to:

1. Strengthening the influence of industry on the educational process through: participation in the formation and correction of the request for training in specific specialties and qualifications; participation in the permanent renewal of educational standards and educational programs; the development of the system of branch faculties, departments in the territory of basic enterprises; creation of joint SECs.
2. Strengthening the material and technical basis of BMSTU educational structures, conducting production practices in modern enterprises, correlating the topics of scientific research projects, course and diploma projects with the actual tasks of industry.

3. Based on the complex of SECs, deepering research cooperation between BMSTU and enterprises through the implementation of research.
4. Ensuring mutual scientific and training integration between BMSTU and enterprises through co-operation within the framework of mixed small creative research groups.
5. Ensuring the upgrading of qualifications, retraining, probation, second higher education for employees of enterprises and holding trainings for BMSTU teachers at the enterprises.

The symbiosis of technologies and methods of advanced education, a powerful production base of the SEC complex, modern information technologies and the "Russian method" in training allows to obtain a synergetic effect, which is an innovative advantage in comparison with other educational systems due to the formation of new competencies and creative skills, acceleration and more qualitative mastery of knowledge, increased motivation of students and specialists. The advanced training system allows to realize the function of habilitation, which provides comprehensive preparation of the students for professional work in new, increasingly complicated conditions, as well as his professional self-realization.

REFERENCES

Aleksandrov, A. A., Proletarsky, A. V., Ke, F., & Neusypin, K. A. (2012). Conception of complex continuous education with innovation technologies. In *Proceedings of the 2012 2nd international Conference on Education and Education Management*. Hong Kong: Information Engineering Research Institute.

Gorelov, V. I. (2007). *Management of the society development*. Moscow: Ekon-Inform Press.

Kazinik, M. S. (2010). *Secrets of the geniuses*. Moscow: Legan Press.

Kim, W. C., & Mauborgne, R. (2005). *Blue ocean strategy: How to create uncontested market space and make competition irrelevant*. Boston: Harvard Business School Press.

Knyaginin, V. N., Meshkov, N. A., & Utolin, K. V. (2016). Advanced education in the information society. *International Review of Management and Marketing*, *6*(3), 89–99.

Kovalenko, I. V. (2012). Theoretical bases of lifelong education research as an innovation system. *Izvestija of the Tula state University*, *1*, 281-287.

Russell, L. A. (1981). *Creating the corporate future: Plan or be planned for*. Wiley.

Tsibizova, T. Yu., & Neusypin, K. A. (2012). Some restructuring aspects of the modern educational-scientific centers control system. *Automation and modern technologies*, *1*, 30-34.

Ugin, E. G., & Demihof, K. E. (2005). *Founders of the scientific school of Bauman Moscow State Technical University*. Moscow: BMSTU Press.

Zinchenko, G. P. (1988). *Continuous education – command of time*. Moscow: Knowledge Press.

KEY TERMS AND DEFINITIONS

Avionics: The electronic systems used on aircraft, artificial satellites, and spacecraft. The term avionics is a portmanteau of the words aviation and electronics.

Education: The process of facilitating learning, or the acquisition of knowledge, skills, values, beliefs, and habits.

Information Society: A society where the creation, distribution, use, integration, and manipulation of information is a significant economic, political, and cultural activity.

Information Technology: The application of computers to store, study, retrieve, transmit, and manipulate data or information.

Management: The administration of an organization, whether it be a business, a not-for-profit organization, or government body.

Management System: The framework of policies, processes and procedures used by an organization to ensure that it can fulfill all the tasks required to achieve its objectives.

Organization: An entity comprising multiple people, such as an institution or an association, that has a collective goal and is linked to an external environment.

APPENDIX

At BMSTU, the concept of training specialists for enterprises has been developed and successfully implemented. Teaching methods with the use of modern information technologies that make it possible to intensify the process of obtaining and assimilating knowledge, the use of the "Russian method" in the process of teaching makes it possible to consolidate theoretical knowledge, to form skills and abilities in each phase of the continuous educational process. A generalized model has been developed at BMSTU for realizing scientific research activities at all levels of training (Figure 2), which was tested in the system of "school - university - enterprises".

The illustrated generalized unified model includes several basic functional blocks: analytical, informative, management and diagnostics, evaluative and correcting. The analytical unit is associated with the attraction, identification and selection of students for the implementation of research work in accordance with the requirements for training, direction, volume and specific knowledge required to participate in research. The informative block includes a set of individual educational programs established for the successful implementation of the research activities of the trainee, as well as the very research work of the student as part of a mixed small creative group. In the management and diagnostic unit, the educational process is monitored in respect of profiling disciplines, including the evaluation of interim results, as well as the process of performing scientific research work. The evaluative unit includes an evaluation system for key results, it analyzes the results of the training (protection of research work, etc.). Corrective block is designed to compare the requirements of legislation, educational standards, requirements for the educational process and the performance of research activities with the requirements of enterprises. Based on the analysis of these requirements, the educational programs and methodology of the research process are systematically corrected, and timely measures are taken to improve the advanced training of future specialists.

Figure 2. A generalized unified model for the implementation of research activities

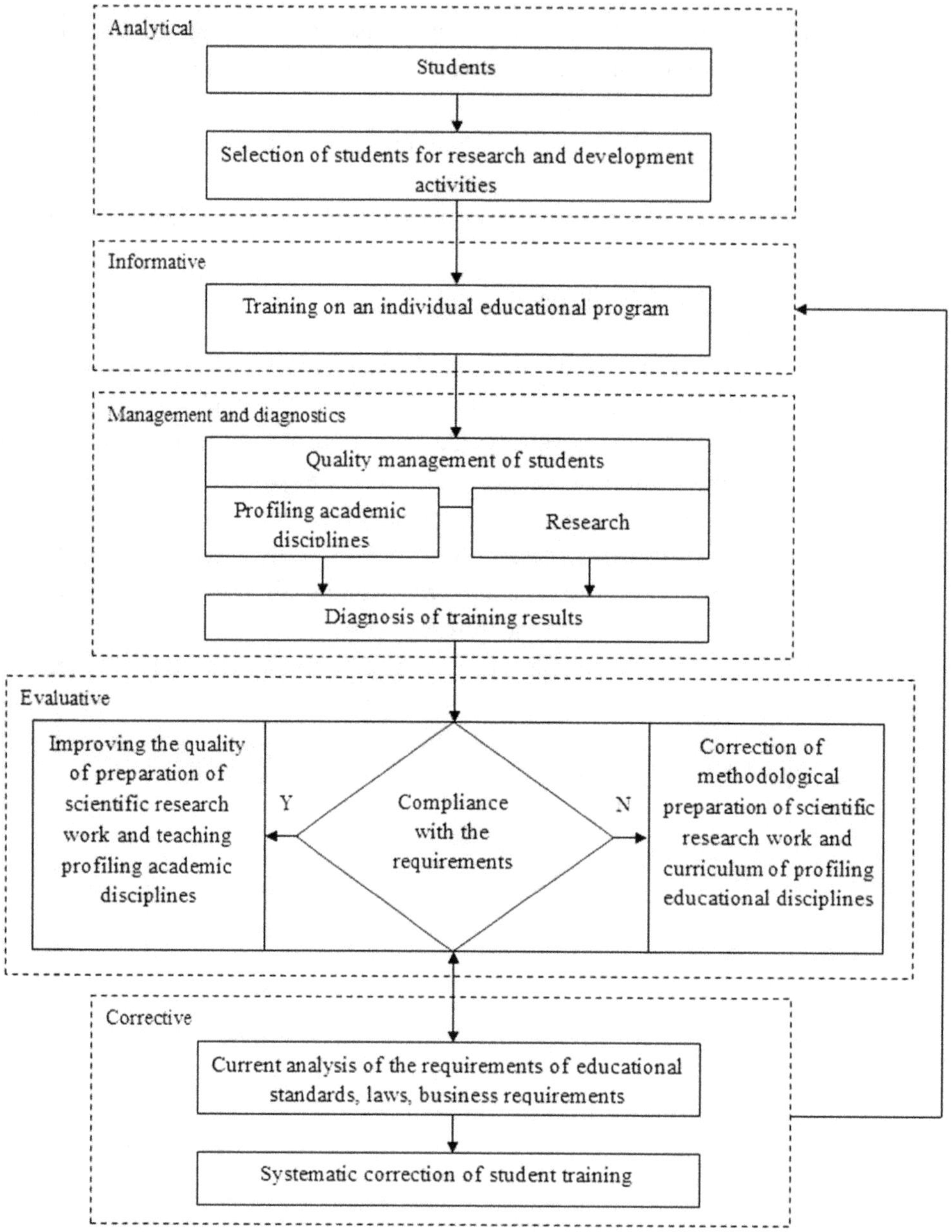

Chapter 10
Concept of Automated Support to Problem:
Modular Vocational Training

Andrey Igorevich Vlasov
Bauman Moscow State Technical University, Russia

Ludmila Vasilievna Juravleva
Bauman Moscow State Technical University, Russia

Vadim Anatolievich Shakhnov
Bauman Moscow State Technical University, Russia

ABSTRACT

The chapter deals with the concept of training for technicians, able to adjust to ongoing changes in the manufacturing environment. It is proposed to apply the systematic approach with a targeted use of collected methods and acmeological planning in a learning process when extra time is found owing to interdisciplinary integration and increased self-learning. Such planning is done with crosscutting design and problem-modular training with cross-rated instruction materials and learning outcomes. Education information systems are recommended as tools (like those based on modular object-oriented dynamic learning environment according to shareable content object reference model standard) together with automated information systems for self-learning, business and role plays with methods to rate career guidance and multimedia aids as learning tools.

INTRODUCTION

Technician training requires developed skills, abilities in project activities, teamwork, quick decision-making, and strict adherence to made decisions. This chapter introduces acmeological planning in the learning process to trigger hidden reserves. The authors search for reserves of training time at the expense of interdisciplinary integration (introducing the concept of the consolidated integrated interdisciplinary task) and increased self-learning (Zhuravleva, 2012). Planning is achieved through crosscutting designs and problem-modular trainings using cross-rated teaching aids and learning outcomes (Kolesnikova, 2009).

DOI: 10.4018/978-1-5225-3395-5.ch010

To intensify the learning process within the same class, teams are formed during workshops, laboratory classes, and business games. Research and development (R&D) teams, which include students from different years and groups, allow undergraduates and junior students to share experiences from scientific and practical activities (Choshanov, 1996). At creative workshop sessions, teams are made within a brainstormed production task depending on student skill level. A level of solved tasks varies from local learning tasks to development of engineering systems following competitive tasks. The project-based preparation to competitions (for example, EUROBOT) has proven an advantage. Within an academic year, project teams use knowledge from different engineering courses to design robotics system under EUROBOT competition terms and conditions. Then, they defend their solution at college, regional, and international competitions (Vlasov& Yudin, 2011).

PROBLEM-MODULAR ENGINEERING TRAINING

Analysis of Processes and Methods of Transfer of Knowledge

Against the background of profound social change, the education system in Russia is in search of new ways to improve the learning process through fundamental and holistic student training. The learning process takes place during sessions in classrooms, laboratories, workshops, and extramural self-learning. Professors act as trainers, tutors, and supervisors students are active subjects within the learning process. Training facilities (or instruction materials) require different tools to present and visualize knowledge and information (Yudin, Kolesnikov, Vlasov, & Salmina, 2017). For the generalized scheme of encapsulated objects in training and education, see Figure 1).

Figure 1. Relationship between types of training and education

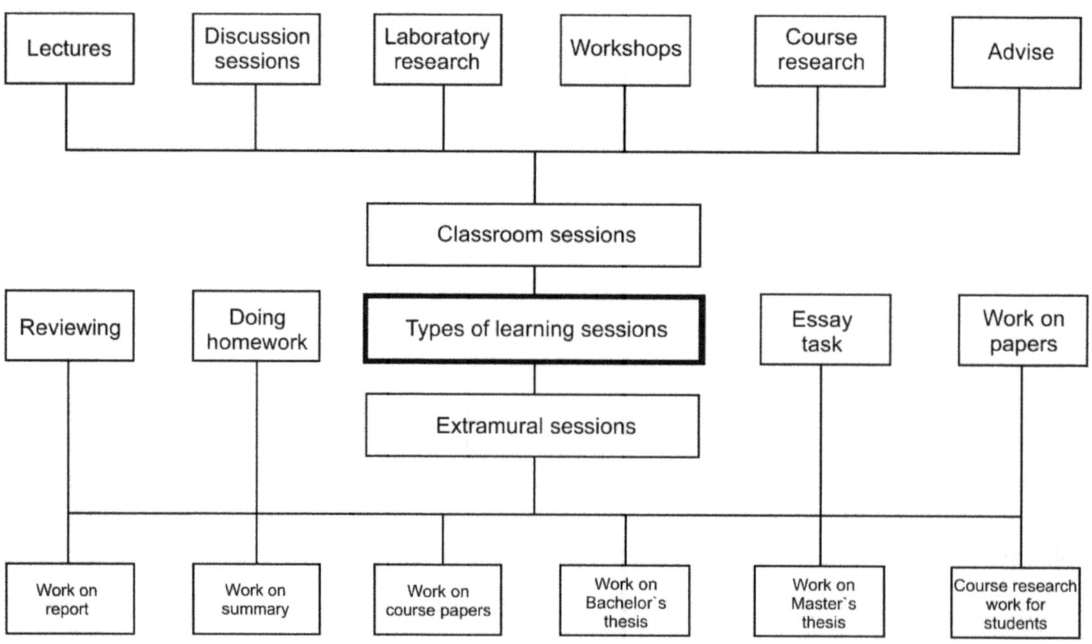

In training competent experts at the college level, improvements to the training quality are possible with the applied new educational technologies, developed learning routes, process plants (public-private partnership), and experienced professionals from process plants involved in the learning process. Other ways include launching art workshops under joint supervision of college teaching staff and leading professionals from core enterprises and establishing the social learning environment close to the working environment. These environments developstudents' abilities to work as a team, make independent decisions, and solve management problems in individual work.

Tools for learning methods and approaches include:

- Project-based approach,
- Problem-based learning,
- Crosscutting design,
- Problem-modular training,
- Cognitive learning,
- Competence-based approach,
- Profession and activity-related approach,
- Value and activity-related approach,
- Process approach,
- Developmental teaching,
- Distance learning,
- Integrative and innovative learning,
- Person-centered learning,
- Qualitative approach,
- Context-sensitive approach, and
- Elkonin-Davydov system.

Analysis of Components of Problem-Modular Training

To improve the learning process, it is necessary to apply information systems to support and administer documents, as well as the learning process (Vlasov, Zhuravleva, Sharipov, & Sharipova, 2012).In training technicians (who adjust themselves to rapidly changing working environments), people often use methods of problem-based and problem-modular training (Archan, 2005). The problem-based learning concept has been widely shared. However, there have been several approaches to its interpretation (Brockmann, Clarke, Mehaut, & Winch, 2008).

Problem-based learning is developmental teaching as a set of activities, including:

- Generating problematic cases,
- Problem statement,
- Support to students in problem solving,
- Solution checks,
- Administration in systematization and consolidation of acquired knowledge to develop creativity,
- Efficient thinking,

- Imagination,
- Cognitive motivation, and
- Mental emotions (see Figure2).

When using problem-based learning, users simulate the real creative process. This makes a problematic case and facilitates a search for a solution. Levels of problem-based learning depend on the content of the instruction material (possibility to generate problematic cases that vary in difficulty) and type of students' self-learning. The method of problem-modular training mirrors the problem-based method (Belous, Bobrovsky, Dobrjkov, Karpenko, & Smirnova, 2012). Problem-modular training has gained popularityat colleges in the United States, Germany, and England. The problem-modular training technique assumes compression of training information, modularity, and rating of knowledge and skills.

Training information compression refers to cases of knowledge compilation, consolidation, systematization, and generalization using the achievements of knowledge engineering. A module may include several modular units; each includes a complete description of a separate operation or technique. Depending on requirements of a professional activity, the modular units may be extended and complement a content of the module (Aleksandrov, Fang, Proletarsky, & Neusypin, 2012).Effectively proved modules with functions use syntactic text analysis to estimate educational task difficulty and complexity (Naumov & Vykhovanets, 2016).

Figure 2. Components in problem-modular training

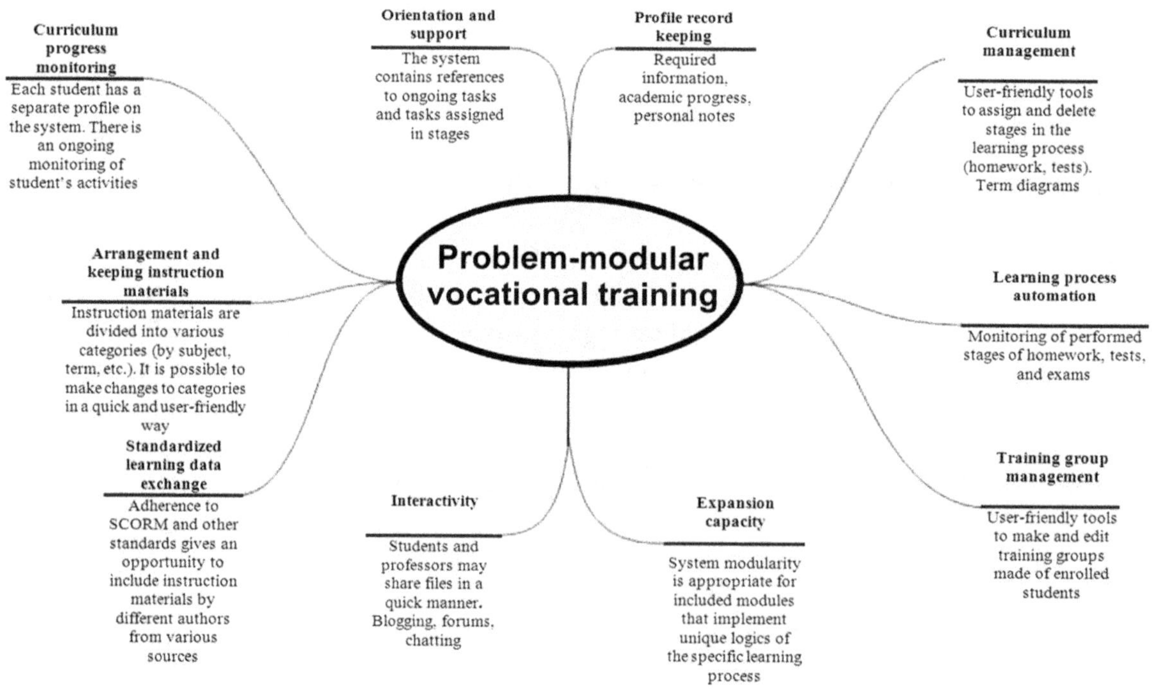

APPROACHES TO AUTOMATED SUPPORT IN PROBLEM-MODULAR VOCATIONAL TRAINING

Dynamic Learning Environment: Open Learning Management System (LMS)

Today's learning process is successfully integrated with computer techniques. These include:

- e-Textbooks,
- Problem books,
- Simulator stands,
- Laboratory workshops,
- Tutorials,
- Syntactic text analysis,
- Virtual enterprises, Visualized thematic lexicons (Shakhnov, Zhuravleva, & Vlasov, 2013),
- Cognitively-structured training, and
- Testing systems (Medvedev& Dobriakov, 2013).

The modular object-oriented dynamic learning environment system (Moodle) has a philosophy of social constructivist teaching. It is primarily focused on establishing professor-student interaction. It is also suitable for an arrangement of traditional distance-learning courses and as a support to full-time studies.

The modular object-oriented learning system produces e-learning courses combining various teaching methods. It is an appropriate structure and method to display teaching aids in an electronic format. Training modules vary depending on their application. They provide support to customized object environments within the framework of the person-centered principle to training management.

Thus, using Moodle, a professor can benefit from broad opportunities to run a course and develop separate training elements. In the teaching and learning process, systematized distance courses can be used for selected aspects or a discipline. This is also true for separate elements and resources. It should be mentioned that Moodle course development does not require special knowledge from the professor surrounding modern information technology. Moodle has a comprehensive set of e-learning aids through efficient standard tools (i.e., text processors, Web forms, etc.). It presents an opportunity to edit and manage learning aids in real time.

When professionals establish e-learning (open education) systems, a crucial point is to select a system to store, manage, and update the content of the training resources. It is possible to achieve these features with a set of standard tools, including regular mail, e-mail, online classifieds, spreadsheets, and databases. Developers usually achieve special-purpose software systems like LMS, learning content management system (LCMS), and development tools for learning environments or authoring software (AS).

Training providers use LMS and virtual learning environments (VLEs) to introduce automation and informational support in administrative procedures related to storage and management of information on training courses, certification learning outcomes, student's personal data, and a scheduled control of the learning process (IEEE 1484.11.2 Standard for Learning Technology, 2003). LCMS create, store, manage, and provide a student with content in an information academic course. Creating content, which is a labor-intensive process, requires regular updates in accordance with changing conditions of the learning process and requirements to students. Tools for course development and AS are designed for teachers or specialists in teaching methods who have no IT working experience. With these tools, they cancreate information and academic content of courses (or content). Training materials developed with the help of such tools are subject to requirements of strict compliance with certain standards and specifications developed by international associations and consortia.

There are three ways to introduce AS: (1) use commercial software (of the shelf solution);(2) commission software development following one's own specifications from third-party development companies (a tailor-made solution); and (3) use in-house design (a home-grown solution).

Commercial software speeds up an implementation process. However, the cost of such software is high. Upon a product purchase, the developments in the field of teaching methods will be restricted to existing formats. The main problem is a need in the learning process redesign to absorb other people's learning methods embedded in the software.

The main disadvantage of commissioning from third-party companies is support and upgrades. Unfortunately, many third-party companies have a short lifecycle. The support ends upon their closure or a change of an activity field. In addition, the product lifetime is large enough.

The third option, in-house design, is the best option for AS implementation at an engineering college. In this option, developers of this class of systems and opportunities for future operational support and improvement is available through professors and students.

Transitioning to a qualitatively new level of the learning process requires a revised architecture of current information systems, as well as replacement and optimization of elements in the data analysis by the complexity criterion. At the current level of society development, education and science has been an object of informatization. Consequences from this process, which are supported by governmental and nongovernmental organizations, are agreements and standards. This application in the open e-learning environment is now becoming mandatory. Due to this, informatization of engineering training—with its specific learning content, used processing methods, and research methodologies—requires a careful study of possible data sources in this field (i.e., material, data from unique equipment, etc.). It also requires developed extensions to standards and agreements in the field of distance learning systems for a possibility of wide remote copying of these data by means of the Web.

It is reasonable to build an up-to-date open learning system following the sharable content object reference model(SCORM) standard. It includes main parts of other important standards for learning process management systems and is progressive because leading automated environments use it (for example, Moodle).

SCORM is a set of guidelines to organize work with teaching materials to shape a pattern of instructions (SCORM 2004 4th Edition Run-Time Environment Version 1.0., 2009). The pattern determines which kinds of auxiliary functions should be used to solve a task and arrange learning materials. It also determines which interrelations exist between these functions and how people can use them.

The SCORM set integrates standards and specifications into a set of technological volumes. Nearly all the specifications and guidelines come from third parties. So far, all the technological volumes are divided by three subjects: (1)SCORM content aggregation model (CAM); (2) SCORM run-time environment (RTE);and (3) SCORM sequencing and navigation (SN).

SCORM combines technologies of companies into a single model. These include IMS, AICC, ARIADNE, and IEEE LTSC. The model, which is usable by all e-learning providers, creates a consistent guide to meet the standard(IMS Content Packaging Information Model, Version 1.1.4 Final Specification, 2004).

To increase a pace with which the standard is introduced and with which they implement the model, SCORM conceptually defines common functionalities that should be available in the all e-learning systems like SCORM. There are the following functionalities:

- accessibility is an opportunity to identify and access learning components from a single remote point and deliver them to many others;
- flexibility is an opportunity to have tailor-made training to individuals and companies.
- cost-effectiveness is an opportunity to increase efficiency and performance by reducing the price and the time required to provide customers with distance learning;
- long life-cycle is an opportunity to withstand the technological progress and changes without expenditures for redesigning, reconfiguring and changes to a source code;
- compatibility is an opportunity to take an educational component developed in one system with a set of tools and platforms and use it in another system with another set of tools and platforms;
- reusability is an opportunity to reuse, flexibility, the opportunity to add the educational component many times to various applications and contexts.

LMS, which is a comprehensive concept in SCORM, refers to a set of functionalities designed for transferring, tracking, reporting, and management of learning objects, students' academic progress, and learners' actions. LMS can refer to basic course management systems or complex distributed environments installed at an enterprise. Many people use the term LMS rather than computer-aided instruction (CMI) to discuss new functionalities and capabilities that are not historically related to previous systems (Computer Managed Instruction Guidelines for Interoperability [CMI001] Version 3.5., 2001). They include links to other information systems, sophisticated tracking and reporting systems for student activities, centralized registration, networking, and adaptive delivery of learning objects. SCORM guidelines review the organization of work using instruction materials to create a model of instructions. It determines which kinds of auxiliary functions should be presented to solve a task of training content organization, possible interrelations between these functions, and possible applications.

There is a wide range of descriptions for LMS. SCORM focuses on an interface between academic content and LMS environments. It does not discuss specific features and opportunities within the LMS. This allows individual sellers to provide users with a variety of training and management services, competitive alternatives maintaining an important aim of SCORM, and capacity for interaction.

SCORM assumes that LMS is a client-server environment with integrated functions of control and delivery to a student. According to SCORM, LMS determines what is to be delivered and at what time. It also tracks user progress and academic performance during the training. SCORM assumes that training resources contain internal logic to solve learning tasks. Such a resource can refer to itself as branching (depending on student actions). Branching is sufficient and locally significant for a training resource. Internal branching should not refer to external training resources in different structures of the content. Due to its importance, a content developer should pay it special attention when choosing training resources and layout.

Analysis of Information Aggregation (IA) Problems

IA can be used as both a specific action-event and to describe conceptual information. It describes an action or process to create a functional set associated with learning objects. This makes it possible to use the set in the learning process. During the SCORM data model choice, IA describes information created during the process or action. IA, which can be a shortened form to describe a training kit, can be used to deliver content as an established structure for the organization distributed between systems and stored in a repository.

A representation of the learning content organization structure in accordance with SCORM requirements includes components aimed at identifying various aspects of its arrangement:

- **Content Hierarchy:** This tree representation looks like a table of contents and represents a logical organization of training resources or events using learning resources. In some cases, a student can go through this hierarchical tree representation in a specific default order designed by a course author.
- **Metadata:** These optional descriptive data are context specific and describe content structure or organization. Such metadata might describe the integration of training resources into a system (for example, students' knowledge or ambitions faced as they use training resources in a learning event).
- **Content Sequencing, Adaptive Content Sequencing, and Navigation:** These optional instructions are used when a course author wants to control training resources presented to students during course navigation. By default, if there are no instructions on content sequencing and navigation, the student can choose any point of the instruction content. By adding specific guidelines, users can change the standard learning behavior. More complex adaptive sequencing of the content can be based on whether the student has completed a learning course or more complex calculations of the user's preferences or evaluation results.

The standard provides for a wide variety of approaches to content aggregation. Aggregation is possible from the smallest learning resources, so small that they will be comparable to a couple of lines of HTML code or a short video, to highly organized learning resources that have got feedback features and are tracked by the LMS.

When a training resource is created, it can be easily described as metadata. It is important to understand that not every resource will require metadata because the training resource can create a single use. In some cases, training resources are made for applications in a specific context. Metadata allow users to find training resources stored in a repository. Training resources may be suitable for repeated use. By reviewing metadata, it is possible to determine whether it is necessary to reuse the resource without a need to review the training resource. Such metadata are considered context-independent because they describe the training resource regardless its location and its application in a specific learning strategy.

Metadata that belong to a specific learning strategy are called context-sensitive. For example, metadata can include an explanation of reasons for a learning event added at a given point in the learning process.

Context-independent metadata usually refer to immutable metadata that describe digital objects, objects of the instruction material, or a set of objects. Context-sensitive metadata, on the other hand, usually refer to metadata that only make sense in the context of a specific context organization.

Using metadata for developing learning resources is a newer concept. A successful practice of this still needs to be developed. In some cases, a conceptual goal of metadata provides users with an opportunity to review several instruction materials. Other cases are strictly informative as they provide an author with information about a model and objectives of a learning object.

Content coherence and SCORM navigation provide an opportunity to find highly adaptable coherence between learning events. The process of standardization for sequencing and navigation through content has proven difficult because of complex approaches required for effective teaching of certain skills or training for challenging jobs and responsibilities. Previous SCORM versions did not provide users with specific sophisticated sequencing capabilities. Users were unable to effectively organize free learning processes because the capabilities were complex and integrated tasks requiring more time. The community has many requirements to learning process development; these are often contradictory. Researchers have not found an approach to cover the necessary applications. However, the flexible approach adopted in SCORM and based on IMS SS Specification provides users with a variety of approaches to training organization.

Based on Moodle in accordance with SCORM standard requirements, the open engineering training system is a part of a large information and instruction system, including:

- **Distance Learning System:** This includes MediaLearning, which is a management system for instruction materials and learning processes (meeting international standard SCORM), a video conferencing system, and an online library. The system follows a modular basis with an option to include additional modules provided by the community of developers for Moodle training systems.
- **Online Laboratory:** This includes hardware and software for experiments with subsequent processing of results through a special-purpose server. The results of each experiment are saved in the experiment database. Therefore, it is possible to create a library of experiments and incorporate results of experiments into instruction materials for courses made in the distance learning system.

This chapter functionally distinguishes the following user roles in the information and instruction system:

- **Laboratory Assistant/Moderator:** This role works with experimental equipment and an application server. Duties include equipment setup, synchronizing work with a data storage and processing server, and correct term nation of a set of research on completed experiments.
- **Course Developer-User:** This role analyzes and arranges materials received from teachers. A server experiments' library is organized into a single and coherent training course in accordance with SCORM. This person possesses skills to work with a special-purpose editor of SCORM courses, a Web-interface of the course developer in MediaLearning instruction materials, and learning process management system. Duties also include working closely with professors.

The professor is an organizer of the learning process and a source of instruction materials. The user must have skills to work with the Web interface used by the professor in MediaLearning.

Student-users who go through an entire learning route developed by the professor will learn materials integrated by course developers and take part in supervisory events monitored in MediaLearning. The user must correctly utilize MediaLearning Web-interface toolkits through all stages of the learning route.

The developed methodology for problem-modular engineering training and the education information environment powered by Moodle ensure integrated work experience in distance learning, processing of research data using remote access equipment, and processing of data with remote access. This chapter presents a methodology to organize the learning process using research findings with remote access. Based on an in-depth analysis of valid standards for distance learning, there is a comparative analysis between SCORM and other standards applied to produce distance learning systems. There is also a concept for the distance learning system implementation based on SCORM. There are results of the designed management system for the learning process. BMSTU (www.bmstu.ru) has introduced the system in the learning process, there are standard routes for teaching and remote access experimenting.

Formalization and structuring are required for lecture materials when estimating the amounts of information and other properties consolidated in the automated training support system. Hierarchical structuring is the most appropriate way to present elements in the formal presentation of system properties (structural complexity) for elements of lecture materials that include different types of information (Shakhnov, Vlasov, Rezchikova, & Zinchenko, 2013).

Based on the hierarchical approach, this chapter identified measures to evaluate structural complexity and formal quality ratings to describe objects with varying physical natures in their quantitative, pragmatic, and semantic sides (Zhuravleva& Shakhnov, 1998). Hierarchical structuring of lecture materials is based on the content hierarchical arrangement (see Figure 3).

In engineering training, hierarchical structuring is directly associated with the crosscutting design methodology. Targeted methods and tools are applied for structuring, visualization, automation, and presentation at each stage of the crosscutting design (Zhuravleva & Shakhnov, 1995).

Figure 3. Hierarchically structured content

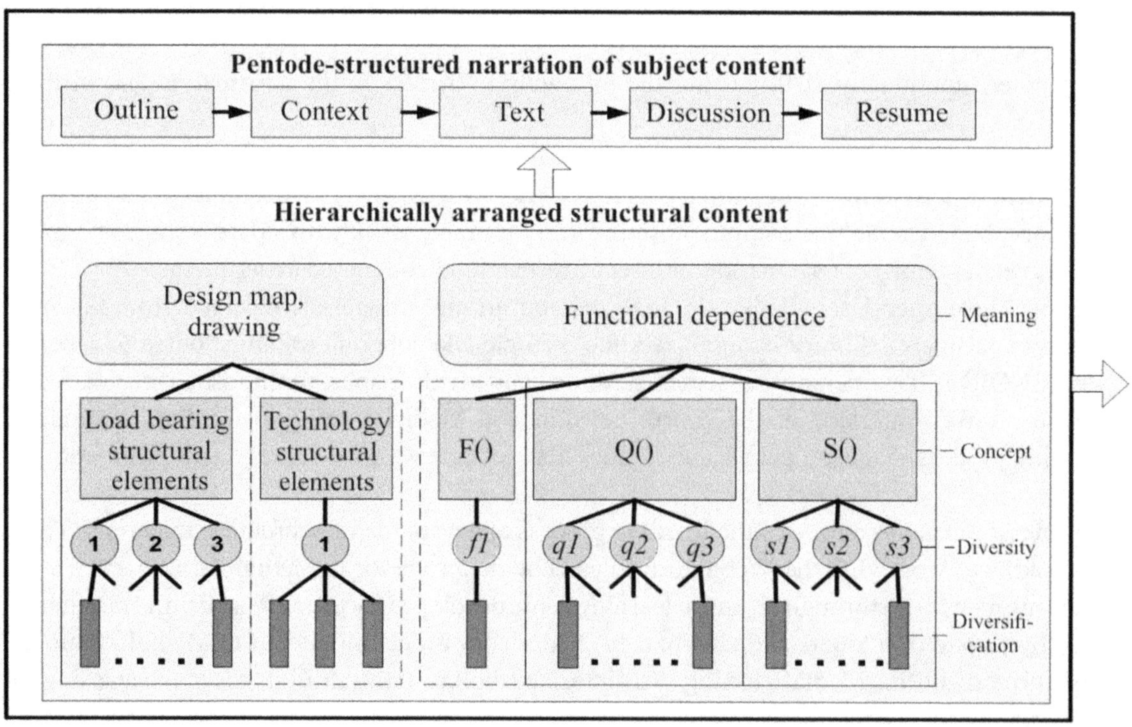

FUTURE RESEARCH DIRECTIONS

As noted above, the problem-modular training techniques are based on compression of training information, modularity, and rating of knowledge and skills acquired by students. At that, training information compression refers to compilation, consolidation, systematization, and generalization of knowledge using the achievements of knowledge engineering. This tendency is expected to increase in the future, the forming databases of knowledge and instruction materials on achieving a set didactic purpose will be subjected to even greater "compression", and new forms of knowledge formalization and transfer will be applied. The concept of "one picture speaks more than 1000 words, and one scheme speaks more than 1000 pictures" will be growing in topicality. Mechanisms of dynamic semantic conceptual search for information, CMAPs and MIND MAPs, etc. will be in greater demand. Developers of the systems for the learning process electronic support will concentrate their efforts on creating interactive tools for dynamic management (including reverse management) of the learning process, as well as introducing multimedia aids and conceptual navigation taking into account individual preferences of a student.

CONCLUSION

With the modular training system, students may find independence or self-sufficiency in their work. This proposed individual training program includes a targeted action plan, data bank, and guidance to achieve a set didactic purpose. At the same time, professors' roles vary depending on a student's level of competence.

ACKNOWLEDGMENT

The work was performed as part of the project of the state order Ministry of Education of Russia: project number N° 2.4176.2017/46.

REFERENCES

Aleksandrov, A. A., Fang, K., Proletarsky, A. V., & Neusypin, K. A. (2012). Conception of complex continuous education with innovative information technologies. In *Proceedings of the 2nd International Conference on Education and Education Management. S. "Advances in Education Research"*(pp. 374-378). Academic Press.

Archan, S. (2005). Modularisierung: ein Weg zur Steigerung der Attraktivität der Lehre in Österreich [Modularisation: A Pathway to increase the attraction of apprenticeship training in Austria]. *BWP Berufsbildung in Wissenschaft und Praxis*, *4*, 23–25.

Belous, V. V., Bobrovsky, A. V., Dobrjkov, A. A., Karpenko, A. P., & Smirnova, E. V. (2012). Multi-criterion integral alternatives' estimation: Mentally-structured approach to education. In *Proceedings of the 2nd International Conference on Education and Education Management. S. "Advances in Education Research"* (pp. 215-224). Academic Press.

Brockmann, M., Clarke, L., Mehaut, P., & Winch, C. (2008). Competence-based vocational education and training (VET): The cases of England and France in a European perspective. *Vocations and Learning, 1*(3), 227–244. doi:10.100712186-008-9013-2

Choshanov, M. A. (1996). *Flexible technology of problem-modular training*. Public Education.

Computer Managed Instruction Guidelines for Interoperability (CMI001) Version 3.5. (2001). Aviation Industry CBT Committee (AICC).

IEEE 1484.11.2 Standard for Learning Technology. (2003). *ECMA Script Application Programming Interface for Content to Runtime Services Communication.*

IMS Content Packaging Information Model, Version 1.1.4 Final Specification. (2004). IMS Global Learning Consortium, Inc.

Kolesnikova, I. A. (2009). *Theory and practice of modular transformation of learning environment at educational institution.* Saint-Petersburg.A. I. Herzen.

Medvedev, V. E., & Dobriakov, A. A. (2013). Conceptual clauses and peculiarities of the cognitively-structured training of engineering elite. In *Proceedings of the International Conference on Interactive Collaborative Learning, (ICL)* (pp. 203-207). Academic Press.

Naumov, I.S., &Vykhovanets, V.S. (2016). Using syntactic text analysis to estimate educational tasks' difficulty and complexity. *Automation and Remote Control, 77*(1), 159-178.

SCORM 2004 4th Edition Run-Time Environment Version 1.0. (2009). Advanced distributed learning.

Shakhnov, V., Vlasov, A., Rezchikova, E., & Zinchenko, L. (2013). Visual learning environment in electronic engineering education. In *Proceedings International Conference on Interactive Collaborative Learning (ICL)* (pp. 389-398). Academic Press. 10.1109/ICL.2013.6644605

Shakhnov, V.A., Zhuravleva, L.V., & Vlasov, A.I. (2013). Nanoengineering Ontology. *International Research Journal, 12*(19), 50-67.

Vlasov, A. I., & Yudin, A. V. (2011). Distributed control system in mobile robot application: General approach, realization, and usage. *Communications in Computer and Information Science, 156*, 180–192. doi:10.1007/978-3-642-27272-1_16

Vlasov, A. I., Zhuravleva, L. V., Sharipov, N. R., & Sharipova, A. F. (2012). Architecture of adaptive multiservice information and training systems. *Scientific Review (Singapore), 6*, 152–154.

Yudin, A., Kolesnikov, M., Vlasov, A., & Salmina, M. (2017). Project oriented approach in educational robotics: From robotic competition to practical appliance. In M. Merdan, W. Lepuschitz, G. Koppensteiner, & R. Balogh (Eds.), *Robotics in education. Advances in intelligent systems and computing* (Vol. 457, pp. 83–94). Springer. doi:10.1007/978-3-319-42975-5_8

Zhuravleva, L., & Shakhnov, V. (1995). Fachspile als Bestandteil der Ingenierpadagogik. Ingenieurausbildung and Strukturveranderungen am Arbeitsplatz des des Ausgehenden 20. Jahrhunders. In *Proceedings 24. Internationalen Symposiums. Ingenieurpadagogik, 95*, 244–247.

Zhuravleva, L., & Shakhnov, V. (1998). Features of business games in engineering education. In Proc. Padagogische Probleme in der Ingenieurausbildung (pp.287-288). Academic Press.

Zhuravleva, L. V. (2012). Formalization of lecture material. In *Proceedings Tenth International Symposium "Intellectual systems"* (pp. 576-580). Academic Press.

ADDITIONAL READING

Aleksandrov, A. A., Neusipin, K. A., Proletarsky, A. V., & Fang, K. (2012) Innovation development trends of modern management systems of educational organizations. *Proc. Proceeding of 2012 International Conference on Information Management, Innovation Management and Industrial Engineering.* ICIII 2012. (pp.187-189). 10.1109/ICIII.2012.6339951

Berduygina, O. N., Vlasov, A. I., & Kuzmin, E. A. (2017) Investment capacity of the economy during the implementation of projects of public-private partnership. Investment Management and Financial Innovations. T. 14. N° 3. (pp.189-198).

de Bruijn, E., & Volman, M. (2000). Changes in occupational structure and occupational practice. a challenge to education. *European Journal of Women's Studies. T., 7*(4), 455–474. doi:10.1177/135050680000700414

Gavrilina, E. A., Zakharov, M. A., Karpenko, A. P., Smirnova, E. V., & Sokolov, A. P. (2016) Software system meta-3 for quantitative evaluation of student's meta-competencies on the basis of analysis of his or her behavior in social networking services. *Proc. Procedia Computer Science 12th. "12th International Symposium Intelligent Systems, INTELS 2016"*. (pp.432-438).

Ibragimov, G. I., Ibragimova, E. M., & Bakulina, L. T. (2016) The state and prospects of development of problem-based learning in higher education. IEJME: Mathematics Education. T. 11. N° 4. (pp.881-889).

Orekhovskaya, N. A., Lavrentiev, S. Y., Khairullina, E. R., Yevgrafova, O. G., Sakhipova, Z. M., Strakhova, I. V., ... Vishnevskaya, M. N. (2016). Management of young professionals in the labor market. *International Review of Management and Marketing, 6*(2), 254–269. URL http://www.econjournals.com/index.php/irmm/article/view/2098/pdf

Shpak, M. A., Smirnova, E. V., Karpenko, A. P., & Proletarsky, A. V. (2016) Mathematical models of learning materials estimation based on subject ontology. Advances in Intelligent Systems and Computing. T. 450. (pp.271-276).

Yudin, A., Kolesnikov, M., Vlasov, A., & Salmina, M. (2017) Project oriented approach in educational robotics: from robotic competition to practical appliance. Advances in Intelligent Systems and Computing. T. 457. (pp.83-94).

KEY TERMS AND DEFINITIONS

BMSTU: Bauman Moscow State Technical University.

CAM: SCORM content aggregation model.

CMI: Computer-aided instruction.

Commercial Software: Developed software for commercial purposes and that a price a license.

Data Mining: A computer process that uses artificial and machine learning methods to discover hidden information or relations in large amounts of data.

EUROBOT: An international amateur robotics contest for teams of young people. This is organized through student projects or independent clubs.

LCMS: Learning content management system.

LMS: Learning management system.

R&D: Research and development.

RTE: SCORM run-time environment.

SN: SCORM sequencing and navigation.

VLE: Virtual learning environments.

Chapter 11
CSRP:
System Design Technology of Training Information Support of Competent Professionals

Vitaly Vladimirovich Martynov
Ufa State Aviation Technical University, Russia

Peter Sakál
Slovak University of Technology in Bratislava, Slovakia

Alexey Skuratov
Directorate of Scientific and Technical Programs, Russia

Elena Ivanovna Filosova
Ufa State Aviation Technical University, Russia

Alena Alekseevna Zaytseva
Ufa State Aviation Technical University, Russia

Elena Shavkatovna Zakieva
Ufa State Aviation Technical University, Russia

ABSTRACT

This chapter proposes a new model of managing educational institutions' activities to provide staffing needs: customer synchronized resource planning (CSRP). It describes a technology that rebuilds the learning process in order to reduce the time needed to prepare staff adequately with the competencies required by employers as requested by the economy sector. At present the development of an open system for the educational institution is being carried out. This system is able not only to create an educational program dynamically, which allows us to get the right number of specialists with the desired competencies in the minimum period, but also to rebuild the agency's management system for new tasks: to generate the necessary training materials, make changes in the timetable, and rebuild the educational portal by adding new data.

DOI: 10.4018/978-1-5225-3395-5.ch011

INTRODUCTION

Over the past decade or two advancements in educational technology have taken place so swiftly that it threatens to revolutionize the education system (Linda Van Ryneveld, 2015). At present specialists training in higher educational institutions need to meet the rapidly changing conditions of the information society and economy development. Today there is a lack of qualified personnel in a number of professions, and the demand is not fully in line with the supply provided by educational institutions. In addition, employers may believe that a large number of graduates from educational institutions cannot apply practically the obtained knowledge and skills, they have a low motivation to work, and they often do not have the competencies required fora particular job. This new approach is especially important in the training of engineers. The flexibility in using new technologies in HEIsis reflecting the socio-economic changes affecting the needs of the student population across the world(Narduzzi & Campbell, 2015)

It is possible to increase the quality and effectiveness of the training by analyzing the requirements of those who are interested in their competitiveness (the educational institution, the state, and the employer), automating the process of learning and evaluating specialists' competitive capability (Martynov, Filosova, & Guzairov, 2014).

Training should be carried out with a focus on the criteria that assess the training quality (Bogoslovskiy et al., 2007). When training specialists in a technical university, the following criteria are of high priority:

- Degree to which graduates have acquired certain competencies
- Intellectual, personal, social, and psychological characteristics of a future specialist (such as communication skills, creative thinking, etc.)
- Focus on future professional activities (desire to work according to a specialty and high motivation to work)

In addition to the common forms of improving the quality of specialists' education, operational training of a specified number of trainees with the competencies required by employers is increasingly becoming an urgent task. This chapter focuses on the introduction of such educational platforms that effectively train specialists at any level of education with a set of desired requirements (in the form of competencies) for his or her level of knowledge. According to its purpose and functional properties, such an information system is the closest one to the CSRP systems in industry (Martynov, & Filosova, 2014).For innovation to take place in higher education, the organizational culture of institutions of higher education and leadership should support such initiatives (Zhu, 2015).

USING CSRP TECHNOLOGIES IN EDUCATION

CSRP systems solve the problem of the individual urgent orders. Such systems presuppose continued possibilities to control the external elements of the production chain. In the CSRP systems with typical enterprise resource planning(ERP)system functions, the function of customer lifecycle support has been added. In the education system, such customers are primarily the representatives of the real sector of the economy, who make individual orders for specialists with the competencies required by the given employer.

To implement the CSRP systems in an educational institution, it is necessary:

- To optimize the activities of the educational institutions, building an effective information infrastructure
- To integrate the representatives of the real sector of the economy into the divisions of the organization, aimed at servicing these customers who plan and conduct training
- To create the technological infrastructure by implementing open technologies capable of supporting the integration of employers, educational institutions, and training control applications.

Successful application of CSRP systems in industry improves the quality and value of the goods or services for the customer, reduces the delivery time and operating costs, creates the infrastructure to meet the needs of the buyer, and improves feedback with the buyer.

Today open CSRP systems are becoming increasingly popular. They are able to integrate a variety of technologies and applications, to collect individual applications produced by different vendors, and to get a single unified application for the manufacturing management. At present the development of an open system for the educational institution is being carried out. This system is able not only to create an educational program dynamically, which allows us to get the right number of specialists with the desired competencies in the minimum period, but also to rebuild the agency's management system for new tasks: to generate the necessary training materials, make changes in the timetable, and rebuild the educational portal by adding new data.

PRINCIPLES OF DEVELOPING CSRPSYSTEMS IN EDUCATION

In addition to the common forms of improving the quality of specialists' education, the operational training of a specified number of trainees with the competencies required by employers is becoming an increasingly urgent task.

This problem can be solved by forming dynamically changing educational programs up to and including personal training planning that will take into account a combination of the acquired knowledge, with its practical application corresponding to vocational interests of a future bachelor's or master's degree. Also, formation of the dynamic learning plans can be useful when improving the skills of workers in different industries to master the missing competencies. The process of forming educational programs promptly allows us to prepare a predetermined number of trainees with the competencies required by the employers. This can be shown in the form of a value-added chain diagram (Figure 1) (Martynov, Tikhonova, Filosova, & Cherkasov, 2013).

Let us consider some of the features of the formation of such a system in detail.

The first stage of CSRP system construction in education is formation and formalization of the employer requirements to the necessary specialists. That is why one of the tasks is to create a domain thesaurus and construct and use a mechanism of determining the necessary competencies.

The competency is understood as a formal system characteristic, which is described as a set of requirements for the knowledge, skills and qualities of the employee for a function, position or role in the organization. Competency is associated with the subject area. If the specific system-wide signs of the knowledge, experience and skill of the specialist coincide with the requirements of the employer for a particular position or the performance of a specific project, then the relevant work should be performed by such a specialist qualitatively and effectively.

Figure 1. Value-added chain diagram model of the process of dynamically forming educational programs

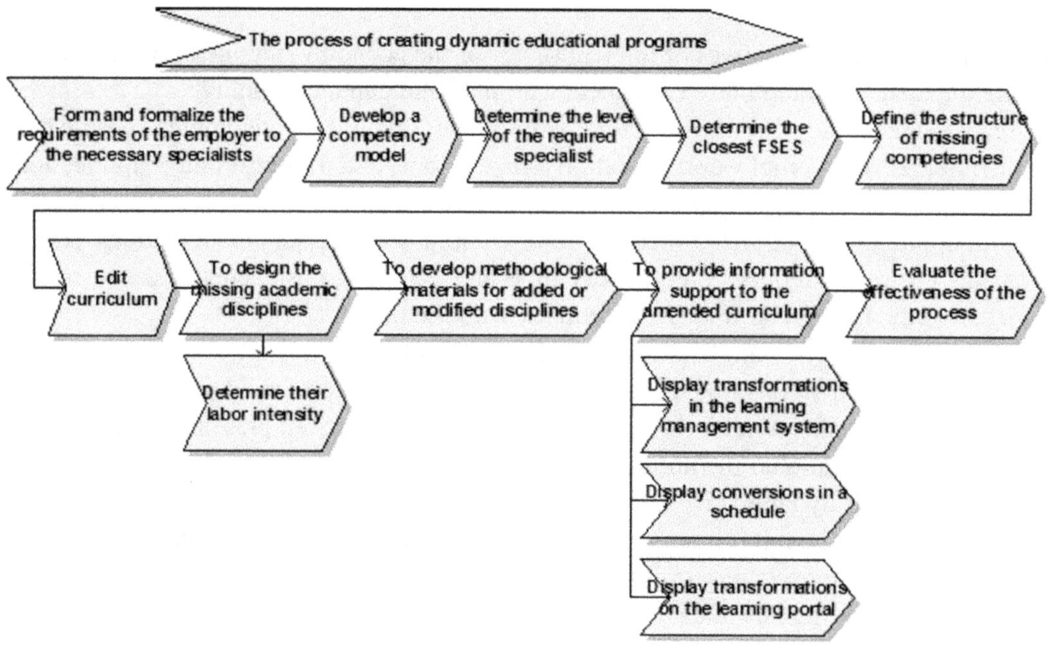

In terms of testology, "competency" is convenient to interpret as the name of the scale, and "competence" - as the level on the scale(Bogoslovskiy et al., 2007). This scale is based on a model of mastering knowledge and skills, according to which workers gradually acquire more experience as they progress through the career ladder. Each specific type of competence is divided into levels, starting from zero, where the employee does not need to have any competence to perform work, up to the maximum level where the highest competence is required. The competence scale for one competence may vary in accordance with the requirements imposed on the employee of this specialty.

In general, the formation of coherent requirements to the specialist competence includes the following steps:

- Construction of the production processes' business model
- Identification of the key professional competencies
- Harmonization of the requirements with the government education fund (GEF)
- Analysis and formation of the hierarchical structure of requirements
- Approval of a set of requirements by the parties concerned
- Creation of the sequence in the form of a training program
- Objective analysis of the educational program's feasibility

Professional standards have already been developed for a number of fields, and there are plans to implement them for another 800 jobs (Decree of the President of the Russian Federation vol. 577 of May 7, 2012). One of the problems concerning the selection of specialists by the employer is the failure of qualification criteria described in these standards or job descriptions to conform to the competencies described in the educational standards, if the standard has not been developed yet.

Therefore, the first task in reaching an agreement is to form the object domain thesaurus and then construct and use the mechanism to determine the required competencies. This thesaurus can be formed as an ontological database, and general requirements are further generated on the basis of the developed ontology (Martynov, Filosova, &Filosova, 2015). The developed ontology was the basis of the established automated information system (AIS), making it possible to choose the federal state standard of higher professional education (FSES), which will be the closest one to the requirements with respect to the structure and content of the basic competencies put forward by the employer.

The employer in the AIS generates the requirements based on the FSES and professional standard. These requirements can be clarified at the request of the university. On the basis of the clarified requirements, the FSES is selected and the employer gets a list of the most suitable areas of training, and the percentage of compliance to the requirements is indicated.

Formation of the employer's requirements proceeds in several stages. First, the employer chooses a profession from the professional standards; if there is no such profession, the employer enters it himself. Further, the employer determines the required competencies on the basis of the professional standards. The last step is to add the competencies associated with the requirements of a particular employer. If a set of competencies and those of the third-generation FSES overlap, the specialist's training area is chosen.

The algorithm determines the direction of specialist training closest to the developed competence model according to the structure and content of the basic competencies, which satisfies a set of required competencies and minimally differs from the competencies formed on the basis of the training standard. This algorithm is presented in the form of the model shown in Figure 2 and is described in Martynov et al. (2013) and Martynov, Filosova, Sharonova, &Shiryaev (2015).

There are still a lot of unfulfilled competencies (Figure 3) (which are not at present covered by FSES), the formation of which is done by filling a variable part of the educational program, consisting of the national-regional component (NRC) and the subjects as may be selected by the students (SSC). Alternatively, they can be formed within the scope of additional education.

Next, it is necessary to make a list of subjects that will allow us to form all the necessary specialist competencies according to the employer's requirements. To design or change the curriculum of the training areas closest to the competence formed, qualification criteria are decomposed up to the competence level and then to the discipline level. Based on the previous decomposition, the compliance matrix of competencies and the components of the educational program (disciplines, training units) are designed. Further, the curriculum in this area is created according to the compliance matrix.

On the basis of the competencies list that is applicable to a specialist but not formed according to the selected area of study, the design of changeable and replaceable disciplines is performed. Then it is necessary to develop a set of training materials and methodological support for these disciplines.

Currently, many universities conduct research focused on the development of an automated system to design a training package (TP). The analysis showed that most of them are directed to the automation of accessing the developed materials, as well as to providing paperwork support. The disadvantage of the proposed solutions is the lack of support of the development of educational materials and their automated search for the relevant disciplines and formed competencies.

To solve these problems, the development of the automated system for the creation and maintenance of the TP is under way. The implementation of this subsystem will help form the TP in accordance with the curriculum in the shortest time. A reduced period of development is achieved through the automatic formation of a set of necessary documents according to the curriculum and the list of competencies that should be covered.

Figure 2. The process of the nearest standard selection

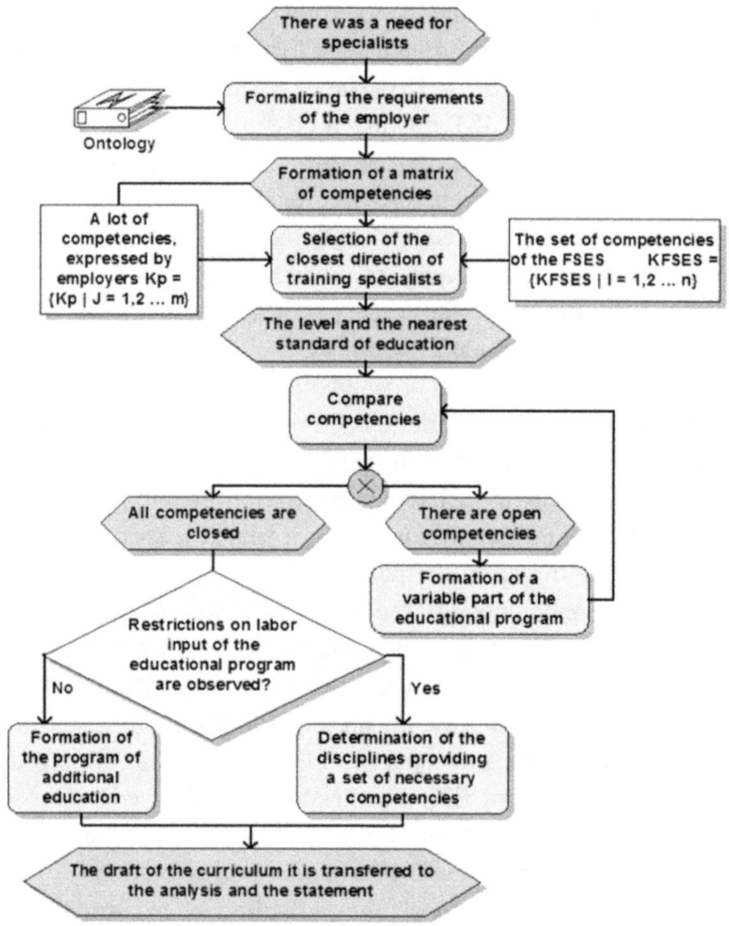

TP paperwork will also be implemented automatically. Coordination of documents will be carried out remotely in the style of modern document management systems, thus saving the time of TP coordination. This system will help the faculty and students make better use of information resources when preparing for the disciplines due to the clarity and completeness of the provided materials (Martynov, & Zaytseva, 2015).

The mathematical model of TP selection from the knowledge base can be presented as follows:

$$EMC_j = \bigcup_{i=1}^{n} E_{ji}, \quad (1)$$

where EMC_j–a set of teaching materials for a specific educational plan (*EP*), j – a number of subjects in the *EP*, $j \in (1;m)$, m – a number of subjects in the EP, E_{ji}– the ith element of the CMD for the jth discipline and n – a number of items of the CMD (the composition is determined by the university training package).

Each TP element consists of a document template, the information from the curriculum, and the content:

Figure 3. Unfulfilled competencies

Training direction	Percentage of realization	Unfulfilled competencies
230100	27.17%	The ability to carry out selection of the source data for the design of reports, articles and papers on the scientific and technical conferences The ability to conduct modeling of processes and systems The ability to carry out the calculation of health and safety conditions Willingness to participate in the work on the final design and development of information technologies in the course of implementation and operation of information systems The ability to use technology development facilities of professional activity in different areas The ability to participate in the formulation and conduct of experimental studies The ability to execute operating results obtained in the form of presentations The willingness to ensure the security and integrity of information systems and data technology
230400	23.52%	Develop components of software systems and databases using modem tools and programming techniques Participate in setting up and commissioning of hardware -software complexes Install hardware and software for information and automated systems

$$E_{ji} = P_{strji} \cup !_{EPji} \cup !_{jiq}, \tag{2}$$

where P_{strji}– the structure of the developed element (template), C_{EPji}– information contained in the EP (labor input, sequence of disciplines, etc.), C_{jiq}– informative material on the subject developed by a teacher on the basis of information sources and personal experience.

The educational plan includes a range of disciplines:

$$EP = \bigcup_{j=1}^{m} Disc_j, \tag{3}$$

where *EP*– educational plan, $Disc_j$– j^{th} discipline of the *EP*.

TP is being developed for each discipline:

$$Disc_j \rightarrow EMC_j \tag{4}$$

FSES provides competencies that should be formed within the scope of this direction. A competence model is complemented by regional components as well. The program must generate a number of competencies (K_l) of the student, as follows:

$$EP \rightarrow \bigcup_{l=1}^{a} K_l, \tag{5}$$

where a – the total number of competencies generated by the student as part of the EP.

The materials (EM_{lf}) should be created with an aim to developing the student's necessary competencies, and these materials are placed based on the teaching material knowledge:

$$K_l = \bigcup_{f=0}^{m} EM_{lf}, \tag{6}$$

where EM_{lf}– educational material contributing to the formation of a competence.

Later, during the formation of new CMD in this database, materials relevant to competencies are selected and formed as part of the discipline:

$$KB = \bigcup_{l=1}^{a}\bigcup_{f=0}^{m} EM_{lf}, \tag{7}$$

where KB– knowledge base of educational materials.

The educational material may be formed by different resource materials:

$$EM_l = Int \cup Lib \cup KB, \tag{8}$$

where Int– material from Internet resources, Lib– materials from the library.

Approval of the training package will be done remotely in the style of modern document management systems, thus saving time when coordinating the TP. This system will help the academic staff and students use information resources and prepare the subjects effectively due to the clarity and completeness of the provided materials.

FUTURE RESEARCH DIRECTIONS

Dynamic design and formation of educational-methodical support of realization of educational programs can be used to solve the problem of increasing the effectiveness of the training demanded by employers graduates through in-depth analysis of the requirements of all stakeholders in their competitiveness (University – state employer), automation of build process of the curriculum and training. In future the development of this system, the formation of a holistic educational environment supporting the educational process at all its stages. It is also planned to develop tools to support the formation of a substantial part of the educational materials using the technology of the semantic web. More about how to develop CSRP systems can be found in the article Martynov, V.V., &Filosova, E.I. (2014). The ontological approach in educational activities is observed in article Martynov, Filosova, Sharonova, &Shiryaev(2015).

CONCLUSION

To provide information support for changing the curriculum, it is necessary to reflect the changes not only in the learning management system in which the new curricula should appear, but also on the university site and the schedule of classes. For communication between the different information systems of any company (as well as the university), it is necessary to build an optimum information architecture.

Evaluation of the effectiveness of forming the necessary professional competencies is often performed on the basis of statistical data on the status and results of the students' individual achievements (Filosova, & Rykova, 2010). This estimation aims to check the knowledge availability in specific subjects taught according to the curriculum and skills to be used in typical situations. To this effect, the procedures of the educational establishment's licensing, psycho-diagnostic tests, physiological monitoring, monitoring of the development of the regional system of education, and examination and certification procedures are used.

The use of the objective approach to structuring knowledge, competence approach, and modern information technology makes it possible to improve the efficiency of training engineers to meet the requirements of fast-growing sectors of the economy. There is an opportunity to optimize the learning process by building flexible, individualized learning paths and improving the quality of the obtained competencies and current changes in the educational process.

ACKNOWLEDGMENT

This contribution is based on the results of project APVV number LPP-0384-09: "HCS 3E Concept Model vs. Concept of Corporate Social Responsibility (CSR)" and project KEGA number 037STU-4/2012 "Implementation of the subject 'Corporate Social Responsibility Entrepreneurship' into the Study Programme Industrial Management in the Second Degree at MTF STU Trnava."

REFERENCES

Bogoslovskiy, V. A., Karavaeva, E. V., Kovtun, E. N., Melehova, O. P., Rodionova, S. E., Tarlyikov, V.A., & Shehonin, A.A. (2007). Guidelines for the design of assessment tools for the implementation of multi-level educational programs HPE with competent approach. In Higher education in Russia (vol. 10, pp. 3-10). Moscow: Moscow Polytechnic University Press.

Campbell, J. D., & Narduzzi, J. L. (2015). The More Things Change: Reflections on the State of Marketing in Continuing Higher Education. In Centennial Conversations: Essential Essays in Professional, Continuing, and Online Education (pp. 309-17). Washington, DC: UPCEA.

Filosova, E. I., & Rykova, O. V. (2010). The use of ontologies in the object learning technologies. In *Proceedings of the Seventh International Scientific Conference "New Educational Technologies in High School"* (*vol. 2*, pp. 123-127). Ekaterinburg, Russia: Federal State Autonomous Educational Institution of Higher Education "Ural Federal University named after the first President of Russia B.N. Yeltsin".

Martynov, V. V., & Filosova, E. I. (2014). Methods for developing CSRP-system in education for human capacity SMART-region. In *Proceedings of the Workshop on Computer Science and Information Technologies (CSIT2014)* (*vol. 2*, pp. 180-184). Ufa State Aviation Technical University Press.

Martynov, V. V., Filosova, E. I., & Filosova, V. K. (2015). The use of ontologies in the task of drawing up training plans for the requirements of employers' information technology. In *Proceedings of the International Scientific-Practical Conference "Problems and Solutions"* (vol. 1, pp. 242-246). Publishing House "East Press".

Martynov, V. V., Filosova, E. I., & Guzairov, M. B. (2014). Improving the efficiency of training requirements for employers. In *Proceedings of the International Scientific-Practical Conference "Innovative Information Technologies"* (pp. 376-381). Moscow: National Research University Higher School of Economics.

Martynov, V. V., Filosova, E. I., Sharonova, J. V., & Shiryaev, O. V. (2015). The technology of ontological analysis in educational activities. In *Proceedings of the Workshop on Computer Science and Information Technologies (CSIT'2015)* (*vol. 1*, pp. 173-178). Ufa State Aviation Technical University Press.

Martynov, V. V., & Zaytseva, A. I. (2015). Educational methodical complex development support system. In *Proceedings of the 2015 Workshop on Computer Science and Information Technologies (CSIT2015)* (*vol. 1*, pp. 183-185). Ufa State Aviation Technical University Press.

On measures to implement state social policy. (2012, May 7). Decree of the President of the Russian Federation vol. 577.

Van Ryneveld. (2015). Introducing Educational Technology into the Higher Education Environment: A Professional Development Framework. In Innovative Professional Development Methods and Strategies for STEM Education (pp. 126-136). Hasan Kalyoncu University.

Zhu. (2015). Organisational culture and technology-enhanced innovation in higher education. *Technology, Pedagogy and Education, 24*(1), 65-79. doi:10.1080/1475939X.2013.822414

ADDITIONAL READING

Intel World Ahead Program. (2009). The Positive Impact of eLearning. Retrieved from https://www.intel.la/content/dam/www/program/education/us/en/documents/positive-impact-of-elearning.pdf Khan, B. (2005). Managing E-Learning Strategies: Design, Delivery, Implementation and Evaluation. Hershey, PA: IGI Global; doi: Learning Technologies. (2017). E-Learning Awards Categories and Criteria, Retrieve from http://www. learningtechnologies.co.uk/enter-an-award/award-categories10.4018/978-1-59140-634-1

Ning, W. (2014). The Latent Appreciation Effect of Interactive Design in Internet Communication, IEEE 11th Intl Conf. on Ubiquitous Intelligence and Computing and IEEE 11th Intl Conf. on Autonomic and Trusted Computing and IEEE 14th Intl Conf. on Scalable Computing and Communications and Its Associated Workshops (pp. 723-726). doi:10.1109/UIC-ATC-ScalCom.2014.65

Technologies, L. (2017). E-Learning Awards Categories and Criteria, Retrieve from http://www. learningtechnologies.co.uk/enter-an-award/award-categories

Weinreich, H., Obendorf, H., & Mayer, M. (2008). Not Quite the Average: An Empirical Study of Web Use University of Hamburg. In *Proceedings of the 15th International Conference on the World Wide Web* (Vol. 2). 10.1145/1326561.1326566

Zualkernan, I. A. (2006). A framework and a methodology for developing authentic constructivist e-Learning environments. *Journal of Educational Technology & Society*, *9*(2), 198–212. Retrieved from http://www.ifets. info/journals/9_2/16.pdf

KEY TERMS AND DEFINITIONS

Competence: It is the possession of a specific competence (i.e., knowledge and experience of their own activity), allowing to make objective judgments and to make the correct decisions.

Competency: The ability to act successfully on the basis of practical experience, skill, and knowledge in solving professional problems. Is understood as a formal system characteristic, which is described as a set of requirements for the knowledge, skills and qualities of the employee for a function, position or role in the organization.

CSRP Systems: Resource planning synchronized with the customer. CSRP includes a complete cycle from designing the future product to customer requirements, to warranty and after sales service. The essence of CSRP is to integrate the buyer into the enterprise management system. In this case, not the sales department, but the buyer places an order for the manufacture of products, he is responsible for the correctness of its execution and, if necessary, monitors compliance with the terms of production and delivery. The enterprise can very accurately track the demand trends for its products.

ERP: Organizational strategy for integrating production and operations, managing human resources, financial management, and asset management focused on continuous balancing and optimization of enterprise resources through a specialized integrated software package providing a common data and process model for all areas of activity.

Federal State Standard of Higher Professional Education: A set of mandatory requirements for the formation of a certain level and (or) the profession, specialty and direction of training, approved by the federal executive body, which exercises the functions of developing public policy and regulatory legal regulation in the field of education.

Object Approach: A programming paradigm in which the basic concepts are the concepts of objects and classes (or, in the less well-known version of prototyping languages, prototypes).

Training Package: The standard name for the totality of educational and methodological documentation, teaching and control tools developed in the higher school of the Russian Federation for each discipline. The TP should include complete information sufficient for the passage of discipline. TPs are designed to ensure the openness of the educational process and should be available to anyone who wants.

Chapter 12
The Usage of GIS in Realizing Engineering Education Quality

Aleksandr Kolesenkov
Ryazan State Radio Engineering University, Russia

Aleksandr Taganov
Ryazan State Radio Engineering University, Russia

ABSTRACT

The chapter has considered research and instructional methodology aspects for development of methodological, informational, and instrumental, ensuring of the education quality management system which are necessary to be taken into account in modern conditions. Mathematical bases of the geoinformation system application for monitoring of the education process realization quality have been developed. Model, method, and algorithm for quality assessment of the educational process realization in institutions have been unfolded. A way of representing some fuzzy production rules in solving application tasks of fuzzy modeling and executing the process of approximate reasoning on educational risks has been introduced. A fuzzy production system of educational risk analysis on the basis of using modified fuzzy Petri nets has been realized. Analysis of possibilities to apply suggested approaches for monitoring of institutions at various levels has been conducted.

INTRODUCTION

Process of monitoring and quality assessment of the education service provision is important for contemporary educational systems. It is necessary to develop a monitoring system realizing specific functions of management and fitting the general system of education (Gurov, Koryachko, Taganov, Moiseenko, & Taganov, 2010) for qualitative realization of educational programs in institutions. Such system is necessary at all levels of the education management for informational support of the management decision procedure to organize, optimize, modernize and increase the educational program realization quality in institutions (Heyneman, & Lee, 2014).

DOI: 10.4018/978-1-5225-3395-5.ch012

Let's mark out specific functions of the education management system (Xin, Li, & Li, 2007):

- Prediction of quality dynamics of the educational process realization;
- Organization of the education quality management process, distribution, assignment and realization of functions;
- Control of the education process realization quality according to characteristics of its potential, current state and result (Brown, 1999);
- Regulation of the educational process realization quality, provision of related indicators;
- Quality assessment of the educational process realization and possibilities of its improvement;
- Quality research of the educational process realization in the area of historical aspects of its formation, limitations, priorities, negative influences and critical factors (Taganov & Taganov, 2006);
- Motivation for actions directed to increase the education quality.

Let's mark out main specific functions of the education quality management (Tan, 2012):

- Professional and teaching staff quality management;
- Student teaching quality management;
- Educational process realization technology quality management;
- Informational and methodological provision quality management;
- Material and technical support quality management;
- Institution infrastructure quality management.

Function composition characterizes a specific nature of the management object and real problems of its functioning and development (Glebova, & Kuznetsova, 2012).

PROCEDURE

Assessment of the education quality is suggested to be accomplished by estimating and monitoring quality of educational program realization in institutions, so a crucial task is to develop a system executing a collection, processing and analysis of information on state of educational programs in order to monitor educational programs (EP) periodically (Dohmen, 1999).

Monitoring of EP refers to a collection, processing and provision of data in relation to set criteria and indices aimed at ensuring efficiency and increase of the educational process quality. Taking into account feedback on the basis of intermediate results for corresponding criteria allows making alterations in educational programs promptly (Attfield, & Vu, 2013).

Criteria and Function of the Education Quality Assessment

All criteria of EP realization quality assessment can be divided into 2 groups (Kolesenkov, & Taganov, 2015):

- Criteria $\{Z_i, i = \overline{1,N}\}$, for which the best value is minimal;
- Criteria $\{R_j, j = \overline{1,M}\}$, for which the best value is maximum.

Model of the educational program monitoring process is suggested to be represented as following:

$$f_m = f(K,T), \tag{1}$$

where K – a set of constant and variable criteria of educational programs which change their values in time t and under influence of various factors; T – a periodicity of fixation of the object state change.

$$K = \{Q, R\} \tag{2}$$

Object state means a set R consisting of sets of attributes $Q = \{Q_i, i = \overline{1,N}\}$ and $R = \{R_j, j = \overline{1,M}\}$.

For assessment of EP the following effectiveness function is suggested to be used:

$$\xi = \frac{M}{N} \cdot \sum_{i=1}^{N} Q_i \setminus \sum_{j=1}^{M} R_j, \tag{3}$$

where Q_i - a normalized value of each from i - cost criteria; R_j - a value of each from i - result criteria.

Development of the function ξ according to years of education will allow to reveal dynamics of the educational program realization quality and execute as well rating assessment of departments responsible for the educational program realization (Sarkisyan, 1977).

Application of Geoinformation Technologies

The use of geographical informational technologies (GIS) will ensure interconnection of attributive and spatio-temporal data taking into account geographic reference to the cartographic base from departments of institutions. It will readily simulating processes and estimating realization of EP (Ninomiya, & Urabe, 2011).

On the basis of GIS-technologies we suggest to execute:

- analysis of the EP current state;
- prediction of dynamics of the EP state;
- prediction of dynamics of a set of criteria;
- analysis of demand for graduates who learnt EP successfully;
- assessment of efficiency of distribution of institutions realizing EP;
- assessment of mobility of graduates who learnt EP successfully;
- assessment of scientific potential of graduates who learnt EP successfully;
- simulation of various processes in institutions realizing EP.

GIS structure represents a multilevel relational model containing a set of layers and objects including geographical data, their connections and attributive information (Kostrov, & Baranchikov, 2014). Each layer includes a table containing information on elements of the cartographic base. Besides, this GIS platform allows to quickly extract information requested by a customer from the data base in the required format and also visualizing it on the map.

ALGORITHM FOR MONITORING OF THE INSTITUTION EDUCATIONAL PROCESS

The following algorithm is suggested for constant monitoring of EP in institutions (Kamens, & McNeely, 2009).

- Choice of criteria and methods for acquisition of their normalized values.
- Calculation of statistical indicators.
- Development and initialization of a specialized set of layers and objects.
- Integration of attributive information.
- Acquisition of data from departments of institutions realizing EP.
- Calculation of the EP effectiveness function according to formula (3).
- Prediction of dynamics of set criteria.
- GIS-simulation of processes in institutions realizing EP.
- Development of the assessment rating of departments by means of GIS facilities according to values of statistical characteristics.

GIS Structure

Task of data storage in GIS is offered to be carried out by putting additional information contained in tables of the integrated or internal data base in correspondence with each graphical object (point, line, polygon) (Dinham, 2013).

Appeal to data from the data base is suggested to by realized be means of SQL-requests, which nature and complexity will be determined by a type of information contained in tables. Application of such approach will let filter data by set parameters, combining tables, sorting and generalizing data.

The following modules are meant to be included into GIS for monitoring of online-learning educational programs (Figure 1):

- Data collection module.
- Data transformation module.
- Data analysis module.
- Module for monitoring of online-learning educational programs.
- Module for assessment of online-learning educational program state.
- Module for simulation of processes in institutions realizing a program of online-learning.
- Module for prediction of dynamics of criteria and indices of the online-learning educational program efficiency.

Executed theoretical researches have shown that application of geoinformation technologies for monitoring of online-learning educational programs in institutions of higher education are reasonable and efficient (Pegat, 2013). It was found out that a part of parameters can have a fuzzy nature that requires an application of a fuzzy logic device in algorithms of data processing. The suggested conceptual technology to be fulfilled in the geoinformation media ArcGIS 10.3.

Suggested model for assessment of efficiency of online-learning educational programs in institutions of higher education will allow considering this process completely since not only economical but also pedagogical and social parameters are taken into account. Such model is also applicable for assessment of efficiency of the educational program realization by all forms of education in institutions of various levels.

Current state of geo information technologies allows developing and implementing a system which is innovative in its analytical possibilities, realizing models, methods and algorithms for monitoring of online-learning educational programs. Practical usage of such specialized GIS will ensure an increase of the management decision-making efficiency in the organization, realization and modernization due to acquisition of reliable information, its processing and analysis.

Modern theories and practice of geoinformation system development for educational risk analysis and monitoring under conditions of miscellaneous given geodata require developing some new effective approaches and algorithms to support decision making on educational risks under fuzzy conditions.

ANALYSIS AND MONITORING OF THE EDUCATIONAL RISKS

Along with classical approaches to structural and functional construction of GIS analysis and monitoring of educational risks the present paper has considered a new approach to organization of the process for analysis and monitoring of educational risks under conditions of fuzziness based on application of contemporary theory and practice for analysis and monitoring of risks under sophisticated software projects. In the suggested approach the process for analysis and monitoring of educational risks as part of GIS includes an algorithm consisting of the following steps (Wheeler, & Bailer, 2009):

- **Identification of Educational Risks:** Determination which risks can influence the quality of the educational project implementation of the region and documentary preparation of their characteristics.

Figure 1. Composition of GIS for the educational program monitoring

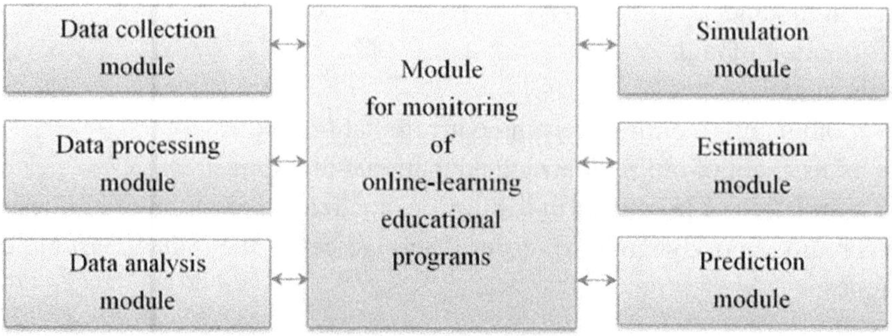

- **Qualitative Analysis of Educational Risks:** Arrangement of risks due to their priority for following analysis or processing by estimation and summation of a possibility of their occurrence and influence of the educational project implementation quality.
- **Quantitative Analysis of Educational Risks:** Quantitative analysis of potential influence of identified risks on general purposes to maintain the quality of the educational project implementation.
- **Educational Risk Response Planning:** Development of possible variants and actions contributing to increase of favorable possibilities and decrease of hazards.
- **Monitoring and Control of Risks:** Tracking of identified risks, monitoring of residual risks, identification of new risks, execution of plans to react to risks and estimation of their effectiveness within life cycle of the quality of the educational project implementation.

An important step of the present algorithm is a qualitative analysis of educational risks executed by expert methods. As a result of this analysis experts form a scaled list of educational risks grouped in categories:

- A list of risks requiring an immediate response;
- A list of risks for additional analysis and response;
- A list of risks of low priority requiring observation.

A formed list of educational risks is a base for execution of the algorithm following stages for system analysis and monitoring of educational risks. Besides, the process of the qualitative analysis of educational risks is a rather labor-intensive process. In order to formalize the process current methods for support of expert decision making under conditions of fuzziness are suggested to be used.

The existing fuzzy production systems of decision support are meant for the realization of a fuzzy inference process and are a conceptual framework of modern fuzzy logic.

Such were the gains in the application of these systems for solving a wide range of control tasks that conditioned choosing the mathematical tools for the formal characterization of the process of educational risk analysis and reduction on the basis of using models, methods and algorithms of the theory of fuzzy sets and fuzzy Petri nets.

Petri nets and their numerous modifications are one of the model classes, a certain advantage of which is an adequate representation possibility of not only the structure of complex organizational and technical systems and complexes but also logic-time specific features of their operation processes. Petri nets are a mathematical model for representing the structure and analysis of the system operation dynamics in terms of «event condition». It can be used for analyzing risk events and revealing potential risks of the educational profile.

An important type of Petri nets is fuzzy Petri nets allowing to constructively solve tasks of fuzzy modeling and fuzzy control in which fuzziness is of subjective nature. In this regard some certain prospects of studying the possibilities of fuzzy Petri nets application for the description and formal characterization of risk control processes and appears as well as the purposes of risk analysis under educational data fuzziness conditions.

On the basis of using modified fuzzy Petri nets can be formulated an experimental version of a computer-aided system of decision support on educational risks (Leonenkov, 2005). This system is in fact an expert system showing the fuzzy logic of input values interrelation – the expert analysis of the risk and output values – the truth degrees of possible educational risks. Scanning of the Fuzzy Petri Net can be effected using an interface represented in Figure 2. Positions of fuzzy Petri nets corresponding to input and output data are marked by color.

PRACTICAL RESULTS

As a result of the theoretical research execution the following points have been developed:

- Model of the educational program geoinformation monitoring process;
- Algorithm for calculation of the educational program realization effectiveness function;
- Functional system of the geoinformation system for educational program monitoring.

Experimental researches represent software realization of the suggested approaches and algorithms in GIS-media ArcGIS 10.3. According to their results the following conclusions can be made:

- Application of GIS for EP monitoring is reasonable and effective from the point of view of prompt informational support for decision-making by institution management;
- Since values of some criteria can belong to fuzzy sets it is necessary to develop technologies for correct operation in conditions of uncertainty for assessment of EP quality.
- Calculations of efficiency criteria for principal educational programs (PEP) have been fulfilled by the example of Ryazan State Radio Engineering University (Figure 3).

Figure 2. Form for scanning of the Fuzzy Petri Net

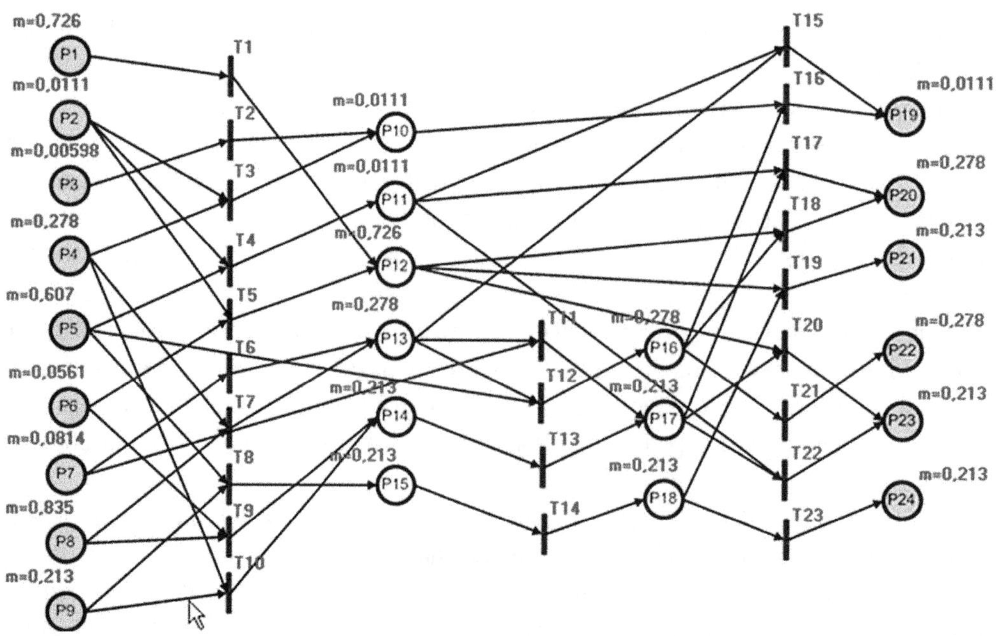

Figure 3. Results of calculation of PEP effectiveness function of the institution

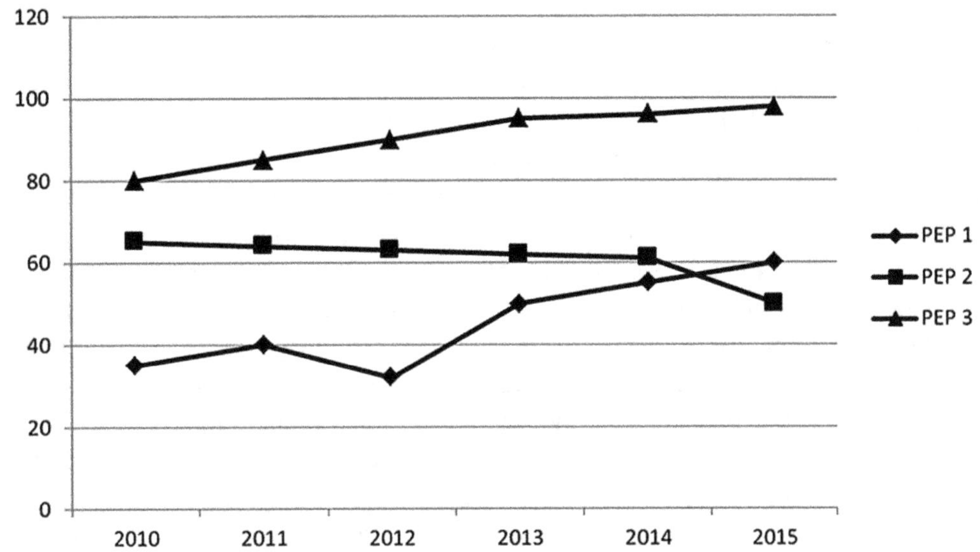

Timely feedback from persons successfully learning EP online will allow to make changes in the educational process at once and that will positively influence quality of services provided by institutions.

FUTURE RESEARCH DIRECTIONS

Several promising future directions of research were identified.

The effect obtained by a person and society is difficult to quantify. This is the main feature of educational services, leading to the emergence of risk. Therefore, the development of technologies for management of educational risk based on modern approaches and systems is an urgent task. The development of the presented technology gives an opportunity to assess the educational risk in educational institutions.

According to the results of the experimental researches, the proposed approach to monitoring educational risk is effective and applicable for informational support of the managerial decision-making procedure in the field of higher education management.

Another research area is the quality monitoring of large enterprises and organizations that have a network of distributed branches.

The program implementation of the characteristics evaluation for the higher education institutions directions is realized as an independent library. It can be integrated into any automated system, where similar technologies are used. Currently, the library contains a classification method that takes on the input attributes of analytics (points, enrollments, places, trained professions, paid training), and at the output gives out the leaves of the tree.

The usage of the presented technology in mobile devices and application based on IOS/Android allows to increase the quality of the results obtained, as well as the analyzed characteristics number.

CONCLUSION

Scientific approach to assessment of the educational process quality leads to significant generalization in the area of educational program realization result. Stable conception of competence has been developing for a long time and includes not only monitoring of the education quality but also assessment of possibility of this quality realization for professional activity.

Modern integrated GIS for the educational process realization quality management becomes a comfortable tool for managers at all levels. Usage of GIS allows not only to increase management efficiency due to automation of collection, processing and analysis of data on realization of educational processes but also improve a level of its management culture. GIS integrated with other systems for the institutional education quality management is a basic integrated system of management in current conditions and should be based on the integrated field of knowledge containing structural and semantic representations of different models and actual data and also mechanisms for their processing.

Suggested model of the EP efficiency assessment in institutions will allow to completely estimate processes occurring in departments since pedagogical, economic and social aspects are taken into account. It is scalable and can be applied for assessment of EP realization efficiency for intramural, extramural and intra-extramural forms of study in institutions of secondary and higher education.

At the macro-level suggested facilities can be used for assessment of institution work in the region by various indices with following visualization of results on intuitive maps-diagrams.

As a result of software realization of the algorithm for analysis of educational risks using an mathematical instrument fuzzy Petri nets an effective and usable software product has been obtained for application in practice both independently and as a part of industrial GIS. The considered way of representing the rules of fuzzy productions in a fuzzy production system of educational risk analysis and reducing based on applying fuzzy Petri nets allows broadening the existing notions of the models, methods and means of the analysis of the educational risks. It opens up new possibilities for applied researches and practical realization of the formal procedures of the decision support within the fuzzy production GIS for analyzing and monitoring educational risks.

Modern GIS technologies provide a possibility to develop and integrate a system having innovative analytical functional and realizing new effective technologies, methods and algorithms for tasks of EP monitoring. Practical application of the developed technologies and system will help increase efficiency of decision-making for the institution management due to prompt processing of great volumes of data.

ACKNOWLEDGMENT

The research is carried out in the Grant of the President of the Russian Federation N° SP-553.2015.3.

REFERENCES

Attfield, I., & Vu, B. T. (2013). A rising tide of primary school standards: The role of data systems in improving equitable access for all to quality education in Vietnam. *International Journal of Educational Development*, *33*(1), 74–87. doi:10.1016/j.ijedudev.2012.02.003

Brown, G. (1999). The Quality of Learning of Learning and How to Assess it. *Lifelong Learning in Europe., IV*(99), 47–54.

Dinham, S. (2013). The quality teaching movement in Australia encounters difficult terrain: A personal perspective. *Australian Journal of Education, 57*(2), 91–106. doi:10.1177/0004944113485840

Dohmen, G. (1999). Lifelong Learning for All – innovative perspectives of continuing education. *Lifelong Learning in Europe., 4*(99), 154–158.

Glebova, L. N., & Kuznetsova, M. D. (2012). Monitoring kachestva vysshego pedagogicheskogo obrazovaniya [Monitoring of quality of higher pedagogical education]. Logos.

Gurov, V. S., Koryachko, V. P., Taganov, A. I., Moiseenko, V. P., & Taganov, R. A. (2010). Opyt sozdaniya i primeneniya resursov elektronnoy informatsionno-obrazovatel'noy sredy po napravleniyu CALS i CASE tekhnologii [Experience of development and application of resources of the electron informational and educational media in the direction CALS and CASE-technologies]. In *Papers of the VII All-Russian scientific and methodical conference Telematics-2010.* Academic Press.

Heyneman, S. P., & Lee, B. (2014). *The impact of international studies of academic achievement on policy and research. In Handbook of international large-scale assessment: Background, technical issues and methods of data analysis* (pp. 37–72). Boca Raton, FL: Taylor and Francis Group.

Kamens, D.H., & McNeely, C.L. (2009). Globalization and the growth of international educational testing and national assessment. *Comparative Education Review, 54*(1), 5–25.

Kolesenkov, A.N., & Taganov A.I. (2015). Kontseptsiya geoinformatsionnoy tekhnologii monitoringa obrazovatel'nykh programm onlayn-obucheniya [Conception of geoinformation technology for monitoring of online-learning education programs]. In *Open and online education.* Tomsk: Publishing House of Tomsk State University.

Kostrov, B.V., & Baranchikov, A.I. (2013) Teoriya i metody issledovaniya modeley i algoritmov predstavleniya dannykh dlya predmetnykh oblastey s ranzhiruyemymi atributami [Theory and methods of researches of models and algorithms for representation of data for subject areas with ranged attributes]. *RSREU Vestnik, 47*, 59-64.

Leonenkov, A. V. (2005). Nechetkoye modelirovaniye v MATLAB i fuzzyTECH [Fuzzy modeling in MATLAB and fuzzyTECH]. St. Petersburg: Academic Press.

Ninomiya, A., & Urabe, M. (2011). Impact of PISA on education policy: The case of Japan. *Pacific-Asian Education, 23*(1), 23–30.

Pegat, A. (2013) Nechetkoye modelirovaniye i upravleniye [Fuzzy simulation and management]. Binom.

Sarkisyan, S. A. (1977). Teoriya prognozirovaniya i prinyatiya resheniy [Theory of prediction and decision making]. Higher school.

Taganov, A. I., & Taganov, R. A. (2006). *Metodicheskiye osnovy sozdaniya informatsionnykh sistem sfery obrazovaniya* [Methodical fundamentals for development of information systems for education: Tutorial.]. Ryazan: RSREU.

Tan, C. (2012). The culture of education policymaking: Curriculum reform in Shanghai. *Critical Studies in Education, 53*(2), 153–167.

Wheeler, M. W., & Bailer, A. J. (2009). Benchmark dose estimation incorporating multiple data sources. *Risk Analysis, 29*(2), 249–256. doi:10.1111/j.1539-6924.2008.01144.x PMID:19000080

Xin, T., Li, F., & Li, L. (2007). An International Comparison of Elementary Education Quality Assessment. *Journal of Beijing Normal University*, 44–47.

ADDITIONAL READING

Alavi, M., & Leidner, D. E. (2001). Review: Knowledge management and knowledge management systems: Conceptual foundations and research issues. *Management Information Systems Quarterly, 25*(1), 107–136. doi:10.2307/3250961

Arthur, J. B., & Aiman-Smith, L. (2001). Gainsharing and organizational learning: An analysis of employee suggestions over time. *Academy of Management Journal, 44*(4), 737–754. doi:10.2307/3069413

Brown, J. S., & Duguid, P. (2001). Knowledge and organization: A social-practice perspective. *Organization Science, 12*(2), 198–213. doi:10.1287/orsc.12.2.198.10116

Cummings, J. N. (2004). Work groups, structural diversity, and knowledge sharing in a global organization. *Management Science, 50*(3), 352–364. doi:10.1287/mnsc.1030.0134

Goodman, P. S., & Darr, E. D. (1998). Computer-aided systems and communities: Mechanisms for organizational learning in distributed environments. *Management Information Systems Quarterly, 22*(4), 417–440. doi:10.2307/249550

Liao, L.-F. (2008). Impact of manager's social power on R&D employees' knowledge-sharing behaviour. *International Journal of Technology Management, 41*(1/2), 169–182. doi:10.1504/IJTM.2008.015990

Lin, H.-F. (2007). Effects of extrinsic and intrinsic motivation on employee knowledge sharing intentions. *Journal of Information Science, 33*(2), 135–149. doi:10.1177/0165551506068174

Mohammed, S., & Dumville, B. C. (2001). Team mental models in a team knowledge framework: Expanding theory and measurement across disciplinary boundaries. *Journal of Organizational Behavior, 22*(2), 89–106. doi:10.1002/job.86

Parise, S. (2007). Knowledge management and human resource development: An application in social network analysis methods. *Advances in Developing Human Resources, 9*(3), 359–383. doi:10.1177/1523422307304106

Reagans, R., & McEvily, B. (2003). Network structure and knowledge transfer: The effects of cohesion and range. *Administrative Science Quarterly, 48*(2), 240–267. doi:10.2307/3556658

KEY TERMS AND DEFINITIONS

ArcGIS: Series of geographical information software products of American company ESRI.

Education Quality: Conformity of education to the needs and interests of the individual, society, state.

Educational Program: The complex of the main educational characteristics, organizational pedagogical conditions, and appraisal forms.

Educational Risk: An element of events uncontrolled development probability with the realization of educational services because of objective exist uncertainty.

Fuzzy Petri Nets: Mathematical instrument for dynamic discrete systems modeling in fuzzy environment.

Geographical Information System: The system of collecting, storing, analysis and graphical visualization of orientation reference data and information connected with them.

Geographical Information Technologies: Information technologies of geographical organized information processing.

Geoinformation System: The system of collecting, storing, analysis and graphical visualization of orientation reference data and information connected with them.

GIS: Geographical information system.

Management: Coordinated activities for the management and management of the organization.

Monitoring: Continuous process of object parameters control and registration in comparison with specified criterions.

Quality: The totality of the characteristics of an object related to its ability to satisfy stated and perceived needs.

Risk: An activity, connected with uncertainty overcoming in a stark choice situation, where there is a possibility of numerical and qualitative estimation of supposed result achievement probability, failure, and divagations from the purpose.

University: Higher educational institution.

Chapter 13
An Approach to Improvement of Master's and PhD Studies in Data Processing and Management Systems

Artem A. Sukhobokov
Bauman Moscow State Technical University, Russia

Vitaliy Baklikov
Optimal Management LLC, UK

Dmitry S. Lakhvich
Bauman Moscow State Technical University, Russia

Andrey V. Sukhobokov
Optimal Management LLC, Russia

Ilya V. Tikhonov
Bauman Moscow State Technical University, Russia

ABSTRACT

To make Master's and PhD theses more influential on the evolution of the software industry, and to make them even more effective and successful, the authors propose that they be directed towards developing cognitive systems and expanding the functionality of integrated big data platforms. Within this field, the authors propose that their themes be organized in the form of the following dual model: for each thesis that develops an emerging new cognitive component inside a core big data platform, there is another thesis that develops a corresponding cognitive component/module within the application layer that uses that component of the core big data platform.

DOI: 10.4018/978-1-5225-3395-5.ch013

INTRODUCTION

In the segment of software industry that deals with the development of business applications, there is a clear trend towards the adoption of new and innovative approaches to business processes. IoT, Blockchain, methods of AI, Big Data analytics, and the use of social media data – all have a significant impact on certain business functions and all, in some cases, transform business models completely.

Universities play a major role in the process of the development and integration of innovative solutions. The enterprise software market is growing by about 6-8 percent annually (Gartner, 2017a; Pang, 2016; Allied Market Research, 2015). The accelerating dynamics will be preserved if more universities, those which were not previously known for their breakthroughs or for outstanding alumni, could take leading positions in modern research and in professional training. In this paper, the authors offer a general approach to these challenges in the area of Data Processing and Management Systems.

If the authors are to take a look at Ph.D. programs, then it becomes evident that the majority of time is spent on research, preparing papers, and preparing the thesis, which ends with the defense. It exceeds the amount of time spent studying various courses, engaged in one's teaching practice, etc. Examples demonstrating such proportions are given in (University of Minnesota, 2017; Massachusetts Institute of Technology, 2017). For Ph.D. programs, such proportions are reasonable—but the same is true, to a large extent, for master's programs. This is because during the two-year term, students are involved in scientific work, prepare publications, and gather practical experience; thereby gathering the material for the thesis and writing the thesis itself. In other words, the execution of a research project, the preparation of papers, and the preparation of the thesis plays a key role in the quality of the professional education of the student. Herewith, the authors mean the quality of professional education as a tolerably-limited set of student indicators:

- The possession of knowledge about a set of interrelated disciplines that are present in the educational program;
- The ability to use existing knowledge to solve professional problems;
- The ability to solve professional problems;
- The ability to learn within the problem-solving process;
- The ability to conduct research in a specific professional field;
- The possession of a set of knowledge, skills, experience, and aptitudes that employers find attractive.

To increase the quality of education, topics of theses must be formulated in such a way that the research is conducted on the most relevant topics within the software industry and modern science. Only then will they have the necessary rationale and lie at the forefront of scientific novelty. Successful identification of such topics for theses requires high qualification, wide professional erudition, and some intuition regarding how things are going to be progressing. One of the signs of the relevance of the topics is if it belongs to the Innovation Trigger area of the Gartner Hype Cycle for Emerging Technologies (Gartner, 2016).

Meanwhile, despite the novelty, all of these theses' topics must be such that Master's and Ph.D. students could finish their research in a specified or reasonable amount of time.

In this paper, the authors aim to offer a certain approach for forming master's and Ph.D. thesis topics in the Data Processing and Management Systems area for universities at various stages of educational maturity, which would allow them to enter new heights on the world stage.

BACKGROUND

Organizational Background

The authors of the paper have developed and delivered lectures and laboratory practicums for the "Big Data Technologies" and "Tools of Business Analytics" courses offered in 2016-2017 within the Data Processing and Management Systems graduate program of Bauman Moscow State Technical University.

The "Big Data Technologies" course was conducted during the 2016 spring semester and included the description of an architecture and functionality of:

- Hadoop distributions;
- Application suites delivered alongside the Cloudera, Hortonworks, and MapR Hadoop distributions;
- More than 20 applications within the Hadoop ecosystem;
- Several Big Data SQL DBMS;
- More than 10 Big Data NoSQL DBMS;
- Open-source integrated Big Data platforms—Apache Spark, Apache Flink, and Apache Apex (in various stages of maturity);
- Proprietary Integrated Big Data platforms—SAP HANA and IBM Bluemix;
- Several sets of services from the Microsoft Azure platform—Microsoft Cognitive Services, Microsoft R Server, Microsoft Azure Machine Learning, and Microsoft Graph Engine.

The "Tools of Business Analytics" course was conducted during the 2017 spring semester and included:

- Business Intelligence and Analytics Platforms;
- Predictive Analytics and Machine Learning Platforms;
- Data Warehouse and Data Management Solutions for Analytics;
- Corporate Performance Management Solutions.

At the end of the course, three lectures were given on topics that are currently three of the fastest-growing fields in Big Data and Cognitive Systems: IoT, Blockchain, and Deep Learning.

Next year, based on this material, the authors plan to develop two new courses: "Promising Applications of Big Data & Machine Learning" and "BI in Corporate Applications". The first course will dive deeply into the applications and use cases, and the second course will concentrate on the adoptability of new approaches towards existing corporate systems such as ERP, EAM, CRM, HCM, SCM, etc.

The students, while taking either or both courses, expressed a great interest in the subject and the ability to capture the latest developments as they emerge. Big Data and Machine Learning technologies, and the tools to which these courses are devoted, are currently undergoing explosive developments. In some cases, the course programs were adapted on the fly to accommodate for the new releases of the software that were released during the term. Students enthusiastically readjusted to the new versions of the software that usually offered a richer feature set. For example, this was evident during the transition from SAP HANA SPS 11 to SAP HANA SPS 12 for using Graph Engine as well as with Microsoft Cognitive Services for using new services.

The courses were conducted for a parallel stream of 60 students (3 groups). At the end of the master's program, 12 students entered the postgraduate program in 2017. Such interest in continuing education may partly be due to students' interest in modern trends of information technologies, which was raised by the authors' who conducted these courses. Witnessing such an interest, the authors think that the education process should be fine-tuned to benefit all three parties involved in the process – the students, the university, and the software industry (the products and technologies of which form the foundation for each course of the master's program considered in the next section).

Master's Program in "Data Processing and Management Systems"

The master's program in "Data Processing and Management Systems" is focused on preparing professionals in corporate applications:

- Consultants, software architects and developers that integrate packaged and preexisting solutions;
- Consultants and software architects for pre-sales of IT systems;
- Consultants, software architects, and developers extending and maintaining existing products or developing new products;
- Consultants, architects, and developers for bespoke development;
- Consultants in tech support roles;
- IT specialists in non-IT companies planning the implementation and development of new applications and interacting with contractors.

It is assumed that the prospective students for this master's program had already passed the following courses:

- Mathematical analysis;
- Linear algebra and analytical geometry;
- Graph theory;
- Mathematical logic;
- Probability theory and statistics;
- Methods of optimization;
- Numerical analysis;
- Programming languages;
- Imitational modeling;
- Databases;
- Internet Technologies;
- Operating Systems.

The authors propose a well-balanced plan for the "Data Processing and Management Systems" master's program, focusing on teaching specialist competencies for the corporate world. In the meantime, it has a rather flexible structure that could be fine-tuned for other programs in the area of software development. The plan is shown in Figure 1.

Figure 1. The proposed plan for the master's program in the direction of data processing and management systems

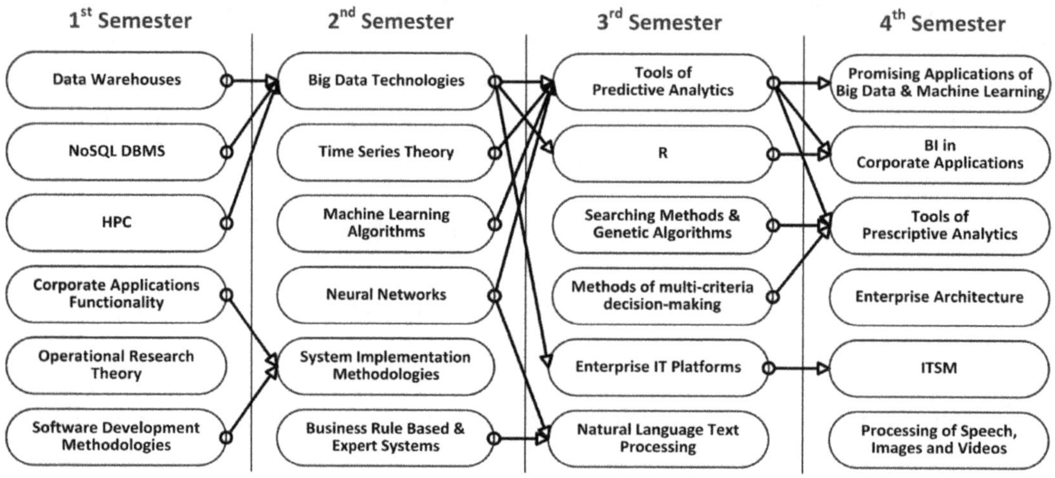

Figure 1 fully depicts the relationships only between those two courses that were prepared by the authors and which formed the foundation for this article. Other relationships are depicted in fragmented form, so as not to overcomplicate the organizational plan. The "Tools of Business Analytics" course, which was developed and conducted by the authors, should in future be renamed as "Tools of Predictive Analytics" as this name better reflects the content. Alongside this course, a "Tools of Prescriptive Analytics" course should be offered, in which optimization and simulation tools will be considered to suggest decision options on how to take advantage of a future opportunity or mitigate a future risk as well as to show the implication of each decision option.

FUTURE-ORIENTED AREA OF RESEARCHES

The authors see the development of cognitive systems on top of Big Data platforms as the most promising direction in the evolution of corporate systems and applications. The authors will now review each of the two foundational parts of this synthesized approach.

Integrated Big Data Platforms

The term (Integrated) Big Data Platform emerged 1-2 years after the first platforms actually appeared (Guess, 2011; Lawson, 2011). A strict and straightforward definition of the term has been absent up until now; but vendors, consultants, and market analysts intuitively understand it and widely use the term (Teradata, 2017; Rose Technologies, 2016; Cloud News Daily, 2015). All of them discuss the components and the functionality of such platforms. The authors use this term to highlight a software tool that is intended for Big Data processing and that includes several instruments of various types. Several of these software tools are listed under Data Management Solutions for Analytics as described by Gartner (2017b). The authors changed the list contained there for the following reasons:

- Open Source Integrated Big Data Platforms such as Apache Spark, Apache Flink, and Apache Apex were not considered by Gartner;
- Some Data Management Solutions for Analytics from Gartner's review are not designed for Big Data processing as they cannot distribute the processing of data between the servers of the cluster;
- Some Data Management Solutions for Analytics from Garner review are not designed to include several components and are, by their functions, either regular DBMS or Data Warehouses.

The reason that our choice of integrated platforms lies within the area of Big Data Tools is caused by the fact that this segment is being intensively developed. Spark is one of the most mature Open Source Integrated Big Data Platforms available on the market and has also been the most active Open Source project in Big Data for several years in a row (Zakharia, 2014; Zakharia, 2015; Databricks, 2016). The same situation is observed within proprietary platforms. Although companies do not publicly disclose such data, it is witnessed through frequent updates of the products and the marketing hype around SAP HANA, IBM Bluemix and other products of this class.

Some components of the integrated platforms can be inferior to analogous independent products. For instance, GraphLab Create is very effective in optimizing the transfer of data between nodes of the cluster during intensive computations over large graphs that are widely distributed over multiple nodes (Turi, 2017). Apache Mahout "Samsara" has its own language for developing Machine Learning algorithms (Lyubimov & Palumbo, 2016). Nevertheless, integrated platforms beat independent modules in their popularity and adaptation by the masses (King & Magoulas, 2017, p. 15; Dinsmore, 2017). For the most part, out-of-box compatibility between various components and wide popularity of the product are factors that far outweigh specific functionality, which often requires additional effort for integration.

The analysis performed during preparation of the course materials helped the authors to generalize the process of the development and proliferation of integrated platforms and also helped the authors to establish a clear picture regarding the future:

1. Open Source integrated platforms emerged as a result of the occurrence of multiple independent applications within the Hadoop ecosystem. Integrated platforms were compensating for the shortcomings of HDFS and allowing for more effective utilization of RAM. Their modules were also combining and integrating disparate applications into a single set of tools that were production-ready out of the box.

The majority of proprietary integrated platforms came from cloud computing providers in the form of additional services – which, for the most part, offered more functionality for application's development with using of cloud resources.

It may seem that there is some incommensurability: Apache Spark, which the authors consider to be an Open Source Integrated Platform, could be launched within public clouds such as Amazon Web Services or Microsoft Azure – service offerings which themselves are also considered by the authors to be integrated platforms. But no matter how, and in what form, the services have come to users, if they combine several different capabilities, the authors consider it to be one of the integrated platforms for Big Data processing.

2. Today, the similarity of components amongst various platforms is evident. The components which are presented within at least one of the selected Integrated Big Data Platforms are listed in the right-hand column of Table 1. (SAP, 2017; IBM, 2017; Microsoft, 2017a, 2017b, 2017c, 2017d; Amazon, 2017; Google, 2017a, 2017b; Apache, 2017a, 2017b, 2017c).

The Integrated Big Data platforms market segment is rapidly developing. The data in Table 1 was refined in April 2017 while working on this paper; and by this time, most of the platforms that are also mentioned in Sukhobokov (2016) from December 2016 have been updated or enhanced with additional components. Certainly, from the table above, it is hard to judge the complexity of the specified item – whether it is a simple system or a sophisticated multifunctional platform with developed interfaces. However, it is not that relevant for this publication – the authors only want to determine the direction of the Integrated Big Data platforms.

The authors believe that further development of these platforms will be expanded by new components that would have a widespread applicability across industries. For example, such components will include libraries of parallel numerical methods for discrete and continuous problems (non-linear optimization methods, packages for differential equations solving, and optimal control packages, etc.); systems for imitational modeling; methods of analysis; situational forecasting and decision making; algorithms of business management; and others. In all the aforementioned cases, parallel processing and the mechanism of distributing the calculations offered by the platforms can yield significant benefits.

3. Specific parallel applications – such as aero- and hydro-dynamic modeling systems, CAD, power network modeling systems, and oil and gas reservoirs modeling systems; as well as ERP, CRM, and SCM systems (whether On-Premise or SaaS) – will not be developed within the integrated platforms. Rather, they will form their own application layer built on top of core platforms as it is shown in Figure 2.

The development of the Integrated Big Data platforms is currently one of the main research topics in software engineering and the evolution of the software industry. The shift of master's and Ph.D. research activities towards this segment could have a significant impact on the growth of universities' international reputations as centers of excellence in preparing high-quality professionals in the most modern segments of IT. Due to the high popularity and relevance of this segment, each thesis in this area containing qualitatively new results would be regarded as an important achievement in professional circles as well as a personal scientific achievement of the student.

Cognitive Systems

The concept of cognitive systems is not strictly defined. As the concept of Artificial Intelligence develops and transforms, the understanding of which systems perform cognitive functions is gradually shifting. At present, the area of cognitive systems can be represented as follows:

Table 1. Components of Integrated Big Data Platforms

Components \ Integrated Platforms	SAP HANA[1]	IBM Bluemix[2]	Microsoft Azure[3]	Amazon Web Services[4]	Google Cloud Platform	Apache Spark	Apache Flink	Apache Apex
SQL DBMS	+	+	+	+	+	+	+	
NoSQL DBMS	+	+	+	+	+			
Graph DBMS	+	+	+					
Object Storage		+	+	+	+			
File & Doc Storage	+	+	+	+				
Graph Engine	+		+			+	+	
Search Engine	+	+	+	+	+			
Crawler	+		+					
Master Data Management Tools	+	+						
Streaming Engine	+	+	+	+	+	+	+	+
Event Processing Tools	+				+		+	
Streaming Analytics	+	+	+	+	+			
Machine Learning Library	+	+	+	+	+	+	+	
Machine Learning Automation Tools	+	+						
ETL Tools	+	+	+	+	+			
Optimization Tools		+						
R Engine	+	+	+		+	+		
Spatial Engine	+	+	+		+			
GIS Engine		+						
Business Analytics	+	+	+	+	+			
Social Networks Analytics	+	+	+					
Rapid Application Development Tools	+	+	+	+	+			+
Data Flow Tools	+	+	+	+	+			
Business Rules Engine	+	+	+					
Messages and Queues Manager		+	+	+	+			
Accept Payment Services		+						
E-Commerce services			+		+			
Online Meeting services				+				
Analytics for investment planning		+						
Services to work with bank accounts		+						
Loan Analytics		+						
Personalized hotel search		+						
Car diagnostic services		+						
Game services				+	+			
Digital Marketing Campaigns services			+					
Contact Center services				+				
Web Server for Data	+	+	+					
Email Client		+	+	+				
Email Server				+				
Data Modeling Tools	+	+	+					

[1] Including SAP HANA ecosystem: SAP HANA Vora, SAP BI Suite, SAP Data Services, etc.

[2] Including services from IBM Partners integrated with IBM Bluemix

[3] Including additional services from Microsoft working on top of the Microsoft Azure platform

[4] Without 3rd party commercial software supplied over the AWS Marketplace

Figure 2. An example of an application layer built on top of an integrated big data platform

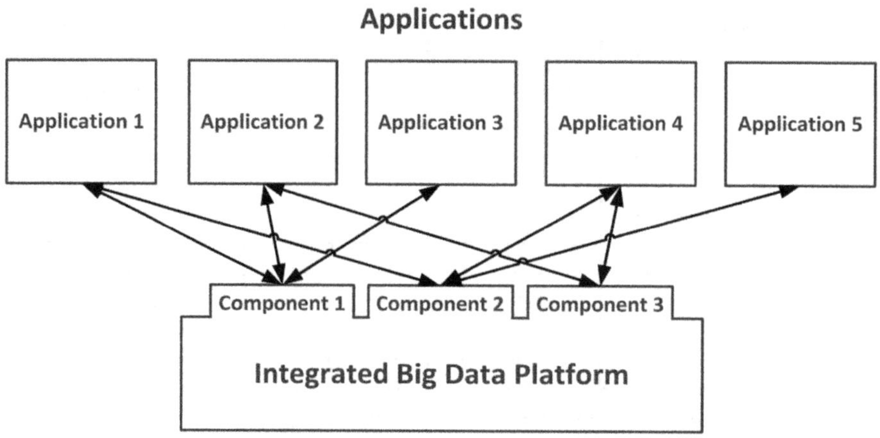

- Predictive analytics systems using the Machine Learning algorithms designed to solve problems of classification, regression, clustering, and association of data; analysis and forecasting of time series (Larose & Larose, 2015, pp. 10-17; Awad & Khanna, 2015, pp. 15-17). These algorithms based on the available data under the conditions of vague definitions of the nature of the system;
- Predictive analytics systems using knowledge of the interactions of individual elements of the system or the rules of its operation. In the first case, simulation systems are used; and in the second case, Inference Mechanisms are used;
- Systems that use neural networks to solve predictive analytics tasks and tasks of text, speech, images, and video processing;
- Adaptive systems of predictive analytics that use a number of Machine Learning models or more complex mechanisms – for example, the Inference Rules set applied to a set of models or an ensemble of parallel algorithms operated by the Machine Learning algorithm of a higher level;
- Adaptive systems using neural networks whose structure changes on the fly under the influence of incoming data. An example of a toolkit for building such networks is Chiner (Hido, 2016);
- Systems of prescriptive analytics that solve optimization problems by exact numerical methods in the presence of a formalized mathematical formulation of the problem, as well as by heuristic numerical methods, often without the formal definition of the problem;
- Systems of prescriptive analytics, constructed as optimization algorithms over the algorithms of Machine Learning or imitation models. The essence of such systems is reduced to solving the problem of selecting the most suitable values of input data or parameters for the simulation model;
- Reinforcement Machine Learning systems, solving the problems of choosing the optimal sequence of actions in a dynamically-changing environment. The basic approaches for solving these problems are divided into passive methods (including adaptive dynamic programming, temporal-difference or "TD" learning, etc.) and active methods (the most common of which is Q-Learning) (Russell & Norvig, 2010, pp. 830-859);

- Systemic Machine Learning systems are partly like the Reinforcement Machine Learning systems because they solve the problems of choosing the optimal sequence of actions in a dynamically-changing environment. But unlike the previous group of systems, they solve these problems in a changing external context. The main approaches to the implementation of Systemic Learning are the Whole-System Learning and Multiperspective Approaches (Kulkarni, 2012, pp. 23-56).

THE APPROACH TO FORMING THESIS TOPICS

To develop and expand the functionality of integrated platforms more effectively, the authors are offering the following approach based on the dual model for master's and Ph.D. theses:

1. For each thesis that addresses an emerging new component inside a core Big Data platform, there is another thesis that develops a corresponding component/module within the application layer that uses the component of the core Big Data platform.
2. Both of the components under development must be cognitive systems. For example, when one thesis develops a new numerical method or a system of imitational modeling that uses certain features of the platform such as parallel computation and/or distributed processing on numerous nodes of the cluster, there is another thesis that develops an application that uses the component from the adjacent thesis. This will promptly demonstrate the new capabilities that are added by the newly-developed component.

Since master's students have less development and scientific experience than Ph.D. students, and since their term lasts 2 years (whereas that of Ph.D. students' studies lasts 4-5 years), when forming the theses topics for master's students, several students could be researching the problem and developing one component of it. Moreover, the development of the second related component also could be undertaken by either a student or a group of students. The key point in dividing up the effort is the synchronization which is achieved by organizing small research groups of students with organizing the focus of their work around a common dual goal and supporting the cooperation of participants. During the parallel workflows, students must almost simultaneously:

- Develop their first working prototype;
- Implement the ability to debug and improve related components;
- Conclude the research in time for submission of the dissertation.

In the case of working within a team, the responsibility of each participant increases and the risk that one of the students will drop the project is reduced.

The development of new components for proprietary platforms within the university is almost impossible without tight connections with the R&D centers of the corporations that author the platforms. The development of such relationships requires long-term contracts and is less likely in the coming years. Such opportunities can exist only within a few universities around the world – for example, in the Hasso Plattner Institute (HPI), which closely interacts with SAP (HPI, 2017) and participated in SAP HANA development (McDermott, 2014, p. 286). Therefore, the authors emphasize the importance of Open

Source platforms. Moreover, the popularity of Open Source platforms within professional environments is not less than, and most likely exceeds that of, the proprietary platforms.

For faculties that do not have any experience in developing the functionality of Open Source platforms and that lack a significant number of contributors to such projects, the most effective way to get noticed could be the invention of new components for the platforms. Amongst such components could be new components that, in our opinion, should appear within Integrated Platforms in the near future, but that could also be the existing components that are currently found in proprietary platforms but which are absent in Open Source platforms.

PRACTICAL IMPLEMENTATION OF THE APPROACH

The discussed approach was used by two students who are co-authors of this paper and who entered the Ph.D. program in the 2015/2016 term. This approach was also used by two master's students who started their programs in the 2015/2016 term. They successfully earned their master's degrees in the summer of 2017, defended their theses (for which each received an excellent grade), and entered the Ph.D. program. These pilot actions offer first proof of the feasibility of the described approach. The authors will continue to use the approach to assess the opportunity to improve the level of scientific achievements of the department.

SOLUTIONS AND RECOMMENDATIONS

At first glimpse, it could appear that the offered approach for the formation of thesis topics for master's and Ph.D. programs in the area of Data Processing and Management Systems could narrow the scope of research subjects in this field. However, this assumption is not true. Cloud, hybrid, and On-Premise clusters that host Big Data solutions become more and more widespread. Multi-node technical platforms – with On-Demand resource provisioning capabilities – are the systems of the new generation (Chapman, Emmerich, Márquez, Clayman, & Galis, 2012). In several years, all applications will be hosted on such platforms (Sukhobokov, 2016). A striking example of this is the process of transferring numerous SAP applications to SAP HANA. Therefore, the approach proposed in the paper does not introduce any limitations, but rather only stimulates the transition of research works to the modern architectures.

Very similar conclusions are valid for cognitive systems. The shift of the whole developments only to cognitive systems will not be a limitation. Cognitive components will become an integral part of future applications within ERP, EAM, CRM, HCM, SCM, and other applied systems. The observed trend in the allocation of smaller services (micro services) in systems and their integration based on a service-oriented architecture will provide deeper cross-use of the entire set of data by means of BI – which, to a large extent, will consist of predictive and descriptive analytics. To interact with all categories of external stakeholders (customers, partners, suppliers, applicants for vacant positions, etc.) as well as with internal staff, bots will be used on a massive scale to build dialogues in natural language. This will improve the interfaces and make it possible to remove most of the work from personnel's interactions with people (Sukhobokov, 2016).

The situation currently being witnessed in the area of Data Processing and Management Systems – in which the barriers to entry into most modern scientific researches, and barriers to the practical implementation of these researches, have been significantly reduced – may not correspond to other scientific areas as well as Ph.D. and graduate programs. For example, the development of engineering education in nuclear technology (Nuclear Energy Agency [NEA], 2012), molecular biology, and other challenging areas requires increasingly sophisticated laboratory equipment. Experts in various areas of science and education should carefully evaluate the applicability of this approach and its components to their respective areas.

FUTURE RESEARCH DIRECTION

Evaluation of the proposed training program requires a review of analogous educational programs and similar courses taught by universities in different countries and regions, as well as courses taught by the company's training centers and distance education providers. This review will require a lot of additional research and involvement of other specialists of the department. We hope that we will be able to extend our team and move forward in this direction.

As of today, only two of the described courses are prepared and conducted. The entire educational program consisting of many such courses is presented as a proposal for discussion. Until it is not implemented in the educational process, it is impossible to assess its quality. The work on the program is continuing however. After the first implementation of the new program, we plan to assess the quality of this program.

Regardless of the chosen approach to forming thesis topics, it requires a review of the state of researchers in leading universities working in scientific field "Data Processing and Management Systems". After increasing the number of students using the approach proposed in the article for carrying out their theses, we want to explore how their opportunities can be used to consolidate information on scientific research conducted by leading universities, to form and support the topical state of such a review.

REFERENCES

Allied Market Research. (2015). *Enterprise application market.* Allied Market Research. Retrieved July 18, 2017 from https://www.alliedmarketresearch.com/enterprise-application-market

Alvesson, M., & Sandberg, J. (2011). Generating research questions through problematization. *Academy of Management Review*, *36*(2), 247–271.

Alvesson, M., & Sandberg, J. (2013). *Constructing Research Questions: Doing Interesting Research.* Los Angeles, CA: SAGE Publications. doi:10.4135/9781446270035

Amazon. (2017). *Amazon Web Services.* Retrieved July 18, 2017 from https://aws.amazon.com/products/?nc1=h_ls

Apache. (2017a). *Spark overview.* Retrieved July 18, 2017 from http://spark.apache.org/docs/ latest/

Apache. (2017b). *Apache Flink documentation.* Retrieved July 18, 2017 from https://ci.apache. org/projects/flink/flink-docs-release-1.3/

Apache. (2017c). *Apache Apex.* Retrieved July 18, 2017 from http://apex.apache.org/docs/apex/

Awad, M., & Khanna, R. (2015). *Efficient learning machines: Theories, concepts, and applications for engineers and system designers.* New York, NY: Apress Media, LLC. doi:10.1007/978-1-4302-5990-9

Chapman, C., Emmerich, W., Márquez, F. G., Clayman, S., & Galis, A. (2012). Software architecture definition for on-demand cloud provisioning. *Cluster Computing, 15*(2), 79-100. Retrieved July 18, 2017 from https://www.ee.ucl.ac.uk/~sclayman/docs/HDPC2010.pdf

Cloud News Daily. (2015). *Guide to Big Data Analytics: Platforms, software, companies['] tools, solutions and Hadoop.* Retrieved July 18, 2017 from http://cloudnewsdaily.com/ big-data-analytics/

Databricks. (2016). *Apache Spark Survey 2016: Report highlights.* Retrieved July 18, 2017 from http://cdn2.hubspot.net/hubfs/438089/DataBricks_Surveys_-_Content/2016_Spark_Survey/2016_Spark_Infographic.pdf

Dinsmore, T. W. (2017). *Spark is the future of analytics.* Retrieved July 18, 2017 from https://thomaswdinsmore.com/

Gartner. (2016). *Gartner's 2016 Hype Cycle for Emerging Technologies identifies three key trends that organizations must track to gain competitive advantage.* Retrieved July 18, 2017 from http://www.gartner.com/newsroom/id/3412017

Gartner. (2017a). *Gartner says worldwide IT spending forecast to grow 1.4 percent in 2017.* Retrieved July 18, 2017 from http://www.gartner.com/ newsroom/id/3672818

Gartner. (2017b). *Magic Quadrant for Data Management Solutions for analytics.* Retrieved July 18, 2017 from https://www.gartner.com/doc/reprints?id=1-3TZLQ0P&ct=170221&st=sb%3f

Google. (2017a). *Google Cloud Platform documentation.* Retrieved July 18, 2017 from https://cloud.google.com/docs/

Google. (2017b). *Google Cloud Platform solutions.* Retrieved July 18, 2017 from https://cloud. google.com/solutions/

Guess, A. R. (2011). *5 key issues any Big Data integration platform should address.* Retrieved July 18, 2017 from http://www. dataversity.net/5-key-issues-any-big-data-integration-platform-should-address/

Hido, S. (2016). *Complex neural networks made easy by Chainer.* O'Reilly Media, Inc. Retrieved July 18, 2017 from https://www.oreilly.com/learning/complex-neural-networks-made-easy-by-chainer

HPI. (2017). *The Hasso Plattner Institute.* Retrieved July 18, 2017 from https://hpi.de/en/the-hpi/overview.html

IBM. (2017). *IBM Bluemix catalog.* Retrieved July 18, 2017 from https://console.ng.bluemix.net/catalog/

Jeanes, E., & Huzzard, T. (Eds.). (2014). *Critical Management Research: Reflections from the Field.* Los Angeles, CA: SAGE Publications. doi:10.4135/9781446288610

King, J., & Magoulas, R. (2017). *2017 European Data Science Salary Survey.* O'Reilly Media, Inc. Retrieved July 18, 2017 from http://www.oreilly.com/data/free/files/2017-european-data-science-salary-survey.pdf

Kulkarni, P. (2012). *Reinforcement and systemic machine learning for decision making*. Hoboken, NJ: John Wiley & Sons, Inc. doi:10.1002/9781118266502

Larose, D. T., & Larose, C. D. (2015). *Data mining and predictive analytics* (2nd ed.). Hoboken, NJ: John Wiley & Sons, Inc.

Lawson, L. (2011). *Big Data platform should support data exploration.* ITBusinessEdge. Retrieved July 18, 2017 from http://www.itbusinessedge.com/cm/blogs/lawson/big-data-platform-should-support-data-exploration/?cs=48159

Lyubimov, D., & Palumbo, A. (2016). *Apache Mahout: Beyond MapReduce. Distributed algorithm design*. CreateSpace Independent Publishing Platform.

Massachusetts Institute of Technology. (2017). *PhD program in social & engineering systems.* Retrieved July 18, 2017 from https://idss.mit.edu/academics/ses_doc/

McDermott, B. (2014). *Winners dream: A journey from corner store to corner office*. New York, NY: Simon & Schuster.

Microsoft. (2017a). *Microsoft Azure solutions.* Retrieved July 18, 2017 from https://azure.microsoft.com/en-us/solutions/

Microsoft. (2017b). *Introduction to R Server and open-source R capabilities on HDInsight*. Retrieved July 18, 2017 from https://docs.microsoft.com/en-us/azure/hdinsight/hdinsight-hadoop-r-server-overview

Microsoft. (2017c). *Graph Engine. Serving big graphs in real-time.* Retrieved July 18, 2017 from https://www.graphengine.io/

Microsoft. (2017d). *Microsoft Azure Machine Learning Studio.* Retrieved July 18, 2017 from https://studio.azureml.net/

Moeini, S. (2014). *6 (very useful!) Approaches to identify research gaps and generate research questions.* Retrieved July 18, 2017 from https://www.linkedin.com/pulse/20140912150946-275561203-6-very-useful-approaches-to-identify-research-gaps-and-generate-research-questions/

NEA. (2012). *Nuclear education and training: From concern to capability.* Organization for Economic Co-operation and Development. Nuclear Energy Agency. Retrieved July 18, 2017 from http://www.oecd-nea.org/ndd/pubs/2012/6979-nuclear-education.pdf

Pang, A. (2016). *Worldwide enterprise applications market to hit $208B by 2020 as replatforming intensifies.* Apps Run The World. Retrieved July 18, 2017 from https://www.appsruntheworld.com/worldwide-enterprise-applications-market-to-hit-208b-by-2020-as-replatforming-intensifies/

Rose Technologies. (2016). *IBM Big Data Platform.* Retrieved July 18, 2017 from http://www.rosebt.com/blog/ibm-big-data-platform

Russell, S. J., & Norvig, P. (2010). *Artificial Intelligence: A modern approach* (3rd ed.). Upper Saddle River, NJ: Prentice Hall.

SAP. (2017). *SAP HANA Platform.* Retrieved July 18, 2017 from https://help.sap.com/viewer/product/SAP_HANA_PLATFORM/2.0.00/en-US

Sukhobokov, A. V. (2016). Perspektivy razvitiia system obrabotki informatsii i upravleniia v sleduiuschem desiatiletii (2020 – 2030) [Prospects of data processing and management systems development in the next decade (2020 – 2030)]. *Informatsionno-izmeritelnyie i upravliaiuschie sistemy, 2016*(12), 69-78.

Teradata. (2017). *Teradata Integrated Big Data Platform.* Retrieved July 18, 2017 from http:// teradata.ru/products-and-services/integrated-big-data-platform

Turi. (2017). *GraphLab Create.* Retrieved July 18, 2017 from https://turi.com/learn/

University of Minnesota. (2017). *Doctoral students. General degree information.* Retrieved July 18, 2017 from https://www.cs.umn.edu/academics/graduate/phd

Zakharia, M. (2014). *Spark's role in the Big Data ecosystem.* Spark Summit, San Francisco, CA. Retrieved July 18, 2017 from https://spark-summit.org/2014/wp-content/ uploads/2014/07/Sparks-Role-in-the-Big-Data-Ecosystem-Matei-Zaharia1.pdf

Zakharia, M. (2015). *Introduction to Spark.* Databricks Intern Event. Retrieved July 18, 2017 from https://www.slideshare.net/databricks/introduction-to-spark-intern-event-presentation

KEY TERMS AND DEFINITIONS

Cognitive System: The term is applied when a computer system mimics "cognitive" functions associated with human intelligence, such as "learning" and "problem solving."

Data Processing and Management Systems: The direction of Master's education which is oriented towards consulting, architecture, and development in the areas of corporate applications and computer science.

Influence on the Industry: Research activities have a significant impact on the software development industry. The results of scientific research offer new and more effective approaches and methods for solving problems and enable the creation and launching of new products on the market.

Integrated Big Data Platform: A software tool that is intended for big data processing and that includes several various instruments.

Master's and Ph.D. Thesis: A document presenting the author's research and one submitted in support of candidature for an academic Master's or Ph.D. degree.

Master's Program: A set of interrelated mandatory and elective courses studied by Master's students, as well as a list of possible research areas.

Quality of Education: The ability to equip students with the knowledge, skills, and core transferable competences they need to succeed after graduation.

Subject of Thesis: The subject of the thesis lies within the chosen specialty and determines the purpose and subject of the dissertation research. It points as well to the solved scientific problem and the results obtained.

APPENDIX

The process of Master's or PhD thesis topic's forming ideally should be split into stages and formalized. One of the most important parts of this process is identifying gaps and generating research questions. LinkedIn post (Moeini, 2014) contains a brief description of six useful approaches on how it could be done. This post has a link to an article (Alvesson & Sandberg, 2011). It is covered in more detail in the book that was released by the same authors a bit later (Alvesson & Sandberg, 2013).
One of the modern approaches that can be used to manage a set of parallel dissertational studies is CMS (Critical Management Studies). The book (Jeanes & Huzzard, 2014) discusses the practical application of CMS to manage various aspects of scientific research. There is no step-by-step detailed methodology, but many questions can be answered.

Chapter 14
Aiding the Transition of Students From School Into Technical University

Tatiana Tsibizova
Bauman Moscow State Technical University, Russia

ABSTRACT

This chapter is about different aspects of creating university-based professionally orienting environment. Issues of students' professional self-determination in transition from secondary education to high school are considered. The author suggests to arrange resource center as a training and research innovative complex for solution of youth's problems with early professional orientation, their motivation, for recruitment and selection of the most prepared for further study. As a result of the center's usability there is a developing trend towards form and direction diversity in scientific, educational, and industrial integration, growing university penetration into secondary school, and high school scientific research's impact into industry.

INTRODUCTION

Under modern conditions, special status is provided for solutions to the problem of professional development and the training of scientific and scientific-pedagogical personnel in non-stop specialized engineering and technical educational systems based on the integration of science and education (Zelencova, Zelencova, & Zelencov, 2014). Obviously, such solutions should be based on the integration of science, education, and production; and should match the new stage in personnel training in a non-stop preparatory system. Such contemporary attitudes are based on different theories and techniques proven by practice. To illustrate the argument – the trend being reflected in conceptual foundation of educational process taught in Bauman Moscow State Technical University (BMSTU) – a well-known method, "The Russian method of craft training," is used (Tsibizova, 2011). The very same system was notable for its reasonable and multistage approach to education. All the works were scientifically analyzed. The master-craft teachers were absolute authorities in their fields, and (as experts) were experienced enough

DOI: 10.4018/978-1-5225-3395-5.ch014

to be able to see the mistakes of their students and to provide explanations. There were three (3) main components (Antsupova, 2005):

- Serious theoretical preparation equivalent to the same quality preparation offered by traditional universities;
- Practice within real factory conditions;
- Non-stop communication between the school and real factories.

Students' professional self-determination (as a process) is influenced not only by educational learning, professional orientation, and research-and-development components, but also by the necessity of personality development in individual and social directions. Such a process 'evolves' the holistic development of the personality of those students who possess a flair for science, research, and creativity and who are considered to be the subjects of the development of professional and social self-determination. It is characterized by the fostering of a desire for creativity, self-expression, and self-affirmation in professional activity; by stable and dominated motives, views and interests, position to knowledge and acquired know-how, social norms, and values; level of moral and aesthetic culture; and development of self-awareness.

Global society needs the educational system's reconsideration that has manifested itself in the practical implementation of a new educational paradigm – one enabled to transform the educational space of a 'high-school-to-university' transition as an important component of a system of continuing professional educational ; and enabled to emphasize the necessity of research with respect to social movements in the context of educational reforms (Arnove, Torres, & Franz, 2012; Tsibizova, 2012).

The author's understanding that the economic and social terms of societal development influences education directly makes the problem even more relevant in terms of its research impact concerning the formation, development, and current stage of the educational environment in the world, in the country, and in a given society (Ivanov, & Ivanova, 2015; Biggeri, & Santi, 2012).

Modern society needs educative, moral, professionally competent, pragmatic people capable of making decisions and of taking correspondent responsibility for those decisions; people who are capable of cooperating with others; people who are notable by their upward mobility, positive dynamics, and constructive communication resulting from social expertise (Arakcheeva, 2012; Sergeeva, 2016). Society demands that young people in their teens define their professional path. In such circumstances, one of the main tasks of high school is to reveal, teach, engage, and support the youth who are interested in science – to regenerate society's scientific potential.

These goals are aimed at the pre-institute period, as the personality is built up in the secondary school period.

In these circumstances, six of the main purposes of high school are as follows:

- To reveal and develop creative abilities and a flair for science;
- To mold key competences and professionally-important personality features as well as to mold motives towards practical implementation of the knowledge received;
- To provide necessary educational conditions for talented children;
- To guide and help youth in their professional orientation;
- To propagate science; and
- To choose those students who are most prepared and ready for the educational process.

The process of professional self-determination proceeds from the expansion and deepening of creative, socially-significant student activities (such as those related to employment, education, gaming, and communication) and the formation of moral, aesthetic, and scientific cultures. Professional self-determination, as defined by Ya. S. Batyshev, refers to independent professional choice, made after personal analysis of inner resources including the student's set of abilities and its match with professional demands. It is a conscious personality attitude, stipulated by the effects of one's upbringing and education, being foundational to any form of professional self-determination (Batyshev, 1987).

Training based on modern technology is an important developmental factor of educational systems connected with integration within scientific, industrial, and educational entities combining practice training with the study of general subjects (Spillane, & Hopkins, 2013).

For an effective solution to the problem of students' professional self-determination, studying must be closely connected with practice. Therefore, certain conditions are to be created to consolidate cross-disciplinary teams of researchers, tutors, and teachers (Donovan, 2013).

Hence, the creating of a university-based professionally-orienting educational environment means a common process influencing the effectiveness and quality of education. As a result, there is a developing trend towards a diversity of forms and directions in scientific, educational, and industrial integration, growing university penetration into secondary school, high school scientific research's impact on industry that effectively affects modern and high-tech industries.

GOALS AND OBJECTIVES OF PROFESSIONALLY ORIENTING THE ARRANGEMENT OF THE EDUCATIONAL ENVIRONMENT

Scientific and technological achievements dominate life in modern society. Thus, society needs field-specific, highly qualified specialists able to create something new in different areas of professional activity. The opportunity to acquire skills based on modern knowledge and to achieve practical results is of utmost importance for the modern specialist.

In his monograph *Philosophy of Education for the 21st Century*, B.S. Gershunsky creates a chain to describe the stages of the educational process: "literacy – scholarship – professional expertise – culture – mentality" (1998). The result of education should be assessed per certain criteria as well as on the basis of the mental priorities and preferences of the certain societies, considering constant changes in social values (Gershunsky, 1998).

Hence, the aims of creating a professionally-orienting educational environment on the basis of a technical university are:

- Students' engagement in design, research, and innovative activity;
- The development of youth's scientific and research creativity and its infrastructure;
- Recommendations for governmental support and regulatory;
- Methodological, and informational supply for large-scale student involvement in scientific and technical activities for the purpose of providing high-quality training to a new, innovation-oriented generation.

To achieve the goals indicated above, the solutions to the following problems are crucial:

- The detection of any and all creative students;
- The development and enhancement of motivation with regards to professional choice and the furthering of education;
- Quality improvement of professionals' training through arrangement and support of the system preparing young people for research-oriented, intellectual, and creative activity.

This will allow the program to assess students' personal creativity, propensity for research, and professional orientation even before admission to the university; and to create opportunities for development of those very abilities in future while studying at the university – thereby providing generational continuity, the training of specialists highly qualified for the fields of science, engineering, and industry; and a replenishment of teaching staff (Tsibizova, 2012).

The main task in setting up a professionally-oriented educative environment on the basis of a technical university model is to form students' creative, intellectual, and professional self-determination in ways appropriate to the features of their individual personalities and to what society demands of staff and modern specialists.

In this case, the leading goal for a teacher, tutor, or research advisor should be each youth's preparation for a reasonable professional choice.

For these purposes, and in order for professional development in the STEM fields to be effective, teachers need worthwhile experiences that will simultaneously develop their knowledge of content, pedagogy, and understanding of how students learn the content (Polly, Martin, Wang, Lambert, & Pugalee, 2016).

On the basis of the overall strategic objective and the main tasks to be performed, much more short-term, specific targets could also be achieved to equip students with definite knowledge, to help them form some essential skills, to reveal creativity, to cultivate aesthetic consciousness, morality, etc.

The result of the above-mentioned aims defines the target levels of educational, creative, intellectual, and personal qualities; provides for high-school-to-university continuity; will it facilitate the fundamental and integral goals of the educational process, and will solve tasks of professionally-orienting training of future specialists in compliance with university profiles and their chosen professions.

This issue is 'near and dear' to the United States, which is concerned that not preparing a sufficient number of students, teachers, and practitioners in the areas of Science, Technology, Engineering, and Mathematics (STEM) is of issue; so improving learning in STEM education continues to be a priority for American policymakers (Congressional Research Service, 2011). The results of the aforementioned study revealed that when simulation and modeling are used under specific learning conditions, a deeper level of understanding of key scientific and mathematical concepts is observed (Moallem, Morge, Narayan, & Tagliarini, 2015).

To create effective classroom environments for STEM teaching and learning, there is a requirement for the comprehensive integration of technology as a fundamental building block of education in the following areas:

1. to develop proficiency in 21st century skills for students;
2. to support innovative teaching and learning; and
3. to create a robust educational support system for both students and teachers.

American policymakers believe that several technologies including probeware, computer simulations, software applications, programmable instruments, mobile devices, and laptop/notebook computers could be used effectively to impact student learning in STEM subjects. To meet the challenges in teaching STEM subjects, the use of computer-based laboratories and features makes it possible to envision dramatic changes in instructional environments and the creation of unique STEM learning environments – in particular, ones which provide support for STEM teaching strategies and active-learning activities (Srisawasdi, 2015).

These tasks demand the formation of science-education integration processes that are reflected in the development and institutionalization of integrated educational systems that associate educational facilities with scientific and professional social institutions (e.g. in the setting of a high school Resource Center).

RESOURCE CENTER FOR SUPPORTING STUDENTS WITHIN THE CONTEXT OF PROFESSIONAL SELF-DETERMINATION

A Resource Center should be an educational-scientific-innovative establishment where, as for its arrangement, all implementation mechanisms for a competitive selection system, students' education, and support are to be provided. Additionally, as for materials, it ought to supply specific research and design laboratories, multimedia interactive classes for collective usage, a youth business incubator, expertise centers for different training lines, etc., all of which are to be fully equipped. Likewise, as for its methodology, it ought to offer a wide range of variable educational programs; allowing students to take the next step in educational, scientific, and innovative integration with the real economy (with deep penetration into that economy) when the educational process is seamlessly connected with the presentation of science, engineering, and design.

The arrangement of a university-based Resource Center is focused on the following problems:

- The development of a methodical basis for students' innovative and scientific-educational activity by means of the updating and creating of variable educational programs, the development of educational technologies, and quality improvement through new methods and means of theoretical and practical training;
- The development of students' scientific and design works, scientific-innovative activities for youth's professional self-determination and motivated training for further education using Center's material base: lab equipment, technology, metrical and analytical means of measurement, and specialized laboratories;
- Scientific-educational performances and student competitions to select those most prepared for further education and for professional activity; annual performances, and contests among research and scientific students based on their works;
- The development of network infrastructure and a common educational data space used to notify young people of educational opportunities;
- A diversification of functional opportunities of the main university departments as a tool for forming students' professional competences; including the establishment of a youth training center for integrating education, research, and innovation; and
- The arrangement and development of an integrated multilevel and multicomponent system for continued anticipatory youth training.

The Resource Center's structure implies the setting up of educational-research laboratories, specialized classes, workshops, competence centers, etc. Each center's department provides for students' training in one of various priority directions of scientific, engineering, and technological progress in single or several profile disciplines; and offers additional educational programs with the involvement of the main departments as well as strategic university partners – entities from high-tech industry, research, and innovative bodies, establishments, and corporations; which stipulates the development of students' scientific research activities through defined innovative projects and an intended incubation of start-ups based on youth teams.

The Structural-Functional Scheme of a Resource Center arrangement could be as shown in Table 1.

A *research, project, and innovative activity cluster* means an organization of students' education and research; specialized profiling laboratory works; students' project activity within creative workshops; the acquisition of hi-tech funds from the expertise centers of leading companies in particular areas; the arrangement of collective-use centers, youth business hatcheries and constructive bureaus; different scientific, research and creative events to finalize every stage of cluster's educational process.

An *educational activity cluster* includes a detailed study of particular subjects and related disciplines; practical classes of special training; additional teaching per professionally orienting programs; professional lectures and master-classes by top scientists and specialists in particular knowledge areas; and educational contests for the assessment of educational quality and students' level of initiative.

A *technical, organizational, and methodical activity cluster* is about material-technical support of the Resource Center – its material-technical base, equipment, and consumables for lab, research, innovative, and creative works; organizational support of the operation of the two other clusters – including group settings, departmental timetables, etc.; methodical support of educational activity – educational programs, academic plans, guides and manuals, methods of contests' and competitions' arrangements, standard-form reviews and questionnaires, etc.

Such an approach towards professionally-orienting educational areas provides opportunities for:

Table 1. Structural-Functional Scheme of a Resource Center

Cluster	Composition, the Department	Functionality
Research, project, and innovative activity cluster	Educational research laboratories Scientific research activity (business hatchery) Creative workshops Expertise centers	Organization research and design activities Innovation The use of high-tech equipment Scientific, research, and creative events (conferences, exhibitions, and contests)
Technical, organizational, and methodical activity cluster	Material technical base Organizational support Methodical provision	Organizational-methodological support Logistical support Support of technical events
Educational activity cluster	Specialized classes Professionally-orienting educational programs Professional lectures Master class Excursions	The teaching of core subjects An in-depth study of disciplines Educational events (competitions; contests)

- The development of youth's mental, intellectual, and creative skills;
- The early proactive and conscious professional orientation of youth;
- Sustainable motivations for seeking professional knowledge in the chosen area, and based on practice;
- Incentives for personnel reproduction for development of the country's scientific capacity;
- Youth's social security;
- The development of supporting research and professional training of youth, including a wide range of opportunities for scientific and technical creativity;
- An elaboration of new methods and educational technologies for the further development of modernization of education .

This complex outlook with regards to the Resource Center arrangement allows for people:

- To develop the variability of additional educational programs and a practical focus on professional education;
- To combine task-reproductive, productive, and creative activities;
- To create new educational technologies based on a combination of theoretical and practical forms of education;
- To solve the problem of integrating education, science, and production while ensuring the integrity of scientific tasks;
- To implement principles of educational continuity, differentiation, and individualization of the educational process, aimed at personality growth and fulfillment;
- To generate socially-significant qualities specific to the young scientist, the motivating needs of the professional educational area, functional literacy, and upward mobility.

This work should result in qualitative and quantitative changes in the educational process's methods and foundational technologies from the point of its continuity in from the "high-school-to-university" educational process. Continuity between both is considered as being a process – and as being the result of defining the integrity of such courses as education, quality improvement, continuity, anticipating character, diversification, and the arrangement of research, scientific, and business components used in the preparation of a new generation of specialists.

Hence, this should enhance youth's educational, research, scientific, and professional levels and help enable them to solve social and educational problems beyond the secondary (high school) level.

CONCLUSION

The implementation of an integrated multicomponent educational system in technical university provides for the opportunity to have a steady and central effect on professionally orienting students' training; to maximally avoid the lack of information on conferences, contests, exhibitions, and other performances held; and to arrange and streamline the process of conducting scientific and educational events. Such a model should be based on the integration of science, education, and business; and should correspond to updated levels of researchers' and technologists' training – their persistent buildup of growth in a system of non-stop professional development.

Basing the educational format on scientific knowledge, the development of additional awareness and the inner motivation generated thereof greatly improves the efficiency of the educational process.

The early development of such patterns within the secondary educational process will become of utmost importance when students reach the 'university' stage of schooling. This program serves as the basis for a non-stop educational process of development with the goal of building up a creative and socially-active personality able to solve innovative problems in a range of different scientific and technological spheres.

FUTURE RESEARCH DIRECTIONS

There are two research directions for further work. First on consist in the development of the methods of young people professional self-determination. One of methods is modernization of the educational programmes. This modernization of educational programs is carried out by means of complex implementation. It is necessary to include into the programs research and design activities of students. This must be done in full compliance with the existing regulatory framework for the development of continuing education. The second line of research is to develop a model of the implementation of research activities in the continuing education system for training highly qualified specialists.

REFERENCES

Antsupova, G. N. (2005). *MGTU – glazami istorika* [BMSTU – eyes of the historian]. Bauman Moscow State Technical University Publishers.

Arakcheeva, O.E. (2012). Propedevticheskaya podgotovka uchitelya k formirovaniyu sotsial'noy kompetentnosti uchashchikhsya [Propaedutic teacher's preparation for the formation of students' social competence]. *Obrazovanie i Samorazvitie*, *3*, 43-46.

Arnove, R. F., Torres, C. A., & Franz, S. (2012). *Comparative education: The dialectic of the global and the local*. Lanham, MD: Rowman & Littlefield Publishers. Retrieved from https://rowman.com/Page/RLContact

Batyshev, S. Ya. (1987). Reforma professional'noy shkoly: opyt, poisk, puti realizatsii [Reforming professional schooling: Experience, research, and implementation]. Moscow: Vysshaya Shkola Publishers.

Biggeri, M., & Santi, M. (2012). The missing dimensions of children's well-being and well-becoming in education systems: Capabilities and philosophy for children. *Journal of Human Development and Capabilities: A Multi-Disciplinary Journal for People-Centered Development*, *13*(3), 373-395.

Donovan, M. S. (2013). Generating improvement through research and development in education systems. *Science*, *340*(6130), 317–319. doi:10.1126cience.1236180 PMID:23599484

Gershunsky, B. S. (1998). *Filosofiya obrazovaniya dlya XXI veka: V poiskakh praktiko-orientirovannykh kontseptsiy* [Philosophy of education for the 21st century: In search of concepts]. Moscow: Sovershenstvo.

Ivanov, O. B., & Ivanova, S. V. (2015). Special features of education space formation in economic and social conditions inherent in the post-industrial epoch. *Lifelong Learning Proceedings of the 13th International Conference*, 43-47.

Moallem, M., Morge, S.P., Narayan, S., & Tagliarini, G.A. (2015). The power of computational modeling and simulation for learning STEM content in middle and high schools. *Improving K-12 STEM Education Outcomes through Technological Integration*, 135-171.

Polly, D., Martin, C., Wang, C., Lambert, R.G., & Pugalee, D. (2016). Supporting the enactment of standards-based mathematics pedagogies: The cases of the CoDE-I and APLUS projects. *Innovative Professional Development Methods and Strategies for STEM Education*, 137-148.

Sergeeva, M. G. (2016). Kriterial'nye pokazateli sformirovannosti sotsial'noy kompetentnosti obuchayushchikhsya [Criteria indicators for students' formation of social expertise]. *Professional'noe Obrazovanie i Obshchestvo*, *1*(17), 30–37.

Spillane, J. P., & Hopkins, M. (2013). Organizing for instruction in education systems and school organizations: How the subject matters. *Journal of Curriculum Studies*, *45*(6), 721–747. doi:10.1080/00220272.2013.810783

Srisawasdi, N. (2015). Motivating inquiry-based learning through a combination of physical and virtual computer-based laboratory experiments in high school science. *Improving K-12 STEM Education Outcomes through Technological Integration*, 108-134.

Tsibizova, T.Yu. (2011). Podgotovka vysokokvalifitsirovannykh spetsialistov v sisteme nepreryvnogo professional'nogo obrazovaniya (na primere MGTU im. N.E. Baumana) [Training of highly-knowledgeable specialists in a system of non-stop professional development (per the example of Bauman Moscow State Technical University)]. *European Social Science Journal*, *2*(5), 154-159.

Tsibizova, T. Yu. (2012). Profil'noe obuchenie kak komponent sistemy nepreryvnogo professional'nogo obrazovaniya [Profile education as a part of the system of continued professional education] [Profile School]. *Profil'naya Shkola*, *4*, 9–13.

Zelencova, N.F., Zelencova, E.V., & Zelencov, V.V. (2014). Problemy professional'nogo stanovleniya i vosproizvodstva nauchnykh i nauchno-pedagogicheskikh kadrov v nepreryvnoy sisteme profil'nogo inzhenerno-tekhnicheskogo obrazovaniya [Problems of professional growth and scientific-pedagogical staff reproduction in a system of continued profile engineering education]. *Nauka i Obrazovanie: Nauchnoe Izdanie MGTU im. N.E. Baumana*, *11*, 827-839.

ADDITIONAL READING

Aleksandrov, A. A., Neusipin, K. A., Proletarsky, A. V., & Fang, K. (2012). Innovation development trends of modern management systems of educational organizations. *In Proceeding of 2012 International Conference on Information Management, Innovation Management and Industrial Engineering*. (pp. 235-238). Sanya, China. 10.1109/ICIII.2012.6339951

Galikhanov, M. F., Ilyasova, A., Ivanov, V., Gorodetskaya, I. M., & Shageeva, F. T. (2015). Continuous professional education as an instrument for development of industry employees' innovational competences within regional territorial-production cluster. *In Proceedings of 2015 International Conference on Interactive Collaborative Learning*. (pp. 251-255). Firenze, Italy. 10.1109/ICL.2015.7318034

Khairutdinova, R. R., & Fedorova, Y. A. (2016). Pedagogical professional self-determination support for students under conditions of additional education program implementation. *International Journal of Environmental and Science Education*, *11*(9), 2275–2285.

Oreshkina, A.K. (2011). Teoretiko-metodologicheskie aspekty formirovaniya proektnoy deyatel'nosti v obrazovatel'nom protsesse nepreryvnogo obrazovaniya [Theoretical and methodological aspects of project activities in the educational process of continuing education]. *Otechestvennaya i zarubezhnaya pedagogika* [*Domestic and foreign pedagogy*], *1*, 159-164. [Russian].

Oreshkina, A. K., Tsibizova, T. Yu., & Nosova, I. S. (Eds.). (2015). *Formy razvitiya sotsial'nogo prostranstva sistemy nepreryvnogo obrazovaniya* [*Forms of development of social space of lifelong learning system*]. Moscow Region State University Publishers. [Russian]

Shaidullina, A. R., Masalimova, A. R., Vlasova, V. K., Lisitzina, T. B., Korzhanova, A. A., & Tzekhanovich, O. M. (2014). Science and Manufacture Integration Models Features in Continuous Professional Education System. *Life Science Journal*, *11*(8s), 478–485.

Tsibizova, T.Yu. (2011). Integratsiya nauki i obrazovaniya kak element sistemy nepreryvnogo professional'nogo obrazovaniya [Integration of science and education as part of the system of continuous professional education]. *Integratsiya obrazovaniya* [*Integration of education*], 4, 25-29. [Russian].

Tsibizova, T.Yu. (2014). Realizatsiya professional'no-orientiruyushchey obrazovatel'noy podgotovki uchashchikhsya na baze Resursnogo tsentra vuza [The implementation of the vocational orientation of the educational preparation of students on the basis of the Resource center of the University]. *Izvestiya Moskovskogo gosudarstvennogo tekhnicheskogo universiteta MAMI*. [*Proceedings of Moscow State Technical University MAMI*], *Vol. 5, 4(22)*, 209-213. [Russian].

KEY TERMS AND DEFINITIONS

Continuity of Educational Activities: The relationship of education levels, types and forms of educational process, carried out in the development of the educational level of students aimed at the acquisition of knowledge, skills, research and socio-cultural experience, secondary, high and professional education.

Diversification of Education: The principle of the education system structuring, providing the possibility of variability of educational services, educational programs, types of educational institutions, etc.

Integration of Education, Science, and Production: The common usage of educational, scientific and industrial potential of organizations with mutual interests, primarily in the areas of training, advanced training and retraining of personnel, as well as joint scientific research, implementation research, etc.

Professional Orientation: A system of science-based activities aimed at preparing young people to choose of profession, to assist young people in professional self-determination and employment.

Professional Self-Determination: The independent choice of profession, carried out the analysis of a persons' internal resources, including their abilities and their correlation with the requirements of the profession.

Research Activity: A process of students study the scientific knowledge system by specially developed means and methods with the aim of cognitive skills developing, world perception, moral and other qualities of the person, as well as creative powers and abilities developing.

Variability of Education: The quality of the education system describing its ability to create and to provide students with options for educational programs and services for selection according to their changing needs and opportunities.

Section 3
Innovation in Engineering Education

Chapter 15
Math-Related Problems in Russian Engineering Education:
Possible Solutions Based on Best Practices in European and Russian Universities

Ilia Soldatenko
Tver State University, Russia

Irina Zakharova
Tver State University, Russia

Oleg Kuzenkov
Lobachevsky State University of Nizhniy Novgorod, Russia

Alexander Yazenin
Tver State University, Russia

ABSTRACT

Engineering education tends to be more and more attractive to Russian students in response to the growing demands of the labor market in this area. However, there is a serious problem of high percentage of drop-outs during first year of study in STEM courses (science, technology, engineering, mathematics) and mathematical disciplines are the most typical reason for that. This problem is addressed by international TEMPUS project MetaMath whose aim is modernization of the Russian education system in accordance with international trends and Russia's cultural and educational traditions as well as needs of business and industry. The purpose of this chapter is to describe research results and analysis of modernization experience of educational programs based on the produced methodology.

DOI: 10.4018/978-1-5225-3395-5.ch015

INTRODUCTION

Engineering education tends to be more and more attractive to Russian students in response to the growing demands of the labor market in this area. However, nowadays there are several serious problems in this area.

These problems, firstly, include global changes in the world that certainly affect education. The speed with which engineering knowledge and competencies evolve has been steadily increasing; new skills required by engineers constantly emerge while some of the existing ones become obsolete. Sometimes it even happens that some technology becomes outdated before a student completes a four-year bachelor course of study.

This, in turn, complicates the process of learning. Modern student is obliged not only to master a certain amount of knowledge but also to learn how to use it to solve practical problems which were not dealt with explicitly during training and may lie on the intersection of different fields. This requires formation of respective competencies of the student.

Secondly, there is a very serious problem of high percentage of drop-outs during the first year of study in STEM courses. Mathematical disciplines are the most typical reason for that. According to current statistics, the average drop-out rate from engineering specialties because of mathematics in Russian universities is about 20%, for some curricula it reaches 40%.

School graduates, who choose these courses, usually underestimate the role and place of mathematics in their upcoming education. Often, prospective students have this false perception that mathematics is unimportant for a chemist, a physicist or a programmer. Everything is also aggravated by the difference in the level of mathematical training between universities and schools.

At the same time, numerous studies have shown that the level of mathematical knowledge is a major factor determining the success of engineering education. In Russia all university students pursuing this kind of curricula are obliged to take a lot of math-related courses at the beginning of their education. Disciplines of engineering profile for which the mathematical knowledge and skills are essential input requirement appear only during senior years.

Other reasons for the above-mentioned problems in Russian universities include reduction of teaching hours (credits) for math subjects in curricula of some Russian universities and the fact that new information technologies are not used to the full extent in education process. The European experiences and research results have proven that significant improvements in learning outcomes in mathematics can be achieved by applying new Technology-Enhanced Learning (TEL) tools and pedagogic approaches.

Because of these circumstances, much methodological work is required to modernize the system of mathematical education for engineers in Russian universities.

This problem is addressed by international TEMPUS project MetaMath (MetaMath, 2016) which involves five Russian universities (Tver State University, Lobachevsky State University of Nizhniy Novgorod, Kazan National Research Technical University named after A.N.Tupolev, Ogarev Mordovia State University, Saint-Petersburg Electrotechnical University (LETI)), Association for Engineering Education of Russia and four European universities (Tampere University of Technology, Claude Bernard University Lyon 1, Saarland University, Chemnitz University of Technology). This project's aim is modernization of the Russian education system in accordance with international trends, best practices of European universities and Russia's cultural and educational traditions as well as needs of business and industry.

The purpose of this chapter is to describe research results and analysis of modernization experience of educational programs based on the produced methodology and to analyze possible ways of modernization of requirements system for mathematical education in Russia basing on SEFI framework approach (SEFI, 2013).

THE ATTITUDE OF STUDENTS TOWARDS MATHEMATICS

As mentioned above, all students of engineering programs in Russian universities have a significant number of mathematical disciplines in the early years of their education. As part of the research several sociological surveys among the students, teachers, employers and graduates were conducted to determine their attitude towards mathematics.

One of the surveys was devoted to the study of students' attitudes towards mathematics as a discipline of their study and its role in their future profession. There are six responses for each question: 0 - completely disagree, 1 – disagree, 2 - rather disagree, 3 - rather agree, 4 – agree, 5 - completely agree. Table 1 shows the results as a percentage of the total number of survey participants. The survey was conducted among students of 3d and 4th years of study.

Table 1. Results of survey "mathematics in the life of students"

#	Question	0	1	2	3	4	5
1	I think solving math problems is quite boring	5	33	40	12	7	3
2	Solving math problems is fun	4	1	13	48	29	4
3	I am OK with how I am doing at solving math problems	0	11	20	51	15	4
4	I consider solving math problems as an opportunity to prove myself	7	21	29	29	11	3
5	I think mathematics is very interesting	0	5	17	36	29	12
6	I liked working on math problems	0	4	24	49	17	5
7	I find mathematics to be very hard	8	21	39	25	5	1
8	I think I am very good in math	1	3	21	49	23	3
9	It is very important for me to be able solve math problems	3	7	12	40	31	8
10	I think math problems are very boring	1	15	19	28	23	15
11	I like situations, in which I have to prove my skills	0	8	5	32	37	17
12	I try to do my best when solving math problems	1	4	17	56	19	3
13	I like tasks that show what I am good at	0	1	1	36	37	24
14	I prefer easy tasks, in which I know I won't make any mistake	0	4	4	41	33	17
15	I am scared to have to solve exercises that might cause me making mistakes	0	13	29	28	25	4
16	I enjoy theoretical aspects of mathematics more than practical problem solving	8	33	36	19	4	0
17	I think for an engineer, problem solving skills are more important than math theory	0	11	19	36	28	7
18	I think we need fewer mathematics course in university	5	7	47	28	11	3
19	I think for my future studies at this university, mathematics has little importance	8	28	35	15	7	8
20	I think for my future career, mathematics has little importance	8	17	29	20	16	9
21	I think, I need to study math to be successful in the future	3	5	24	40	21	7

One can see from the results that students prefer more practical mathematics, rather than theoretical. At the same time, they consider reduction of credits for mathematical disciplines inappropriate, because they believe that mathematics is important in order to study other disciplines. Second, students consider mathematics to a lesser extent essential for their future lives and careers. Thirdly, the students believe that for an engineer practical problem-solving skills are more important than theoretical understanding of the principles, on which these methods are based.

Polls generally show that students are aware of the importance of mathematics at least for their further education at university.

THE ROLE OF MATHEMATICAL COMPETENCIES IN HIGHER EDUCATION

In 2003 Russia joined Bologna Process by signing the Declaration. In 2011 Russia introduced new federal state educational standards (FSES).

One of the main distinguishing features of the new standards is a competence-based approach. The essence of this approach is that the emphasis of the educational process is transferred from the content of education to learning outcomes, which should be transparent. The learning outcomes are described by the system of competencies, which are a dynamic combination of knowledge, skills, abilities and personal qualities that a student should be able to demonstrate after completion of his or her education.

At the same time, a lot of attention is paid to competencies related to mathematical training and ability to use mathematics in real life applications. These competencies are even more important in the field of information and communication technologies. As part of the project TUNING RUSSIA (Petrova, et al., 2013) the subject area group (SAG) on ICT conducted sociological research aimed at identifying the importance of specific subject competences for graduates in ICT.

Representatives of students, teachers and graduates of some of the leading Russian universities, as well as representatives from leading Russian and international companies in the field of ICT were given a list of competencies and were asked to rate their importance for professional work and estimate the level of their formation in existing Russian system of higher education.

The evaluation was conducted on the four-point scale. Among others, the list included competence for a deep mathematical training: “to apply and develop fundamental and interdisciplinary knowledge, including mathematical and scientific principles, numerical methods ...”. Table 2 shows the results of the survey.

These data clearly demonstrate the need for this competence in the education process (the average level of importance).

Table 2. Result of the survey

Group of Respondents	Importance	Achievement
Academics	3.46	2.62
Employers	3.09	2.75
Students	3.07	2.77
Graduates	3.09	2.7

The fundamental importance of mathematical competencies for professional work in the field of ICT has been reflected in the design of Russian Sectoral Framework of qualifications and competency characteristics for Computer Science degree programs (TEMPUS project INARM (Kuzenkov&Tikhomirov, 2013)). In this framework labor functions were correlated with the areas of fundamental knowledge necessary for their successful acquiring. An example of such correlation is shown in Table 3.

Possession of fundamental mathematical knowledge (optimization methods, operations research, systems analysis, discrete mathematics, etc.) was included to the passports of all groups of labor functions of this framework as an obligatory component.

RESULTS OF COMPARATIVE ANALYSIS AND APPROACHES FOR MODERNIZATION

In order to achieve the goal of upgrading system of mathematical education, the structure and content of educational process in math-related engineering courses in the selected Russian and European universities have been studied, comparative analysis was held and, as a result, recommendations for the best practices implementation in the educational process of Russian universities were elaborated.

Comparative analysis shown that thematic contents and learning outcomes in Russian and European universities are almost identical. The main difference is observed in active use of information technologies and, in particular, e-learning systems in European universities. This allows to take out some of the material for independent study and focus on difficult topics. E-learning systems also allow to automate and, as a result, simplify the knowledge assessment process. One of the goals of modernization was to minimize this difference.

The following ways for courses modernization were developed:

- There are two ways of teaching math: theory-oriented and practice-oriented. We should find a golden mean but make emphasis on practice.
- Give more real-life practical examples in math subjects from the very beginning to justify necessity of math.

Table 3. Correlation of labor functions and fundamental knowledge

Groups of Labor Functions	Labor Functions	Areas of Knowledge as an Instrument of Formation of Learning Outcomes (Based on the CS 2014)	Knowledge Modules for Building Education Programs
A.1. Harmonization of information system (IS) and business strategies	Foresee long-term prospects for business development and determine the IS infrastructure in accordance with organizational policy	AL. Algorithms and Complexity AR. Architecture and Organization HCI. Human Computer Interaction IAS. Information Assurance and Security GV. Graphics and Visualization IM. Information Management	Fundamentals of management; fundamentals of economics; math modelling; systems analysis; optimization methods; operations research; software engineering; Cybernetics basics

- Involve business community in participation in students practice: starting from term papers and finishing with diplomas and industrial practices.
- Increase the role of independent work of students.
- Use project method in teaching. The main purpose of this method is to provide students with possibility of independent knowledge acquiring as well as solving practical problems that require integration of skills from different subject areas.
- Use bridging courses to simplify students' transition from school to university.
- Use ICT tools and technologies more actively to enhance and support education process.

As for requirements system Lobachevsky State University of Nizhniy Novgorod in the framework of TEMPUS project developed maps of mathematical competencies, carried out modernization of programs for mathematical disciplines and created adequate funds of assessment tools in the field of ICT (Zakharova, et al., 2014). Competence map contains description of a set of indicators that show the specific quality aspects of mastering the competence and descriptors that characterize them. SEFI framework was actively used in the process (SEFI, 2013).

Here is an example of competence map fragment for PC3 "The ability to understand and apply in research and applied activities modern mathematical apparatus and the basic laws of science" in preparation of bachelors in the program "Fundamental science and information technologies" (Gugina&Kuzenkov, 2014; Bedny, et al., 2014).

The fragment given in the Table 4 belongs to the first level of mastering – "technology literacy", which corresponds to the first and second year of bachelor program and also corresponds to the first level of SEFI Framework. This fragment contains only material from "Calculus". Groups of competence

Table 4. Fragment of competence card

Indicators	Descriptors				
To know the concept: converging and diverging sequences; continuity of the function; differentiability; smoothness; derivative	lack of knowledge	presence of major errors in the knowledge of basic material	knowledge of basic material with a number of minor errors	knowledge of basic material with a number of notable errors	knowledge of basic material without errors
To be able to: find limits of sequences; find derivatives of complex functions; differentiate inverse functions; differentiate functions defined implicitly; differentiate functions defined parametrically	no ability to solve standard problems	major errors in solving standard tasks	ability to solve standard problems with minor errors	ability to solve all standard problems with minor errors	ability to solve standard and non-standard tasks
To know a variety of methods and ways of calculating limits, methods of differential calculus	lack of skills	lack of a number of important skills	presence of the minimum required skills	presence of most of the basic skills, demonstrated in standard situations	presence of all the skills demonstrated in standard and non-standard situations

indicators consist of knowledge, skills and abilities that correspond to the understanding of a competence as a dynamic entity that unites them.

Thus, SEFI standard is a useful and effective tool that allows adaptation of mathematical education in the Russian state educational standards to international requirements.

Another approach to the modernization of higher education has been undertaken in the framework of the project "Scientific and methodological support for the development of exemplary basic professional educational programs for different fields of education" in which groups of authors developed exemplary basic educational programs aimed at the formation of general and universal competences for combined groups of education programs.

Work (Zakharova, et al., 2016) describes the development of an exemplary basic educational program for group 02.00.00 "Computer and Information Sciences". An analysis of professional standards was conducted with the goal to verify compliance of optimized general professional competences with labor functions, related to the professional activities of graduates of corresponding education programs (Zakharova, et al., 2016; Zakharova, et al., 2015; ECTS, 2009).

4. PRELIMINARY RESULTS OF AN EXPERIMENT

During the project an experiment was conducted and approbation of the developed modernization methodology in the Russian universities was carried out. As part of the experiment, several groups of students in each institution took the modernized courses. Not only the selected disciplines were modernized but also some changes to curricula were made. This work was conducted in accordance with the general process of refining of federal state standards of higher education in Russia, due to the participation of authors of the chapter in the relevant working groups of the Ministry of Education.

At the beginning of the course, testing of input level of students was carried out and at the end – testing of output level. Input and output tests were made in the same framework of competencies on the same level of complexity therefore it helped to reveal the knowledge gain of students. The following tables show test results. The first table shows level of mastering aggregated groups of SEFI competences. The test results are indicated as a percentage of correctly accomplished tasks.

Table 5. The relative level of mastering aggregated groups of SEFI competences at the beginning and end of the course

#	SEFI Competences (Aggregated Group)	1st Group (Traditional Program)		2nd Group (Modernized Program)	
		Pre-test (%)	Post-test (%)	Pre-test (%)	Post-test (%)
1	ordinary differential equations	33,60	40,69	43,33	63,75
2	equations of the first order	63,00	52,59	62,50	73,44
3	second-order equations	57,14	54,19	40,48	67,86
4	tasks associated with eigenvalues	91,20	78,62	63,33	72,50
5	nonlinear optimization	92,00	75,86	83,33	100,00

Table 6. Relative increase / decline in the level of mastering SEFI competencies (knowledge gain = (output – input) / input)

#	SEFI Competencies (Aggregated Group)	1st Group (Traditional Program)	2nd Group (Modernized Program)
1	ordinary differential equations	21,10	47,13
2	equations of the first order	-16,52	17,50
3	second-order equations	-5,16	67,64
4	tasks associated with eigenvalues	-13,79	14,48
5	nonlinear optimization	-17,54	20,00

The results of the test show that for a number of competencies modernized program provides a higher degree of mastering by students in absolute terms. It is even more noticeable that the modernized program gives best results in knowledge gain (Table 6). However, the most interesting thing is that traditional program gives negative knowledge gain for number of competencies, while the modernized program shows no decline anywhere.

This can be explained by the fact that during the process of traditional education not much attention is given to these competences. It is believed that they were mastered during previous courses and there is no need or time to repeat it. As a result, students forget even what they knew before. On the contrary, the modernized program allows students to update their basic.

CONCLUSION

The chapter describes problems in engineering education in Russia. These problems were addressed by international TEMPUS project which involves several Russian and European universities and Association for Engineering Education of Russia.

The chapter analyses the experience of modernization of educational programs based on produced methodology. Obtained results of approbation of the methodology showed that the chosen modernization methods are an effective tool for solving the specified math-related problems in engineering education in Russian universities.

As a part of the project analysis of possible ways of modernization of requirements system for mathematical education in Russia basing on SEFI framework approach was carried out.

ACKNOWLEDGMENT

Tempus project MetaMath has been funded with support from the European Commission [grant number 543851-TEMPUS-1-2013-1-DE-TEMPUS-JPCR]. This chapter reflects the views only of the authors, and the Commission cannot be held responsible for any use that may be made of the information contained therein.

REFERENCES

Bedny, A., Erushkina, L., & Kuzenkov, O. (2014). Modernising educational programmes in ICT based on the Tuning methodology. *Tuning Journal for Higher Education*, *1*(2), 387–404. doi:10.18543/tjhe-1(2)-2014pp387-404

ECTS. (2009). ECTS User's Guide. Brussels, Belgium: European Commission /Luxembourg: Office for Official Publications of the European Communities. doi:10.2766/88064

Gugina, E. V., & Kuzenkov, O. A. (2014). Education standards of Lobachevsky State University of Nizhniy Novgorod. *Herald of Lobachevsky State University of Nizhniy Novgorod*, *3-4*, 39-44. Retrieved from https://elibrary.ru/item.asp?id=22862964

Kuzenkov, O. A., & Tikhomirov, V. V. (2013). Using the methodology of "Tuning" in the development of national ICT competences framework. In *Modern Information Technology and IT Education*. Moscow, Russia: Fund for Promotion of Internet media, IT education, human development League Internet Media. Retrieved from https://elibrary.ru/contents.asp?issueid=1371690

MetaMath. (2016). Objectives. In *MetaMath project official web-site*. Retrieved from http://www.metamath.eu/objectives/

Petrova, I., Zaripova, V., Ishkina, E., Militskaya, S., Malikov, A., Kurmishev, N., ... Zakharova, I. (2013). *Tuning Russia: Reference Points for the Design and Delivery of Degree Programmes in Information and Communication Technologies*. Bilbao, Spain: University of Deusto. Retrieved from http://www.deusto-publicaciones.es/index.php/main/libro/1041/eu

SEFI. (2013). *European Society for Engineering Education (SEFI), Brussels*. A Framework for Mathematics Curricula in Engineering Education. A Report of the Mathematics Working Group. Brussels, Belgium: European Society for Engineering Education (SEFI). Retrieved from http://www.sefi.be/?page_id=125

Zakharova, I. V., Dudakov, S. M., & Yazenin, A. V. (2016). On the development of an exemplary curriculum for UGNS (combined group of specialties) "Computer and Information Sciences" in accordance with professional standards. *Herald of Tver State University. Series: Pedagogy and Psychology*, *2*, 84-100. Retrieved from https://elibrary.ru/item.asp?id=26555960

Zakharova, I. V., Dudakov, S. M., Yazenin, A. V., & Soldatenko, I. S. (2015). On methodological aspects of the development of exemplary educational programs of higher education. *Educational Technology & Society*, *18*(3), 330-354. Retrieved from https://elibrary.ru/item.asp?id=24102764

Zakharova, I. V., Kuzenkov, O. A., & Soldatenko, I. S. (2014). The use of modern educational technologies for the improvement of mathematical education within the engineering courses in Russian universities. *Modern Information Technology and IT Education*, *10*, 159-171. Retrieved from https://elibrary.ru/item.asp?id=23020629

KEY TERMS AND DEFINITIONS

Bologna Process: A series of ministerial meetings and agreements between European countries to ensure comparability in the standards and quality of higher-education qualifications.

FSES: Federal state educational standard – a set of mandatory requirements for education of a certain level and (or) to the profession, specialty and direction of training.

Labor Functions: An activity that is performed by a person occupying a certain position in the staff list of the enterprise and possessing a specific skill level.

SEFI: The largest organization for engineering education in Europe. An acronym for its French name, Société Européenne pour la Formation des Ingénieurs.

STEM: A term that refers to the academic disciplines of science, technology, engineering, and mathematics.

TEMPUS: The TEMPUS (Trans-European Mobility Programme for University Studies) program encourages higher education institutions in the EU Member States and partner countries to engage in structured cooperation.

Chapter 16
The Problem-Oriented Approach in the Basic Mathematical Courses for Engineering Education

Olga Alexandrovna Dotsenko
Tomsk State University, Russia

Andrey Alexandrovich Zhukov
Tomsk State University, Russia

Tatiana Dmitrievna Kochetkova
Tomsk State University, Russia

Elena Gennagyevna Leontyeva
Autonomous University of Barcelona, Spain

ABSTRACT

Problem-based learning takes a well-deserved place in the educational programs of leading universities in the world. Meanwhile it is known that this approach has been well developed for training students of economy and medicine. There are certain difficulties in setting targets as well as in teaching methods in basic technical subjects, in particular in the mathematical courses. The chapter presents an analysis of the peculiar features of problem-based learning in a research university for basic courses of the first two years of study. The discipline "Numerical Methods and Mathematical Modeling" is given as an example of the application of this approach. The main topics are proposed and lesson plans are provided. The information support of the courses is carried out in the learning management systems. The elements of this approach have been put into practice of training course and it was shown that the material was achieved much better.

DOI: 10.4018/978-1-5225-3395-5.ch016

ORGANIZATION BACKGROUND

Tomsk State University (TSU), located in the south of Western Siberia, opened in 1888 as the first university in Russia behind the Urals. There are 22 faculties, three institutes of applied sciences, the scientific library, and the Siberian Botanic Garden. In 2016, TSU was listed as one of four leaders in the Project 5-100 in Russia. In addition, the university continues to improve its position in the world rankings of institutions of higher education.

INTRODUCTION

PBL takes a well-deserved place in the educational programs of leading universities (Barrett, 2010; Cotič, & Zuljan, 2009; Mertins, 2012). It can be a challenge to develop teaching research students' independent learning skills and transformation of reality (Hung, 2011; Schmidt, Rotgans, & Yew, 2011). The creative portion of the learning process includes the research of new facilities, storage of acquired knowledge, and identification of task and method solutions. Students develop and show personal competences, including original thought, problem recognition, quick orientation in new conditions, and intuition in the process of solving creative nonstandard tasks. PBL meets the full requirements of modern engineering education. It can be used as a means of developing trainees' learning activities and independence.

The basic concept of PBL is an educational problematic situation. PBL is a complex theoretical or practical question. It contains a latent contradiction and, upon solving, leads to different (and often opposing) viewpoints. PBL is a mental state (or cothinking interaction) of a student or group of students under the guidance of a teacher.

This chapter will elucidate the peculiarities of applying the problem-oriented approach in teaching mathematical disciplines. It will also describe the use of PBL in the Numerical Technique and Mathematical Modeling course during the second year of the radiophysics faculty engineering specialty (Radio Electronic Systems and Complexes) at National Research TSU. The chapter will provide main topics and lesson plans. It will analyze problems encountered by participants and teachers during the educational process. The conclusion will discuss the implementation of PBL elements.

BACKGROUND

This approach has been successfully developed to educate students in the fields of economy and medicine (Iskrenko & Poulton, 2008; Spencer & Jordan, 1999). A review of the references shows that PBL has been widely used in medicine. Recently, this approach had made spectacular progress in engineering education. However, there are limited methodological developments in educational disciplines, including mathematics (see Figure 1). Figure 1, which is based on data from the international citation database Scopus, shows a nine-year containing publications and key words.

Articles focusing on PBL in both exact sciences and engineering contain, as a rule, a description of the experience of applying this method in a university and for a specific subject. During the last two years, some articles reviewed existing developments and offered generalizations of experiences (Merritt, Lee, Rillero, & Kinach, 2017).

Figure 1. PBL in various disciplines

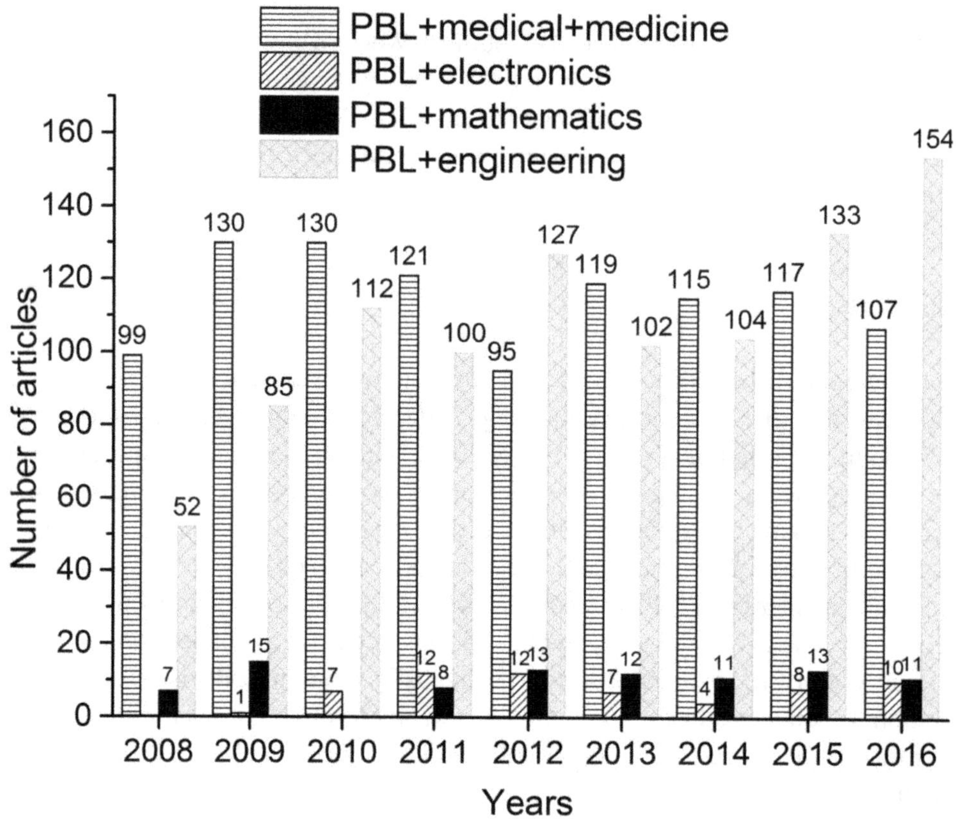

Dionysiou and Ktoridou (2012) reviewed a computer security course for senior students from the Department of Computer Science at the University of Nicosia, Cyprus. Through the application of problem-oriented teaching, students learned how to make decisions and were encouraged to plan continuous self-education.

Bevinakoppa, Ray, and Sabrina (2016) pointed out that pure forms of PBL were used in medical schools, business courses, and economic disciplines. They found that hybrid training in computer science combined the problem-oriented method, traditional lectures, seminars, and laboratory work. The authors used experiences from Australia's University of Sydney and Melbourne Institute of Technology as examples.

El-Khalili (2013), after reviewing the use of the problem-oriented approach in software engineering courses at Petra University in Amman, Jordan, found positive student feedback in end-of-semester surveys.

Valtanen et al. (2012) concluded that the analysis of a real-world problematic situation in training is required to offer standard solutions. It is important to understand the problem's meaning, causes, and features as prescribed by the system analysis. The authors found a lack of research on student opinions surrounding the application of problematic education in higher education, particularly in the field of information technology at the University of Tampere in Finland. Student, teacher, and expert views helped in the development of effective training programs promoting in-depth training. Their view also helped in the understanding of problem scenarios and problem-solving processes.

Figuera et al. (2011) described the experience of applying problem learning for engineers in the field of telecommunications at King Juan Carlos University, Spain. During the project, students were tasked with developing a complex device, including a radio control system for a car. As learners linked knowledge in individual disciplines, they understood the practical applications and interconnections. Learners simultaneously acquired skills in teamwork, planning, and communication. Figuera et al. (2011) argued that this type of innovative approach ensures multidisciplinary training of engineers with business skills.

Fortelius and Akerman (2015) described the implementation experience of a PBL project in the field of materials, food, and chemical engineering at Helsinki Metropolia University of Applied Sciences, Finland. They emphasized that this approach was necessary for the creating of the conceiving, designing, implementing, operation (CDIO) model in engineering education (Crawley, Malmqvist, Östlund, Brodeur, & Edström, 2014). This model was a response to the needs of society in the face of the rapid progression of science and technology.

This chapter will elucidate the peculiarities of applying the problem-oriented approach in teaching mathematical disciplines. It will describe experiences in using PBL for TSU students in the Numerical Technique and Mathematical Modeling course.

CASE DESCRIPTION

The Numerical Technique and Mathematical Modeling course contains the basic methods for solving mathematical problems. These tasks are based on models, which describe the different physical processes. By solving abstract tasks, it is possible to identify a problem in electrical engineering, electronics, electrodynamics, and other disciplines within the basic educational program. These disciplines are studied after the Numerical Technique and Mathematical Modeling course. Problem definition from these special disciplines is the students' problem-based situation because they are unfamiliar with the theory.

In the first lesson, the teacher describes a problematic situation in an area of knowledge or practice familiar to the students. The problem's solution requires the use of numerical methods from the current lesson theme. During the discussion, brainstorming and idea reduction occur under the guidance of a teacher. Algorithms and methods of solution are offered. This technique motivates students to take an interest in the subject and demonstrate their connection to the practice. It also involves independent work outside of the classroom.

To participate in the discussion, students must be theoretically prepared on the current theme. It is assumed that students and teachers analyze the strengths and weaknesses of the methods used to solve the problem. The discussion, which can be in the form of a workshop, should include presentations. During this step, students will understand the use of the methods.

Teachers should allocate time to work with students who arrive unprepared to the workshop. There is also an individual-collective method of materials, in which small groups of students formulate and discuss questions related to a portion of the reading material.

The next stage focuses on solving analysis problems in the individual assignments. This task deals with computing programs implementing a numerical solution. Most of the students are unable to achieve this task without the teacher's assistance or availability of appropriate samples. A control test with theoretical questions is given upon completion of this stage.

A control task relating to group work consists of nonspecific practical problems. During the task, students work independently. They present their decisions during the next lesson. Consultations to address emerging issues are appointed between classes. The eight topics in the Numerical Technique and Mathematical Modeling course contain a task related to future professional activities.

Topic 1: Features of computer arithmetic
Topic 2: Methods for solving nonlinear equations

Task: Identify the lower and upper limits of the passband of the *RCL* filter with the level K_0. To solve the task, form an expression for the transfer coefficient of the filter $K(\omega)$. The student must solve the nonlinear equation $K(\omega) - K_0 = 0$, which has two roots.

Topic 3: Methods for solving linear algebra problems

Task: Locate the currents flowing in a closed circuit using mesh current method.

Topic 4: Data interpolation

Task: The results of measurements of the gain-frequency characteristic of a microstrip resonator are given. It is necessary to construct an interpolation polynomial function and a spline function according to proposed data. Next, compare interpolation error for both cases.

Topic 5: Ordinary least square technique

Task: An experimental data table is provided. This is a result of measuring the permittivity of a certain material (for example, wood, soil, or cement on moisture content). It is required to fit the experimental data by the least squares method and to determine the optimal number of basic functions for the known measurement error of the dielectric constant. Using the inverse interpolation method, determine the value of the moisture content of the sample for a value of the permittivity according to the instruction of the teacher.

Topic 6: Numeric integration methods

Task: A rectangular pulse sequence is provided. Every continuously differentiable periodic function can be expanded in a Fourier series, in which the summands are definite period integrals. Values of amplitude, period, and duty cycle are given by the teacher. Find the amplitude of the second harmonic of the given signal.

Topic 7: Methods for solving ordinary differential equations

Task: A linear circuit is given. It consists of two resistors (R1, R2) and two capacitors (C1, C2). The transient processes in this circuit are described by a second-order differential equation. The initial conditions are $u(0) = 0$, $u'(0) = 0$; $e(t) = U_0 \sin(\omega_1 t)$ is the external signal. It is necessary to determine the value of the voltage on the capacitor C2 at time $t_0 \neq 0$ for set values of resistances, capacitances, and the frequency of the external signal f_1 ($\omega_1 = 2\pi f_1$). Parameters of a circuit, a signal, and a method of numerical solution of a differential equation are given in several variants.

Topic 8: Methods of unconditional optimization of functions

Task: An oscillating circuit containing a resistor, a capacitor, and an inductor is given. It is necessary to find the maximum and minimum values of the complex impedance module at the frequency range from 1 kHz to 100 kHz. The parameters of the circuit are specified in several variants.

"Methods for Solving Linear Algebra Problems" is a task solved by students. The input data for the task is an electrical diagram with two or three closed circuits. The circuit includes resistance values of the resistors and voltage of the power supply. The electrical schematic diagram is also given. The students must find currents flowing in the diagram's electrical loops. To solve this problem in electric engineering, it is necessary to choose an appropriate method to calculate electric circuits (i.e., Kirchhoff's law, mesh current method, nodal analysis, etc.; Ivanov, Lukin, & Soloviev, 2002). Students should be familiar with theoretical material from the recommended literature. Diagrams are selected in such a way that the students must solve a system of linear algebraic equations to calculate the currents. At the end of the work, testing is carried out using an electronic schematic capture and simulation program called MultiSIM. See Figure 2.

Figure 2. Electrical circuit and solution check in MultiSIM

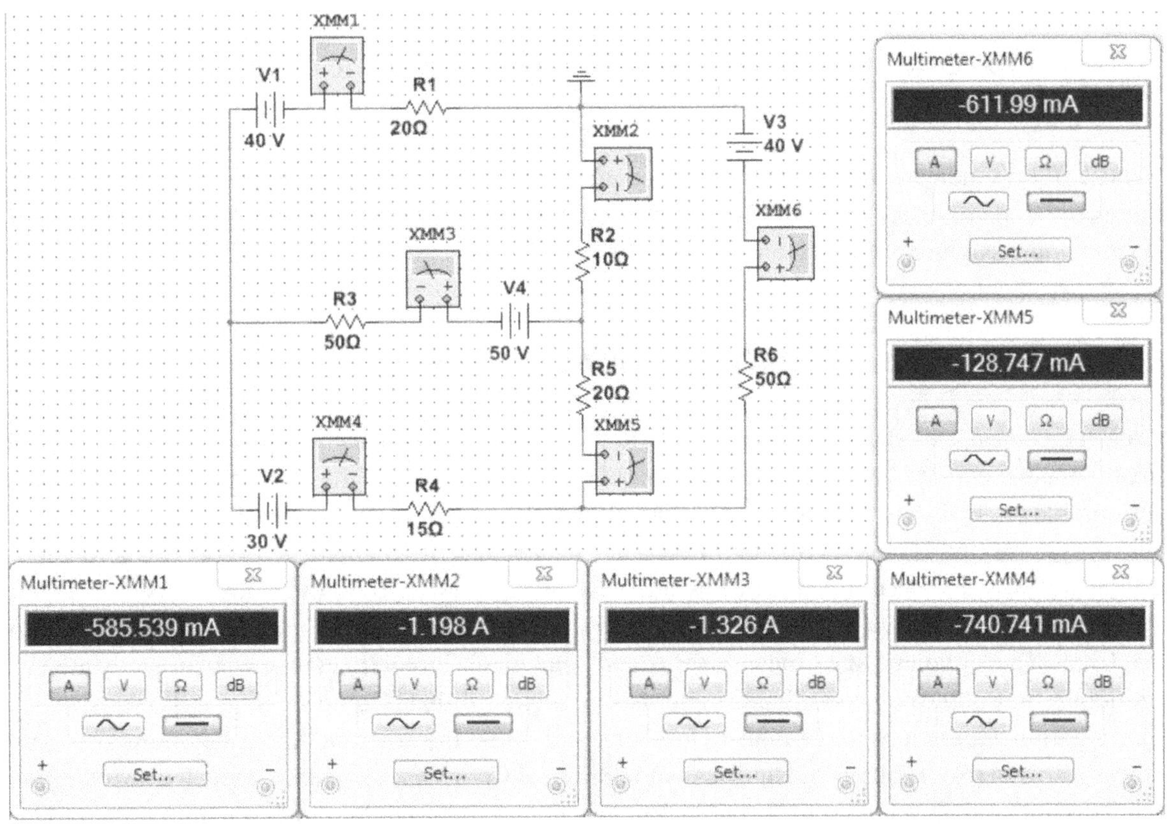

The information support is carried out in the learning management system's modular object-oriented dynamic learning environment (MOODLE). This widespread system established itself as a convenient tool for the organization of students' independent work (Macías, 2012; Martinez, Herrero, & de Pablo, 2011; Regueras, Verdú, Verdú, & de Castro, 2011; Rodriguez, Fuertes, Piera, & de Castro, 2012). Students of the faculty use the system's electronic resources from the first years of study in the educational process (Zhukov & Korotaev, 2015). Theoretical material, rehearsal and control tests, recommended literature, and a glossary are provided for each topic. A set of individual targets was compiled, including detailed examples of solutions in multiple programming environments (i.e., PascalABC, MathCad, Matlab, Oktave).

RESULTS AND DISCUSSION

Results of the study were announced at the conclusion of the 2016 spring semester. Tools for analyzing student activity are provided by MOODLE. The results were as follows:

1. E-learning course attendance in the 2015/2016 academic year increased by one-third in comparison to the previous academic year when studying was carried out with a traditional method. It amounted to more than 120,000 applications with the number of students to 80 people.
2. The number of students who learned the theoretical material online (before studying the topic in class) increased by approximately 30%.
3. In the 2015/2016 academic year, the average grade of the final course test increased significantly (20%).
4. The number of students who were given a credit test according to the results of semester work increased by 20%.

These figures indicate that students have become more active during the semester. By the end of the training, the students improved their knowledge of the subject. Students positively evaluated the seminars and teamwork assignments on the end-of-semester survey.

FEATURES OF PBL FOR MATHEMATICAL COURSES

Setting targets and creating teaching methods in basic technical subjects can be difficult, particularly in mathematical courses. As a rule, it is not problematic to formulate the case study in economic and health problems. PBL has been applied in mathematical, physical, and engineering disciplines. TSU's Control Systems and Radioelectronics (TUSUR) Department of Mathematics proposed a method of teaching students advanced mathematics while considering challenges in the study of special technical disciplines (Magazinnikova, 2015). In order for the students to successfully understand the Foundations of the Chain Theory course, the Linear Algebra and Analytical Geometry course was supplemented with an in-depth study of the polar coordinate system using sections titled "Complex Numbers: Forms of Their Representation, Operations on Complex Numbers, and Some Information About Polynomials." Attention was also given to the topic "Cylindrical and Spherical Coordinate Systems," which was required for special disciplines related to radar, radio navigation, electrodynamics, and propagation of radio waves.

Mathematics teachers simultaneously collaborated with teachers of special disciplines to find and set problem-based cases. Each topic in advanced mathematics was associated with another professional or special discipline. The task of defining frequency transmission coefficients or impedance of electric circuit was chosen for the topic "One-Variable Function." Problems related to the spectral analysis of rectangular pulses were solved in the topic "Fourier Transform."

When elements of PBL were introduced in the educational process, the teachers faced the following problems:

1. Although some students were familiar with PBL, other students needed to be prepared or "woken up."
2. It was difficult to identify problematic situations for solving mathematical and technical tasks to perform with one or two classes.
3. Poor school training limited the number of tasks used in the implementation of problem-oriented approach.
4. Students wanted to solve the problem on the example's model (the stencil).
5. Mathematics called for consistent presentation of items. However, breaking down to smaller problems provided an incomplete knowledge base. High levels of professional training in mathematical subjects cannot occur without generalization and systematization.
6. Thinking inertia of teachers (the habit of traditional relay learning affect) prevented good teaching practice.

Students also faced difficulties that manifested as resistance. Therefore, it was impossible to use this approach during the first two years of study.

1. During the first two years of study, students were not good acquaintances.
2. A large number of the young individuals were poorly motivated for acquiring knowledge, which is necessary for the more intensive independent work.
3. Poor training in school demands students' time and effort to solve problems. Otherwise, an achievement of result becomes impossible.

In science, technology, engineering, and mathematics (STEM) disciplines, it is important to preserve the fundamental component of education. It is necessary to avoid the situation when training is reduced to learning the recipes for solving a certain range of tasks.

FUTURE RESEARCH DIRECTIONS

Engineering education, following the requirements of society, is in constant development and is undergoing continuous changes. This process is as unstoppable as scientific and technological progress. In addition, the search for new methods of teaching, effective and attractive, has become necessary in the conditions of competition of universities. Therefore, an approach such as PBL has good prospects for application in engineering education. The movement towards PBL is a fascinating and interesting process. The application of this approach to learning process at Radiophysics Faculty of TSU will also continue. The next steps will be following:

1. Modification of the course "Materials and Components for Electronics" with application of elements of a problem-oriented approach. The necessary lecture material can be submitted with examples of solving practical problems, thereby understanding and internalizing the course can be improved. For example, practical task: to choose the components of the dielectric material with needed mechanical and electrophysical properties from the existing database of materials and to calculate the effective permittivity of the material for manufacturing the capacitor ceramics. This task shows where and how the theoretical material can be useful, helps to submit it in a logical connection with practice and to remember better.
2. Development of a set of tasks for the course "Basics of Programming" for the specialty "Electronic systems and complexes". That is, the conditions (texts) of tasks will have a practical focus; for example, they contain specific information about experimental data. Currently, there is the possibility for some classes to use the ARDUINO microcontroller, whose software is publicly available on the Internet. This device allows formulating tasks for programming the operation of individual external devices that are compatible with it.
3. Development of a set of cases for practical classes on the course "Basics of designing and production technology of radio electronic systems". Finding a solution to a particular problem, for example, increasing the reliability of a control unit for Christmas tree garland or choosing an electronic device cooling system, helps to get necessary skills faster.
4. Modification of the lecture course "Virtual Instruments LabVIEW" with application of elements of the problem-oriented approach. The lecture material will be supplemented with examples of solving practical problems of data collection and processing, and practical lessons on the course will be supplemented by a set of individual tasks for creating virtual instruments for automating electric engineering measurements. Skills that students will assimilate in the implementation of these assignments will contribute to the successful execution of their course projects and final qualification work.

CONCLUSION

PBL elements have been used in the Numerical Technique and Mathematical Modeling training course. The following conclusions are based on the experience of practical over-notions during the spring semester of 2014/2015 and 2015/2016 academic years:

1. A topic's problem statement creates interest and motivation for learning.
2. Students were more successful at grasping material.
3. It is necessary to combine traditional lectures with practical lessons to use various active methods and approaches. These combinations are especially important for engineering subjects in the mathematical disciplines during the first and the second years of study.

REFERENCES

Barrett, T. (2010). The problem-based learning process as finding and being in flow. *Innovations in Education and Teaching International*, *47*(2), 165–174. doi:10.1080/14703291003718901

Bevinakoppa, S., Ray, B., & Sabrina, F. (2016). Effectiveness of problem-based learning implementation. *International Journal of Quality Assurance in Engineering and Technology Education*, *5*(3), 46–58. doi:10.4018/IJQAETE.2016070104

Cotič, M., & Zuljan, M. V. (2009). Problem-based instruction in mathematics and its impact on the cognitive results of the students and on affective-motivational aspects. *Educational Studies*, *35*(3), 297–310. doi:10.1080/03055690802648085

Crawley, E. F., Malmqvist, J., Östlund, S., Brodeur, D. R., & Edström, K. (2014). *Rethinking engineering education: The CDIO approach* (2nd ed.). New York, NY: Springer International Publishing; doi:10.1007/978-3-319-05561-9

Dionysiou, I., & Ktoridou, D. (2012). Enhancing dynamic-content courses with student-oriented learning strategies: The case of computer security course. *International Journal of Cyber Ethics in Education*, *2*(2), 24–33. doi:10.4018/ijcee.2012040103

El-Khalili, N. H. (2013). Teaching agile software engineering using problem-based learning. *International Journal of Information and Communication Technology Education*, *9*(3), 1–12. doi:10.4018/jicte.2013070101

Figuera, C., Morgado, E., Gutiérrez-Pérez, D., Alonso-Atienza, F., del Arco-Fernández-Cano, E., Caamaño, A. J., ... Requena-Carrión, J. (2011). A multidisciplinary problem based learning experience for telecommunications students. *International Journal of Human Capital and Information Technology Professionals*, *2*(3), 15–28. doi:10.4018/jhcitp.2011070102

Fortelius, C., & Akerman, M.-L. (2015). Project/problem based learning in the field of materials, food, and chemical engineering at Helsinki Metropolia University of Applied Sciences. *International Journal of Quality Assurance in Engineering and Technology Education*, *4*(4), 39–46. doi:10.4018/IJQAETE.2015100103

Hung, W. (2011). Theory to reality: A few issues in implementing problem-based learning. *Educational Technology Research and Development*, *59*(4), 529–552. doi:10.100711423-011-9198-1

Iskrenko, E. V., & Poulton, T. A. (2008). Problemno-orientirovannoe obuchenie: osobennosti metodiki prepodavaniya v Velikobritanii [The specificity of problem based learning: the technique of operation in St. St. George University of London, Great Britain]. Nauchnyie vedomosti – Scientific statements, 10(50), 214-218.

Ivanov, I., Lukin, A. F., & Soloviev, G. I. (2002). *Elektrotehnika. Osnovnyie polozheniya, primeryi i zadachi* [Electrical engineering. The main provisions, examples and exercises]. Sankt-Peterburg, Russia: Lan Publishing house.

Macías, J. A. (2012). Enhancing project-based learning in software engineering lab teaching through an e-portfolio approach. *IEEE Transactions on Education, 4*(55), 502–507. doi:10.1109/TE.2012.2191787

Magazinnikova, A. L. (2015). Ob izmeneniyah kursa matematiki dlya studentov-radiotehnikov [The alteration of mathematics course for students-radio engineers]. *Izvestiya vyisshih uchebnyih zavedeniy. Fizika, 8/3*(58), 324-326.

Martinez, F., Herrero, L. C., & de Pablo, S. (2011). Project-based learning and rubrics in the teaching of power supplies and photovoltaic electricity. *IEEE Transactions on Education, 1*(54), 87–96. doi:10.1109/TE.2010.2044506

Merritt, J., Lee, M. Y., Rillero, P., & Kinach, B. M. (2017). Problem-based learning in K-8 mathematics and science education: A literature review. *Interdisciplinary Journal of Problem-Based Learning, 11*(2). Retrieved online at 10.7771/1541-5015.1674

Mertins, K. V. (2012). Higher liberal education prediction by way of educational technologies. *European Researcher. Series A, 2*(17), 205–209.

Regueras, L. M., Verdú, E., Verdú, M. J., & de Castro, J. P. (2011). Design of a competitive and collaborative learning strategy in a communication networks course. *IEEE Transactions on Education, 2*(54), 302–307. doi:10.1109/TE.2010.2053933

Rodriguez, S. B., Fuertes, M. C., Piera, A. F., Garcia, I. P., & Solsona, F. J. A. (2012). Lessons learned in the use of WIRIS quizzes to upgrade Moodle to solve electrical circuits. *IEEE Transactions on Education, 3*(55), 412–417. doi:10.1109/TE.2011.2181381

Schmidt, H. G., Rotgans, J. I., & Yew, E. H. J. (2011). The process of problem-based learning: What works and why. *Medical Education, 45*(8), 792–806. doi:10.1111/j.1365-2923.2011.04035.x PMID:21752076

Spencer, J., & Jordan, R. K. (1999). Learner centred approaches in medical education. *British Medical Journal, 318*(7193), 1280–1283. doi:10.1136/bmj.318.7193.1280 PMID:10231266

Valtanen, J., Berki, E., Georgiadou, E., Hatzipanagos, S., Ross, M., Stamelos, I. G., & Staples, G. (2012). Features for suitable problems: IT professionals' and IT students' opinions. *International Journal of Human Capital and Information Technology Professionals, 3*(3), 27–41. doi:10.4018/jhcitp.2012070103

Zhukov, A. A., & Korotaev, A. G. (2015). Metodicheskoe i informatsionnoe obespechenie kursa "Osnovyi rabotyi v SDO Moodle" [Methodical and information support of the course "Basics LMS Moodle"]. *Innovatsii na osnove informatsionnyih i kommunikatsionnyih tehnologiy – Innovations based on information and communication technologies, 1*, 46-49.

KEY TERMS AND DEFINITIONS

CDIO: CDIO in engineering education involves teaching future specialists to work throughout the lifecycle of the product using conceive-design-implement-operate.

Multidisciplinary Training: Methods of teaching to reveal the interconnection of sciences in practice, as well as solving complex problems requiring knowledge in various fields of science.

Problem-Based Learning (PBL): Learning based on self-searching for solutions to specific problems under the guidance of a teacher.

Student-Oriented Educational Environment: Training aimed at developing a student's professional skills on an individual trajectory in a creative atmosphere.

Teacher: A teacher is a leader as the independent students work and search for information. The teacher discusses the scientific approach to solving a problem.

Chapter 17
A Project-Oriented Approach to Practicum on Software Engineering Methodology Courses

Tatiana Nikolaevna Romanova
Bauman Moscow State Technical University, Russia

Tatiana Ivanovna Vishnevskaya
Bauman Moscow State Technical University, Russia

Dorjsuren Odselmaa
Mongolian University Science and Technology, Mongolia

ABSTRACT

This chapter suggests a method for practicum on software engineering methodology course using a project-oriented approach. The chapter features basic organization principles of the approach and examples of methodical support for laboratory works based on these principles, and provides recommendations on choice and use of methodologies and technologies of software engineering for the development of distributed information systems. The experience of using this technique for teaching students studying for a Master's degree in Software Engineering in Bauman Moscow State Technical University is presented.

INTRODUCTION

Wide software application in numerous spheres of life calls for dramatically higher standards of software quality, reliability and security. Software engineering is an integral part of all the stages of the production of complex software and software-technical systems starting from the very beginning of software creation to its deployment and maintenance (Gomes, Mendes, & Marcelino, 2015). The main concepts of software engineering are formalized in a set of international standards and are studied in the Methodology of software engineering (MSE), which is a theoretical course (Lipaev, 2006). The course is taught to students studying for a master degree in Software engineering.

DOI: 10.4018/978-1-5225-3395-5.ch017

The course comprises lectures and practical classes. Different universities have different approaches to practical classes. For example, students studying at the department of Software Engineering of the faculty of Computer Science of NRU HSE write a paper on a specified topic as a report on the practicum and build an application using the techniques studied in the course (Saleh, 2016). In Southern Federal University students studying for the master's degree explore existing software tools for program verification within laboratory classes. In BMSTU our target is to organize practicum in such a way, as to not only study the theory of MSE but also use it to ensure the development of quality software. Students studying for the master's degree in Software engineering are to be well versed in the fundamental principles and systematic approaches namely solving issues of development, deployment and maintenance of software systems, described in (Lattanze, 2008), and as well as in methods of their application (Bass, Clements, & Kazman, 2012).

METHOD

Within their professional training master students encounter some difficulties in studying the following disciplines: Distributed systems of information processing (DSIO) and Methodology of software engineering. Thus, a new, distinct from traditional, approach of training was required. A project-oriented approach described in (Zamyatina, & Mozgaleva, 2014) was chosen as such. Implementation of the project-oriented approach features shaping master students' professional skills through completing real engineering tasks. As such, within their laboratory works on MSE, the students are tasked to perform the information systems design, which is also constitutes a part of their course paper on the DSIO discipline.

In addition, the subject area to be analyzed as well as development tools are selected for each master on the basis of his/her personal professional interests, activity areas and experience of relevant software developments. This allows the teacher to work with each master individually or in small teams of two or three students (Pressman, 2009).

The level of competence of an expert is determined by the amount of knowledge and experience gained through his/her own activities in a particular subject area. As the proposed approach to shaping professional competences of master students is based on their own professional experience and activities, such an educational approach can also be classified as a competence-oriented approach. The project-oriented approach has been further complimented by a methodological educational model, which allows to bring together the theory of the methodology of software engineering and practical experience of the master students.

The proposed method objective: To form the professional competence of master students in the process of studying the Methodology of software engineering.

To achieve this goal the following tasks are to be solved:

- To develop the skills of formalization of customer requirements;
- To develop the skills of using modern paradigms, methods and notations of software engineering for building problem domain models;
- To develop the skills of process and data structures modeling ;
- To develop the ability for team-work;

- To develop the ability of application of the international ISO standards and the Standards of the Russian Federation at each stage of software development.

The Model of the proposed methodology is shown in *Figure 1*.

The assignments for master students featured within the presented methodology reflect their future professional environment and requirements and are shaped by the current trends / demands of the time. At the same time, the theoretical information provided in the lecture course in the discipline Methodology of software engineering, should comprise the interdisciplinary connections that play an important role in the formation of the basic competencies that result from a combination of different disciplines. Therefore, the choice of topics for laboratory works is based on topics of lectures within the disciplines Methodology of software engineering and Distributed systems information processing. The master student based on his/her personal programming experience and areas of interest chooses the subject area. In addition, the tasks requirements of the workshop are to use object-oriented approach to describe a system as a series of objects interactions. The system development should start with the analysis of the subject area, definition of the aim and purpose of the future system, its quality requirements and functional characteristics. Per the results of such analysis, a technical specification for a distributed information system and a conceptual model of the designed system are further developed. To describe system models in object-oriented approach it is appropriate to use a visual modeling in a fairly complete UML notation (Unified Modeling Language), which is further expanded during the transition from analysis to design (Schmidt, Hatebur, & Heisel, 2015)

Main Topics of Laboratory Works

1. Analysis Phase: At this stage, the student chooses the topic of the coursework in the Distributed systems of information processing discipline (DSIO) and describes the selected subject area. He outlines the purpose of the proposed system, the existing counterparts, assesses the topicality of development and describes the concept of the information system.
2. Planning Phase: At this stage, the working schedule of individual and collective development is formed; the resources and timing of the project development are assessed. Also the financial and strategic values are estimated and the risk level of the project is described.
3. Composing of the Technical Assignment for the Development of a Distributed Information System: This phase consists of the development of functional system requirements, requirements to functional characteristics, the system topology, the requirements to the technical means and requirements for system reliability.
4. Project development of a distributed system in one of the standard notations (UML, IDEF0, etc.).
5. Development of a structure and architecture of the distributed information system and the distribution of tasks between the project participants.
6. Development of a logical and physical design of the project.
7. Description of Testing phase. The information system should be thoroughly tested. In the report the student must describe the used testing methods and cite the performed tests.
8. Preparation of the report.

Figure 1. Model of method

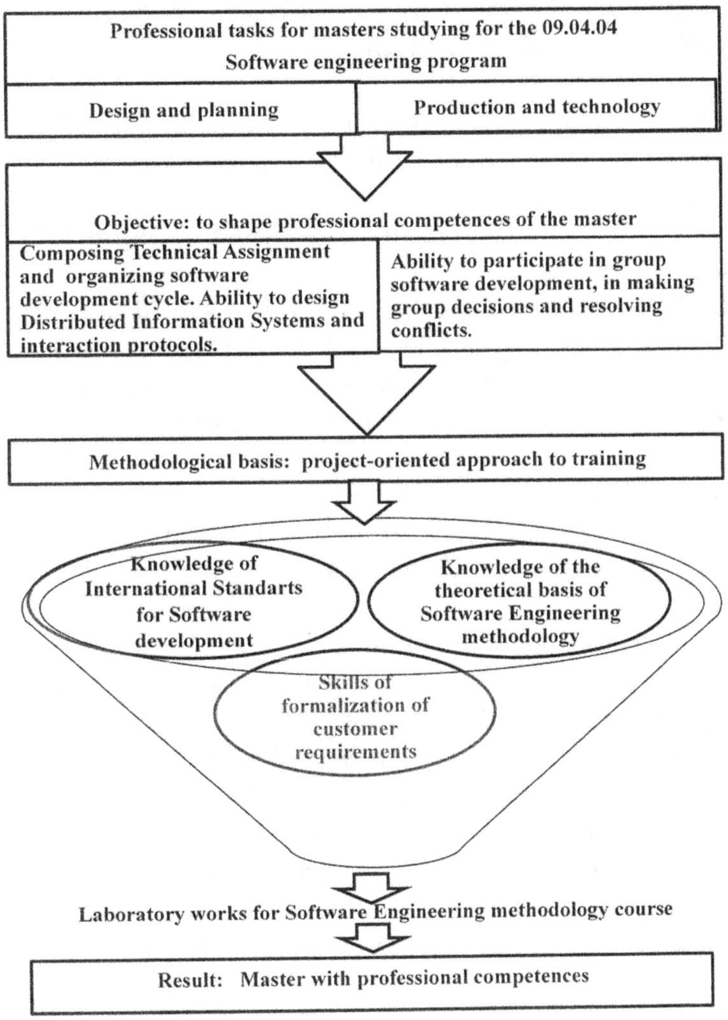

Today many information systems are distributed and heterogeneous. Distributed systems allow us to solve more complex problems and improve efficiency of production processes, but at the same time they are much more complex to implement and support. Design of distributed systems has much in common with the design of regular information systems, but we should take into consideration some specific features (Delgado, 2015). To build correctly the architecture of a distributed information system in the laboratory work within the discipline of *Methodology of software engineering* it is necessary to use the theoretical material the master students obtain in a course of lectures on DSOI. They should know the principles of databases and Front-Ends scaling, the principles of the protocols and methods of ensuring secure data transmission through networks. Students should understand the difference between Service-Oriented Architecture (SOA) and micro services, have the knowledge of the existing approaches for building asynchronous communication Back-Ends and the modern soft engineering approaches to the resolving of irregular situations in the operation of distributed information systems.

EXPERIENCE IN IMPLEMENTING PROJECT-ORIENTED APPROACH IN LABORATORY WORKS FOR SOFTWARE ENGINEERING METODOLOGY COURCE

Following the proposed method, based on project-oriented approach, the training process of students studying for master's degree in Software engineering is implemented according to the following scheme:

- In the first semester master students study the basic course in Distributed systems of information processing (DSIO);
- In the beginning of the second semester they are assigned individual or group tasks for the course project on the development of a specific distributed system of information processing;
- In the second semester within the framework of Software Engineering Methodology laboratory works they design the distributed information systems for their course work (by using the theoretical basis of the Methodology of software engineering and meanwhile developing design skills under the guidance of a tutor);
- At the end of the second semester the master students develop the technological part of the course project on their own (write the code of the DSIO).

Methodological support plays a big role in the performing of laboratory works for the course «Methodology of software engineering». Manuals (Vishnevskaya, & Romanova, 2012, 2017) developed at the Computer Software and Informational Technologies Department describe every step in the development of an information system. As a DSIO design sample the guidelines describe a Web-portal designing process (which is still relevant). Typically, portals use service oriented architecture (SOA) which requires splitting the system into independent subsystems (services). This approach is useful for it provides for the work to be split between team members as well as allows for the use of various techniques and separately scaling services.

All the laboratory works for "Software Engineering Methodology" course are divided into four stages of DSIO development:

1. **The Formalization of Customer Requirements:** Master students must write a general technical assignment for the development of DSIO on the chosen topic, analyze and evaluate risks in the development of a complex set of programs, describe performance requirements toward the quality of the developed information system and write a profile of standards governing the development of a distributed information system in a chosen subject area.
2. **The Design Phase - Designing the Structure and Architecture of DSIO:** Master students must perform the following tasks:
    a. describe the topology of a distributed system;
    b. describe the requirements for implementation of the subsystems in detail;
    c. define the basic communication protocols;
    d. design the architecture of a distributed information system;
    e. develop high-level requirements for services.
3. **The Design Phase - The Construction of Object-Oriented Models of the System Architecture Using UML Modeling Language:** Master students must perform the following tasks within the laboratory works.

   a. Create a conceptual design of the developed system. The conceptual design allows us to view the developed information system from the point of view of users.
   b. Describe the object-oriented models of the system architecture:
      i. static models that describe the structure of the system in terms of object classes and interactions between them;
      ii. dynamic models that describe the structure of the system and the interaction between the objects of the system.
   c. Develop a logical design of DSIO which describes the organization of the elements of the software solution, and how they interact. At this stage we analyze the created system from the point of view of the project team.
   d. Develop a prototype of the user interface.
   e. Develop models of data organization in the system.
   f. Perform the analysis and selection of tools to ensure the life cycle of the software.
4. **Report on DSIO project:** Master students should hand over the scientific-technical report on the software design to be composed of:
   a. The technical assignment for the development of a distributed system of information processing, written in accordance with the standard of the Russian Federation;
   b. The design of the system in the form of diagrams on the UML modeling language
   c. The text providing justifications for the choice of OS, DBMS, language of DSIO component development and framework, as well as technologies, to ensure reliability.

Using these methodological recommendations allows the students to meet the tasks of the laboratory works in accordance with the curriculum of «Software Engineering Methodology» and provides for the shaping of the required professional skills:

- Ability to formalize customer requirements and compose the technical assignment in accordance with the standards for software development;
- The skill of using modern paradigms, methods and notations of software engineering for building problem domain models and modeling processes and data structures;
- Ability to identify the main components within software development, to define the relationships between them, to describe the interface and behavior of all software components;
- Ability to justify the choice of technological platform for the development of DSIO;
- The skill of writing scientific-technical reports on the results of the performed work.

Under previous method, students received separate tasks in DSIO and MSE disciplines and spent a lot of time on designing and implementing a large number of projects. As a result, 30% of students failed to submit the course project on DSIO in time. Application of the proposed method for MSE practicum boosted the motivation of students to perform a high qualified design of the distributed system, the task chosen in accordance of their professional orientation on different classes of tasks.

The extracts from the course project on DSOI by a master student Dorjsuren Odselmaa and entitled *"Communication portal for foreign students"* are given as a demonstration of the proposed methodology. The main purpose of the developed portal is to provide a platform for communication of foreign students studying in Russia. The portal main languages are English and Russian. The particular feature of this platform will be that it will enable foreign students studying in different universities of Russia, to exchange opinions about their universities, find someone to talk to and to obtain detailed answers to all their questions related to studying in different cities of Russia.

For the course of the project DSOI, in the laboratory works on discipline Methodology of software engineering, Dorjsuren Odselmaa developed a detailed technical specification for the development of the portal. It presents the topology of the developed distributed information system (Figure 2) and describes the functionality of all subsystems.

The specification substantiates that for the realization of its basic purpose, the system should consist have front-end and four back-ends.

The front-end should accept requests from users by the means of the HTTP Protocol and analyze them. Based on the analysis, the front-end is to make requests to the back-ends, aggregate them, and send the back-end responses to the user. The front-end must also include a module to gather statistics.

The profile back-end must implement the following functions:

- To add, remove and modify the profile of a particular student;
- To get a list of students with specific filtering criteria.

The location back-end must implement the following functions:

- To add a new location;
- To get a list of locations with the filter conditions;
- To obtain, modify, and delete specific location.

The message back-end must implement the following functions:

- To get the list of message filtering criteria;
- To obtain, modify, and delete a specific message;
- To add a new message.

The session back-end should be responsible for user registration and authentication.

In addition, the specification defines requirements for the system implementation:

- All the back-ends and front-end must be running separately and in isolation from each other.
- The system uses micro services.
- The development of server applications can use different frameworks for different back-ends, since back-ends don't need to be closely related to each other according to the requirement of the micro-service approach to development.
- The front-end should have two interfaces: a web interface and a mobile interface. Interfaces must be accessible via thin client, browser and mobile app.
- The administrator interface should be implemented with the use of the command line (console), enabling the configuration of the nodes system, adding and removing nodes.
- The back-end servers are unavailable for users, which is secured by their location in the internal network.

Figure 2. Topology of system

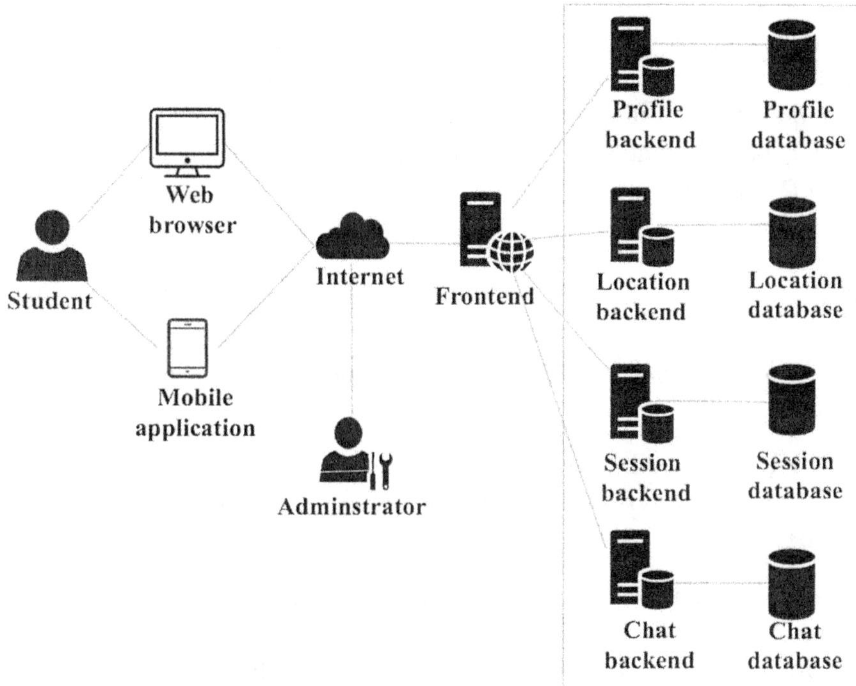

- Validation of input data should be carried out on the user side using JavaScript scripts and on the front-end. Back-ends should not validate input because the user cannot refer to them directly, the back-end is to receive filtered input from the front-end.

In accordance with the presented methodology requirements, Dorjsuren Odselmaa in her DSOI course project in detail designed the system behavior using UML notation, which allowed her to integrate in her distributed information system design such important features as the latent period in the presence of geographically remote sites, and the distribution of information across nodes. Dorjsuren Odselmaa described and justified the choice of the database management system, protocols, and technology for remote data access.

Furthermore this approach allowed to implement the group work of students for the development of information systems. For example, a team of five students using the proposed method developed a portal to automate the control of the educational process: the assignment of educational tasks, automated and manual testing of program codes issued tasks and the reception of reports on all information disciplines. The portal also implements the collection of statistics and provides the possibility of rapid interaction between students and teachers.

In addition, the number of tasks decreased, thus the attention of students concentrated on the solution of larger and more complex projects. This led to higher academic achievements in both disciplines – DSIO and MSE. The application of the proposed method for teaching students studying for the master's degree allowed to bring the number of underperforming students in both disciplines down to 5%. Moreover, the number of excellent marks in the MSE exam amounted to about 80%.

CONCLUSION

The work presents the methodology of practicum conducted within the Software Engineering methodology course and featuring the project-oriented approach and it is unique. Furthermore, this technique is used by students in the design of information systems for master's degree. However, a project-oriented approach can also be applied in the study of other disciplines in the field of IT-education, such as Operating systems and Databases.

REFERENCES

Bass, L., Clements, P., & Kazman, R. (2012). *Software Architecture in Practice*. New York, NY: Addison-Wesley Professional.

Delgado, J. C. (2015). Distributed Interoperability in Heterogeneous Cloud Systems. In S. Bagchi (Ed.), *Emerging Research in Cloud Distributed Computing Systems* (pp. 1–40). IGI Global; doi:10.4018/978-1-4666-8213-9.ch001

Gomes, A. J., Mendes, A. J., & Marcelino, M. J. (2015). Computer Science Education Research: An Overview and Some Proposals. In R. Queiros (Ed.), *Innovative Teaching Strategies and New Leaning Paradigms in Computer Programming* (pp. 1–29). IGI Global; doi:10.4018/978-1-4666-7304-5.ch001

Lattanze, A. J. (2008). *Architecting Software Intensive Systems: A Practitioners Guide*. New York, NY: Auerbach Publications. doi:10.1201/9781420045703

Lipaev, V. V. (2006). Programmnaya engeneria. Metodologicheskie osnovy [Software Engineering Methodological Foundations]. Moscow: TEIS.

Pressman, R. (2009). *Software Engineering: A Practitioner's Approach*. New York, NY: McGraw-Hill Education.

Saleh, H. M. (2016). *Program of discipline "Software Engineering Methodology"*. Retrieved from https://www.hse.ru/data/2016/10/04/1117071415/program-1513156455-EHWApXuS2m.pdf

Schmidt, H., Hatebur, D., & Heisel, M. (2015). Developing Secure Software Using UML Patterns. In V. Diaz, J. Lovelle, & C. Garcia-Bustelo (Eds.), *Handbook of Research on Innovations in Systems and Software Engineering* (pp. 32–70). IGI Global; doi:10.4018/978-1-4666-6359-6.ch002

Vishnevskaya, T. I., & Romanova, T. N. (2012). *Techonologiya programmirovaniya: Metodicheskie ukazaniya k laboratornomu praktikumu – Ch. 3* [Programming Technology: Methodical instructions for laboratory works – P.3]. Moscow: Moscow State Technical University Publishing House. Retrieved from http://www.catalog.inforeg.ru/Inet/GetEzineByID/293408

Vishnevskaya, T. I., & Romanova, T. N. (2017). *Metodologiya programmnoy ingenerii: Metodicheskie ukazaniya k laboratornomu praktikumu* [Methodology of software engineering: Methodical instructions for laboratory works]. Moscow: Moscow State Technical University Publishing House. Retrieved from http://ebooks.bmstu.ru/catalog/199/book1531.html

Zamyatina, O., & Mozgaleva, P. (2014). Competence component of the project-oriented training of elite engineering specialists. In *Proceedings* of *the 2014 IEEE Global Engineering Education Conference.* New York, NY: IEEE. 10.1109/EDUCON.2014.6826077

KEY TERMS AND DEFINITIONS

Conceptual Design of the Developed System: The conceptual design allows to view the developed information system from the point of view of users.

DSIO: Distributed information system is a system in which data storage and processing are not concentrated within one computing machine, but are distributed between multiple computers, which appear to users as a single system.

Dynamic Models of DSIO: Dynamic models that describe the structure of the system and the interaction between the objects of the system.

Front-End and Back-End: Terms in software engineering, which are distinguished according to the principle of shared responsibility between external representation and internal implementation, respectively. Front-end interaction interface between the user and the main software and hardware (back-end). Front-end and back-end can be distributed across one or more systems. In software architecture there may be many levels between the hardware and the end user, each of which may also have front-end and back-end. Front-end is an abstraction that provides a user interface.

Logical Design of DSIO: The developed a logical design of DSIO which describes the organization of the elements of the software solution, and how they interact. At this stage we analyze the created system from the point of view of the project team.

MSE: Methodology of software engineering.

Object-Oriented Models of DSIO: The construction models of the system architecture using_UML modeling language.

Static Models of DSIO: Static models that describe the structure of the system in terms of object classes and interactions between them.

SOA: Service-oriented architecture. A modular approach to software development based on the use of distributed, loose coupling model components, which have standardized interfaces to communicate according to standardized protocols.

Topology of DSIO: Describes the requirements for implementation of the subsystems in detail, defines the basic communication protocols, the architecture design of a distributed information system and the developed high-level requirements for services.

Chapter 18
Application of Interactive Technologies in Engineering Education in the Research University

Gennady Konstantinovich Baryshev
Moscow Engineering Physics Institute, Russia

Aleksandr Vasilyevich Berestov
Moscow Engineering Physics Institute, Russia

Yuri Valentinovich Bozhko
Moscow Engineering Physics Institute, Russia

Nadezhda Aleksandrovna Konashenkova
Moscow Engineering Physics Institute, Russia

ABSTRACT

The chapter describes the problem of application of interactive technologies of engineering education in the contemporary world-class research university: National Research Nuclear University "MEPhI" (Moscow Engineering Physics Institute). The results of the ongoing process of transformation of engineering education in compliance with the CDIO (conceive-design-implement-operate) international and Russian federal national educational standards are discussed. The pilot projects on the modernization of engineering educational programs have demonstrated that interactive technologies are effective in fixing sustainable results of developing and monitoring the students' project-implementational and other engineering skills.

DOI: 10.4018/978-1-5225-3395-5.ch018

INTRODUCTION

Engineering programs aimed at training qualified bachelors, specialists and masters capable of accomplishing personal goals, should be implemented by nuclear, are-earth and other high-tech industries. On the one hand, since its very establishment, the Moscow Engineering Physics Institute has a long-established tradition of training engineers through combining research and developmental activities. This was reflected in the way the educational process was formed. Along with strong fundamental academic training in Mathematics and Physics, it was obligatory for students to be involved in scientific research and development projects (R&D), carried out by the Institute's departments together with industry leaders. Students were also actively engaged in the designing of high-technology devices, appliances and systems that were well ahead of their time; the development and production of pilot models; the perfomance of technology intensive experiments, including, those in reactor conditions (high temperatures, ionizing radiation, mechanical loads).

BACKGROUND

Science constituted the basis for educational process in the second-generation university model (research university) (Fedorov & Medvedev, 2011, Saprykin, 2012, Vladimirov, 2011). However, nowadays, the practical, innovative activity is becoming the key driver of changes in education. Generally, entrepreneurial universities are the leading trend in the world today (Altbach & Salmi, 2011, Clark, 2004, Delbanco, 2012, Grasso & Burkins, 2010, Simonyants, 2014). Considerable efforts are made in the National Research Nuclear University "MEPhI" to modernize basic educational program on nuclear physics and technologies, based on the international engineering educational standards by the Worldwide CDIO Initiative (CDIO, 2017) and other programs taking into account the project-implementational approach and the active role of innovations in the educational process of a modern research university (MEPhI MA Educational programs, 2017).

According to the requirements of the CDIO international standards (Crawley, 2001), relevant laboratory infrastructure is essential for the high-quality engineering education that provides future engineers with necessary knowledge and skills. The practice-oriented approach applied in engineering education cannot be pursued without developing skills of designing and setting up physical experiments in nuclear, aerospace and medical technologies. Such an approach has been historically pursued in MEPhI. It comprises fundamental training and mastering practical skills of working with technical devices and systems through laboratory works in combination with individual designing and development of constructions, devices, installations and systems (Yakovlev et al., 2015). The Department of Engineering Science and Technology of the National Research Nuclear University "MEPhI" employs the above-described approach too. Furthermore, let us examine the examples of the development of these principles of sustainable practice-oriented engineering education in a modern research university with the help of current interactive technologies.

MODERNIZATION OF ENGINEERING MODULES OF MAJOR EDUCATIONAL PROGRAMS WITH THE HELP OF INTERACTIVE TECHNOLOGIES: EXAMPLES

The development of network and interactive technologies allows to reformat the elements of the educational process. Thus, it is now possible to develop and organize distance laboratory works for some disciplines within engineering modules of the core educational programs, which enables students to obtain the skills of working with the basic instruments of computer-aided engineering and designing.

Various computer software solutions are used in engineering projects, namely, Mathcad and Matlab, LabVIEW system for the development of virtual instruments and various CAD systems for engineering devices and systems (Surin et al., 2013). Laboratory works are vital for providing students with the skills of working with these systems. Pilot projects of such laboratory works on Mathcad and LabVIEW basics have been developed within the framework of the project realization. Another example of the modernization of the engineering modules of the core educational programs with the help of interactive technologies is the integration of social media tools into the educational process and individual work management.

Nowadays, social media (Facebook, Vkontakte, Instagram, etc.) are a popular and effective communication tool. Almost all of us have a social media account. The overwhelming majority of students visit the "Vkontakte" social network on a daily basis, communicate with friends, post and comment photos and videos, share information and take part in discussions in special communities. In this regard, it is a long-held fact that students also use social media to do homework together. Vkontakte has a range of useful tools for that purpose, namely, the opportunity to organize a multiparticipant online chat, to share photos, documents, etc.

Nowadays, global smartphones and tablets market size exceeds the personal computers and laptops market size. Consequently, a considerable part of the visits to social media are made via mobile gadgets (phones, smartphones and tablets). The development of modern wireless networks (Wi-Fi, LTE, etc.) makes it possible to use all the technological opportunities of modern devices such as exchangeing not only messages but also multimedia content (photo, audio and video) to maintain communication and to be online. According to the roadmap of the "Education 2035" foresight by the Agency for Strategic Initiatives to Promote New Projects, the "Everybody online" is a trend gaining momentum by 2020 (ASI Foresight "Education 2035", 2017). Its name clearly reflects the fact that a considerable part of communication will move online. That will be a natural process for the modern digital generation. A substantial part of educational processes will obviously move online as well.

Communication in social media eliminates the limitations of space and time coming from people being in different cities, regions and time zones. This circumstance may spread to the educational process, providing students, teachers and other subjects of the process with the opportunity to communicate interactively regardless of the limitations of time, space, etc., inherent to the traditional format of classroom education.

Therefore, the opportunity of using Vkontakte, being the most popular social network in the Russian Federation, the Customs Union and the CIS countries, was quite evident in the context of the organization of education of bachelor and specialist students at the Faculty of Physics and Technology in compliance with international engineering educational standards by the CDIO.

A community named "Basics of Design, Development and Engineering" was created on Vkontakte for the "Machine Elements and the Basics of development" course for the continuous communication between teachers, experts and students, promptly informing, working out any organizational matters, and, most importantly, assisting students in their individual and team work on engineering projects. The community name reflects the prospective changes in the program of the discipline within the framework of the implementation of international engineering educational standards by the CDIO in MEPhI.

The community is private, but visible for potential members. Moderators from among the teachers have been appointed managers (administrators) of the community in order to post publications on the behalf of the community (for all members to see these publications in their newsfeed) and to provide effective technical administration.

There is a possibility to post the community status update with this information appearing in the members' newsfeed, which is useful for posting short important announcements about current activities, schedules, exams, etc. Publications are the main instrument within the community. Administrators, moderators and community members are authorized to post publications. All the community members also can post comments, including those personally addressed to the teachers, students, experts, etc.

The following may be posted in the community:

- Discussions in a forum format;
- Documents (administrative regulations, examples of technical specifications and reports, educational books, etc.);
- Photos and pictures, including technical drawings;
- Videos, including video lectures, workshops, etc.;
- Online voting, polls, getting feedback, etc.

It is noteworthy that the tone of community publications is independent from any regulations or requirements applied to formal letters and is semantically and stylistically closer to the common way of communication between ordinary social media users.

The community instruments make it possible to use various social practices to diversify the process of teaching:

- Discussing the uploaded materials for background studies (science fiction, video lectures on system engineering, TRIZ, system-activity methodology, lectures by leading nuclear industry experts such as S. Kirienko, Y. Adamov and others);
- Publishing the biographies of outstanding engineers and inventors using a special hashtag;
- Organizing small local contests, flash mobs and discussing topical engineering events in Russia and worldwide.

One may upload to the community's photo albums various pictures about engineering of the past, the present and the future (with the ability to discuss them with other members), photos from lectures, seminars, workshops and business games. The community's videos contain useful and entertaining videos and lectures by leading experts and scientists. The community "The Basics of Design, Development and Engineering" has become an effective and useful instrument of providing e-learning and online communication between students and teachers. It enables students to broaden their horizons, worldview and engineering thinking by uploading diverse multimedia content. It is hardly possible within class hours to manage to touch upon and discuss such an abundance of materials aimed at developing the students' personal and project skills. At the same time, these materials do not seem to be an extra work for students because looking through and discussing such content automatically fit into the daily process of students' activity on Vkontakte. Getting new information, considering and discussing it becomes an inherent part of the students' communication in social media. This synergy between educational objectives and sharing and discussing information in social media catalyzes the all-round development of students and the process of providing them with not only knowledge and skills connected with the engineering disciplines of the Department of Engineering Science and Technology, but also with personal and project skills necessary for becoming the engineers of the future.

INTERACTIVE METHODS OF THE ASSESSMENT OF COMPETENCIES IN ENGINEERING EDUCATION

Designing new educational programs in compliance with the international CDIO standards implies the modification of basic engineering disciplines programs. For instance, the "Basics of Development" discipline for the Physical and Technical Faculty undergraduates (third-year students). Designers of the discipline's teaching materials qualify the following as the types of educational activity: lectures, tutorials, practical training, laboratory works, examinations, colloquiums, individual works, research projects, practical training and term projects (term theses). Institution of higher education may introduce other types of educational activity (Baryshev, Berestov & Konashenkova 2016).

According to the "Basics of Development" lectures and seminars are the types of teaching activity. The development and realization of the "Basics of Design, Development and Engineering" course in compliance with the CDIO standards made it possible to introduce the additional elements of educational technologies (business games, workshops, etc.) on the basis of this format, which promotes the development of engineering thinking and training to work in groups on engineering projects.

Due to the specifics and syllabus of the educational format, it is necessary to assess the students' skills by a range of directions:

- **Teamwork and Leadership:** At the beginning of project work the teams were formed on the ground of personal relations between students. The boundary condition was to form a team from classmates. Mixed teams were also permitted without prejudice to educational process. Such cases took place and were managed individually. In general, 35 teams of students were formed with their own leaders, idea providers and executors.
- **Feasibility of an Engineering Solution:** Undoubtedly, developing and solving an engineering challenge is a creative process. That is why the discipline's program engages workshops on design engineering, theory of the resolution of invention-related tasks methods and other instruments that may help students find original, creative, high-tech and quality solutions.
- **Report Design Correctness:** An important assessment criterion is the compliance of reporting documentation with the requirement of the Unified system of engineering drawings and GOST (national standard). As experience has shown, the students should be particularly trained to design reports correctly.

- **Accuracy of Engineering Designs:** Within the framework of developing engineering solutions students designed separate units and elements of constructions in accordance with the "Machine Elements and Basics of Development" discipline. Most computations were performed using automatic means and software solutions such as Mathcad.
- Understanding an engineering project within the context of the product's life cycle.

It is an inherent part of designing a device in compliance with the international CDIO standards. All projects are relevant developments for nuclear and other high-tech industries, implemented within the framework of R&D conducted by the Department of Engineering Science and Technology and "Signal-MIFI" LLC.

For example, a number of engineering tasks was formulated so that to involve the best students with engineering skills in solving high-priority tasks for development of systems of quality control of new-type perspective nuclear fuel elements for increasing reliability and economic effectiveness of nuclear power plants – the R&D performed by National Research Nuclear University MEPhI (the Department of Engineering Science and Technology). They included patent research on the problem as well as development documents for information measuring systems, software and prototypes.

Understanding the product's life cycle is an inherent part of solving an engineering challenge because educators are at the same time the employees of enterprises responsible for larger projects and integrating the students' developments.

The modular structure of the student work on engineering projects created competitional conditions when the best project was chosen. On the other hand, the competitional format should have provided the opportunity to assess the students' skills related to all the above-stated aspects of project work.

An ordinary format of academic presentation does not provide such an opportunity. That is why the Department of Engineering Science and Technology established a group engineering projects contest with experts of the "Constructing Future" research group involved for the organization of the contest and the business game.

The objectives of the contest's steering committee were the following:

- organizing the contest;
- organizing the work of the program committee;
- organizing the work of the contest commission;
- organizing the distribution of awards to the winners and runners-up.

The objectives of the program committee were the following:

- organizing and holding a business game;
- developing the procedures of assessment for the contest commission;
- organizing the conversion of contest results into the students' academic assessment scores at term examinations.

The objectives of the contest commission were the following:

- assessing the students' skills by the criteria developed by the program committee;
- choosing winners and runners-up.

Using social media, particularly the community on Vkontakte, during preliminary selection and the analysis of projects made it possible to increase the efficiency of the work of the experts and to keep timely discussions and cooperation with contestants.

Vkontakte is a Russian social network that provides a number of instruments sufficient for daily online communication. Every student has 24/7 access to its services via their smartphones of tablets. It was considered a much better and more comfortable way for project day-to-day communication between students, teachers and experts, that any other system, including MEPhI's own ICT instruments.

Standard 5 of the CDIO, according to which the curriculum of an educational program should include two or more projects involving the familiarization of a student with project implementation, plays a crucial role in the implementation of the CDIO concept.

A project will be understood to be a training and practical task related to the design and production of products, which are realized with the help of the engineering disciplines complex. By solving the training and practical tasks integrated in the curriculum, the students improve their skills of the design and production of new products and systems, as well as their ability to apply theoretical knowledge in the engineering practice.

The tasks on the design and creation of new products and systems could vary from basic to advanced depending on their intensity, difficulty and place in the curriculum.

The development and creation of products and systems in the context of actual engineering practice allows students to determine the fields of their future professional interests. To implement Standard 5, a team project-based learning methodology was developed by the Department of Engineering Science and Technology. The methodology was successfully tested on the Department's bachelor students. A team project is a hands-on training that is aimed at the development of engineering, designing, technological etc. projects in the real-life environment (Baryshev, 2016).

CONCLUSION

All in all, the pilot projects on the modernization of engineering educational programs have demonstrated that interactive technologies are effective in fixing sustainable results of developing and monitoring the students' project-implementational and other engineering skills in compliance with international engineering educational standards by the Worldwide CDIO Initiative, federal national educational standards and the university's own educational standards (National Research Nuclear University "MEPhI", 2017). More efforts are planned to be made to distribute the successful examples and practices, to enhance the use of interactive technologies in engineering education, and specifically within the framework of competitive growth of MEPhI in the global educational environment.

The future of application of interactive technologies of engineering education in the contemporary world-class research university is connected with competitive engineering educational bachelor, specialist and masters programs, aimed at training skilled engineers for high-tech industries, including companies of State corporation for nuclear energy Rosatom. At the moment, these programs are at the stage of development, such as brand-new programs for training specialists with fundamental knowledge, engineering and management skills required to develop electrophysical and electromechanical equipment and its implementation in high-tech industries of digital economics (MEPhI admission office, 2017).

REFERENCES

Altbach, P. G., & Salmi, J. (Eds.). (2011). *The road to academic excellence: the making of world-class research universities*. Washington, DC: The International Bank for Reconstruction and Development / The World Bank. doi:10.1596/978-0-8213-8805-1

Baryshev, G. K., Berestov, A. V., & Konashenkova, N. A. (2016). The Application of Online Team Project Training in Nuclear Engineering Education. *Communications in Computer and Information Science*, *674*, 370–379. doi:10.1007/978-3-319-49700-6_35

CDIO. (2017). Retrieved from http://cdio.org/about

Clark, B. R. (2004). *Sustaining Change in Universities. Continuities in Case Studies and Concepts*. New York, NY: McGraw-Hill.

Crawley, E. F. (2001). The CDIO syllabus: a statement of goals for undergraduate engineering education. Cambridge, MA: The Department of Aeronautics and Astronautics, Massachusetts Institute of Technology.

Delbanco, A. (2012). *College. What It Was, Is, and Should Be*. Princeton, NJ: Princeton University Press.

Fedorov, I.B., & Medvedev, V.E. (2011). Engineering education: problems and tasks. [Inzhenernoye obrazovaniye: problemy i zadachi]. *High Education in Russia*, *12*.

Foresight, A. S. I. "Education 2035". (2017). *Presentacia forsaita "Obrazovanit 2035"*. etrieved from http://asi.ru/molprof/foresight/12254/ (in Russian)

Grasso, D., & Burkins, M. B. (Eds.). (2010). Holistic Engineering Education. Beyond Technology. Springer Science+Business Media. doi:10.1007/978-1-4419-1393-7

MEPhI MA educational programs. (2017). *Unikal'nyye obrazovatel'nyye programmy dlya nabora v 2017 godu* [Unique Educational Programs for the 2017 Recruitment]. Retrieved from https://admission.mephi.ru/admission-2017/magistracy/unique

National Research Nuclear University. (2017). *Obrazovatelniye standarty vyshevo obrazovaniya Natsionalnovo issledovatelskovo yadernovo universiteta MIFI. Napravleniye podgotovki 14.03.02 Yaderniye fizika I tehnologii* [Educational Standard of Higher Education of National Research Nuclear University "MEPhI". Direction of training 14.03.02 Nuclear Physics and Technologies]. Retrieved from https://mephi.ru/obrdeyat/edst-vo.php

Saprykin, D.L. (2012). Inzenernoe obrazovanie v Rossii: istoriya, kontseptsiya, perspektiva [Engineering education in Russia: history, conception, future trends]. *Vyssheye obrazovaniye v Rossii.*, *1*, 125-137.

Simonyants, R. P. (2014). *Problems of engineering education and their decision involving industry*. Moscow, Russia: BMSTU. (in Russian)

Surin, V. I., Biryukov, A. P., & Dubkov, B. A. (2013). Ispol'zovaniye CAD-sistem v uchebno-tekhnicheskikh uchebnykh kursakh dlya studentov Natsional'nogo issledovatel'skogo yadernogo universiteta MIFI [The use of CAD-systems in educational engineering training courses for the students of the National Research Nuclear University MEPhI]. In *V trudakh Mezhdunarodnoy konferentsii «Sistemy proyektirovaniya, predplanirovaniya i upravleniya zhiznennymi tsiklami promyshlennogo produkta CAD / CAM / PDM 2013»*. Moscow, Russia: V.A. Trapeznikov Institute of Control Sciences of Russian Academy of Sciences.

Vladimirov, A. I. (2011). *Ob injenerno-tekhnicheskom obrazovaniy* [About engineering-technical education]. Moscow, Russia: Nedra Publishing House.

Yakovlev, D., Pryakhin, A., Korolev, S., Shaltaeva, Y., Samotaev, N., Yushkov, E., & Avanesyan, A. (2015). Engineering competitive education using modern network technologies in the NRNU MEPhI. In *Proceedings of the 2015 IEEE Workshop on Environmental, Energy, and Structural Monitoring Systems*. Trento, Italy: IEEE. 10.1109/EESMS.2015.7175849

Chapter 19
The Educational and Academic Innovation of the Avionics Engineering Center

Andrey V. Proletarsky
Bauman Moscow State Technical University, Russia

Konstantin A. Neusypin
Bauman Moscow State Technical University, Russia

Kai Shen
Beijing Institute of Technology, China

ABSTRACT

Research directions for carrying out scientific works are presented within the Avionics Engineering Center at Bauman Moscow State Technical University. The structure of Avionics Engineering Center is illustrated and prospective areas of working are highlighted. Methods on implementation of perspective scientific research and educational programs are developed for innovative development of the Avionics Engineering Center. Symbiosis of new developed programs allows training and getting a set of better quality specialists and innovative technologies in the defense and aerospace industry.

INTRODUCTION

For the education of high-level professionals, the "Russian method of engineering training" was originally developed and successfully implemented at Bauman Moscow State Technical University (BMSTU). Nowadays, for implementation modern innovative technologies (Knyaginin, Meshkov, & Utolin, 2016), the Avionics Engineering Center was built as a joint project of Bauman Moscow State Technical University and Ramenskoye Design Company. In this sense, the project "Avionics Engineering Center" is a practical implementation of the method "Russian method of engineering training" in modern information society.

DOI: 10.4018/978-1-5225-3395-5.ch019

Avionics Engineering Center has been working on the following main research and educational directions:

- Preparation of bachelors, engineers, masters and highly qualified specialists in the priority areas;
- Improvement of training methods, creating a highly effective system for training technical professions;
- Carrying out joint research projects;
- Participation in scientific competitions, grants, awards, including Grants of Ministry of Education and Science of Russian Federation;
- Joint organization of international conferences and symposia;
- International cooperation, for example, international laboratory;
- Organization of joint scientific journal, scientific and methodical bases;
- Joint intellectual property rights (patents);
- Joint participation in competitions, grants, awards and research funds.

THE EQUIPMENT AND RESEARCH WORKS OF THE CENTER

The equipment of the Avionics Engineering Center located in three main areas:

- The multipurpose multifunctional exerciser based on four-generation aircraft cabin with spread spectrum of functionality: simultaneous work for several students, practicing coalition cooperation, advanced training of flight crews under conditions that are close to combat, practicing dueling and group interaction of several aircrafts, and further complication for training purposes;
- The Laboratory of Intelligent Systems at Bauman Moscow State Technical University includes several operational working stations for pilots which are adapted to current and future requirements, such as formation of flight mission during multi-conflict flight situation, coalition, interaction and formation of models of external environment;
- The research is in progress with real navigation and piloting systems and platforms of modeling aerodynamic characteristics.

The research works are carried out in the following directions:

- The research in the field of aerodynamic characteristics, application of nanotechnology and research of aerodynamic properties, creation of models and working out on modeling complex engineering center;
- The work on dynamic formation of flight mission on the basis of predictive models with Genetic Algorithms to meet changing operational conditions of aircraft and conditions of our aircrafts against coalition enemies;
- The development of measurement systems and complex of modern aircraft from the concept of measuring complex synthesis with variable structures which can provide information as accurately as possible under conditions of intensive maneuvering.
- Perspective areas of research works are:

Figure 1. The View of Testing Platform and Navigation Systems at the Avionics Engineering Center

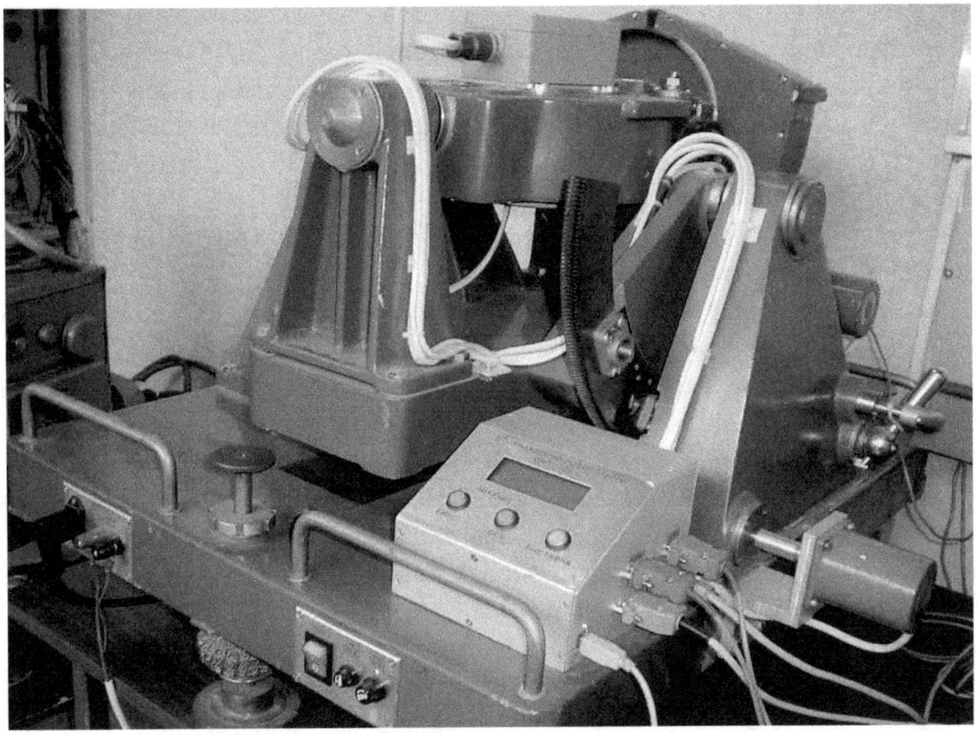

- ◦ Development of the theory of intelligent control systems (Proletarsky, Shen, & Neusypin, 2015);
- ◦ Creation of intelligent systems for space applications;
- ◦ Development and application of intelligent technologies for control systems of atmospheric flight vehicles;
- ◦ Research in the field of control theory;
- ◦ Global aggregation and management of multi-site complexes and coalitions.

THE TASKS OF THE AVIONICS ENGINEERING CENTER

The primary purpose of the Avionics Engineering Center is to provide scientific and engineering school of world level.

The main tasks of Avionics Engineering Center are:

- The involvement of prominent scientists as supervisors of creative groups of young researchers to research current and future problems of equipment, technologies and educational programs through innovative projects;
- The development and coordination of fundamental and applied research works which are carried out in Ramenskoye Design Company and Bauman Moscow State Technical University;

- The increase of level of fundamental education and engineering training of young professionals through several ways:
    - The organization of new lecture courses with the participation of leading scientists;
    - The preparation of new textbooks, monographs, training manuals, new labs using the latest scientific achievements in the course and degree design works of students;
    - The preparation of a new generation of young scientists through the involvement of graduate students to carry out scientific subjects;
    - The active involvement of graduate and undergraduate students to participate in international seminars, conferences, symposia;
    - The training of young scientists and specialists at the world's leading research centers and universities;
- The creation of specialized scientific research stands, simulators equipped with modern equipment, instrumentation, multi-channel high-speed communication devices and information display;
- The development of international cooperation in the field of research and educational activities, the implementation of joint research work with foreign partners in the framework of international projects and on the basis of bilateral agreements and contracts;
- The organization of international scientific and technical conferences, symposiums, competitions, seminars, exhibitions;
- The creation of modern laboratory workshop for students and postgraduates at Bauman Moscow State Technical University;
- The organization of research works for students and graduate students;
- The promotion and popularization of engineering knowledge, advanced information technology, the creation of a joint scientific journal;
- The attraction of additional resources for the development and strengthening of material and technical scientific and methodological base;
- The work on the creation of joint intellectual property right (patents);
- The joint participation in competitions, grants, awards and research funds.

Thus, the establishment and successful operation of the engineering center will help to solve one of the most important tasks of present stage – the training of a new generation of young scientists which are capable of creating breakthrough technologies in the aerospace and defense industry.

SOME SCIENTIFIC METHODS OF THE AVIONICS ENGINEERING CENTER

Vehicles that are capable of sustained motion through air and space are termed flight vehicles, and are generally classified as aircrafts, spacecrafts and rockets. All flight vehicles require manipulation (i.e., control or adjustment) of position, velocity and orientation for efficient flight mission completion based upon measurement information about operation state of flight vehicles. Whereas, measurement information is usually disturbed by external noise in practice. In order to realize effective control of flight vehicles, it is highly required to improve the accuracy of on-board measurement complex, including Inertial Navigation System (INS), Global Navigation Satellite System (GNSS), Radio Navigation System

(RNS) and other kinds of information measurement systems (Brown, 1973). For working concordance of various measurement systems of navigation information above and accuracy improvement of whole measurement complex, some state estimation algorithms must be employed, such as Linear Kalman Filter (LKF) (Brown, 1973; Kalman, 1961).

Based on Linear Kalman Filter, information processing in measurement systems of navigation may be easily fulfilled without high-accuracy. In order to improve the accuracy of navigation information estimation, nonlinear filtering methods (Stepanov, & Toropov, 2010), such as Nonlinear Kalman filter (NKF) (Carvalho, Moral, & Monin, 1997), Bayesian Algorithm (Stepanov, & Vasiliev, 2010), were suggested to apply in on-board measurement complex. Nevertheless, because of the nonlinear properties in dynamical measurement sensors, it also ineluctably appears some unpredictable errors during the process of information estimation and filtering. Hence, adaptive nonlinear estimation algorithms must be designed to adapt to external changing environment.

Self-organization Algorithm (SOA) (Ivachnenko, 1971) as one kind of adaptive estimation algorithm was invented to settle complex nonlinear problems by academician USSR A.G. Ivachnenko. Especially, Self-organization Algorithm with Redundant Trends (Neusypin, Proletarsky, & Shen, 2014) can not only reduce the calculation capacity, but also enable to obtain a better optimal solution during information processing of measurement. During intensive maneuvering of flight vehicles, it is impossible to apply priori models of measurement systems of navigation obtained under laboratory tests. Therefore, it is necessary to reconstruct some high-accuracy and adaptive measurement models during the functioning process of flight vehicles based on more intelligent algorithms. Thus, Neural Networks (NN) and Genetic Algorithms (GA) were proposed to use for constructing novel complex measurement system of navigation information.

In terms of processing and filtering of navigation information, one of the most applied algorithms might be the well-known Kalman Filtering algorithm and its modifications, such as Extended Kalman Filter (EKF), Unscented Kalman Filter (UKF) and etc. Nowadays, research interests on estimation algorithms might be divided into two different orientations. One of the orientations is in-depth study and analysis of Bayes theory, while another way is to apply advanced intelligent algorithms, such as Neural Networks, Self-organization Algorithms, Genetic Algorithms or their combinations, to modify traditional Kalman Filter and design novel measurement systems of navigation information.

The Kalman Filter (KF), also known as linear quadratic estimation (LQE), was firstly proposed by Hungarian-born American mathematician R.E. Kalman in his paper "A new approach to linear filtering and prediction problems" published by Transactions of the ASME in 1960. Kalman Filtering algorithm provides a recursive solution to the linear optimal filtering problem Linear Kalman Filtering algorithm can only solve linear problem. For nonlinear information processing, Nonlinear Kalman filter was suggested to employ in measurement complex

As for Self-organization Algorithms, Group Method of Data Handling algorithm represent as a sorting-out method that are usually employed to process navigation information in complex measurement systems. Whereas, during the full self-organizing processing for updating new states, it is inevitable to cause aging and inbreeding effects that have a negative impact on the accuracy of this algorithm. To some extent, Self-organization Algorithm with Redundant Trends can overcome those negative influences above and thus may be applied to process navigation information with less calculation workload and better optimal solutions via rebuilding real-time nonlinear estimation models by utilizing redundant trends.

What's more, Neural Network is a general mathematical computing paradigm that models the operations of biological neural systems. In 1943, McCulloch and Pitts (1943) published a seminal paper titled "A logical calculus of ideas imminent in nervous activity", which inspired the development of the modern digital computer and the electronic brain. At approximately the same time, Rosenblatt (1957) investigated the computation of the eye, which eventually led to the first generation of Neural Networks, known as the perceptron. Artificial Neural Networks generally might be used as nonlinear filter of a signal for the purposes of noise reduction and signal enhancement. For applying Neural Networks as nonlinear filter, an extensive set of training samples to cover all possible situations must be collected so as to adapt to the given training set.

During the operation of highly dymanic objects, such as combat aircrafts, the mathematic models of studied system are usually employed for control of the aircraft to run as we want to be. When we know nothing about the prior information of studied objects, we can put Neural Networks into use for the sake of building mathematic models during operation. Thanks to Neural Networks, some high accuracy mathematic models can be established after long time training. The training of Neural Networks might be progressed as follows: i) giving initial weights randomly; ii) realizing training epoch; iii) testing the formed models of neural network.

One of the most popular Neural Networks is Backpropagation (BP) algorithm, a method for learning in multi-layer networks. After choosing initial weights of the network randomly, the Backpropagation algorithm can be applied to compute the necessary corrections. The BP algorithm generally is decomposed as the following four steps: i) feed-forward computation; ii) Back propagation to the output layer; iii) Back propagation to the hidden layer; iv) weight updates.

Similarly, Neural Networks may also be used to make prediction of system states, and thus can be implemented for further building linear or nonlinear prediction models that might offer essential compensation information to correct deviations of measurement systems.

Based upon natural selection and natural genetics, Genetic Algorithms were initiated as stochastic search techniques by Holland (1975). Standard genetic algorithm was first described in his monograph "Adaptation in Natural and Artificial Systems" published by the University of Michigan press. Later, variations of basic Genetic Algorithms were proposed, but few of them significantly represent as new methodologies. During the performance of flight vehicles, we may compare estimates INS errors from Kalman Filter with predictions INS errors obtained with GA to correct deviations on the output terminal of measurement complex.

During the function of highly dynamic objects, such as aircrafts, rockets and spacecrafts, the theoretical values or reference values are impossible to be obtained. Therefore, it is highly required to apply some accuracy analysis and evaluation methods to evaluate the efficiency of traditional and advanced intelligent estimation algorithms in complex measurement systems. In this case, the formulae of location errors, functioning as an accuracy evaluation criterion, can be formulated according to the movement model of gyro-stabilizing platform. By applying this accuracy evaluation criterion, the performance and accuracy available of estimation algorithms above can be evaluated during semi-physical modeling of complex measurement systems.

SOME SCIENTIFIC RESEARCH RESULTS OF THE AVIONICS ENGINEERING CENTER

At the Avionics Engineering Center, some perspective scientific research tasks are carrying out based on advance multifunctional experiment platforms.

In the following studies, we would like to select the errors model of navigation systems as the object of our study. The mathematical model of navigation errors has a form as:

$$x_k = \mathbf{\Phi}_{k-1} x_{k-1} + w_{k-1} \tag{1}$$

where $x_k = \begin{bmatrix} \delta V_k \\ \psi_k \\ \varepsilon_k \end{bmatrix}$ is state vector, $w_{k-1} = \begin{bmatrix} 0 \\ 0 \\ \varpi_{k-1} \end{bmatrix}$ is input noise, $\mathbf{\Phi}_k = \begin{bmatrix} 1 & -gT & 0 \\ \frac{T}{R} & 1 & T \\ 0 & 0 & 1-\mu T \end{bmatrix}$ is the transfer matrix, δV is the velocity error, ψ is the misalignment angle, ε is the drift rate, g is the acceleration of gravity, R is the Earth radius, T is the sampling time, ϖ is the Gaussian white noise with zero mean and covariance matrix $\boldsymbol{Q}_k$.

In general, part of the state can be directly measured as:

$$z_k = \boldsymbol{H}_k x_k + v_k \tag{2}$$

Figure 2. Multipurpose multifunctional exerciser based on four-generation aircraft

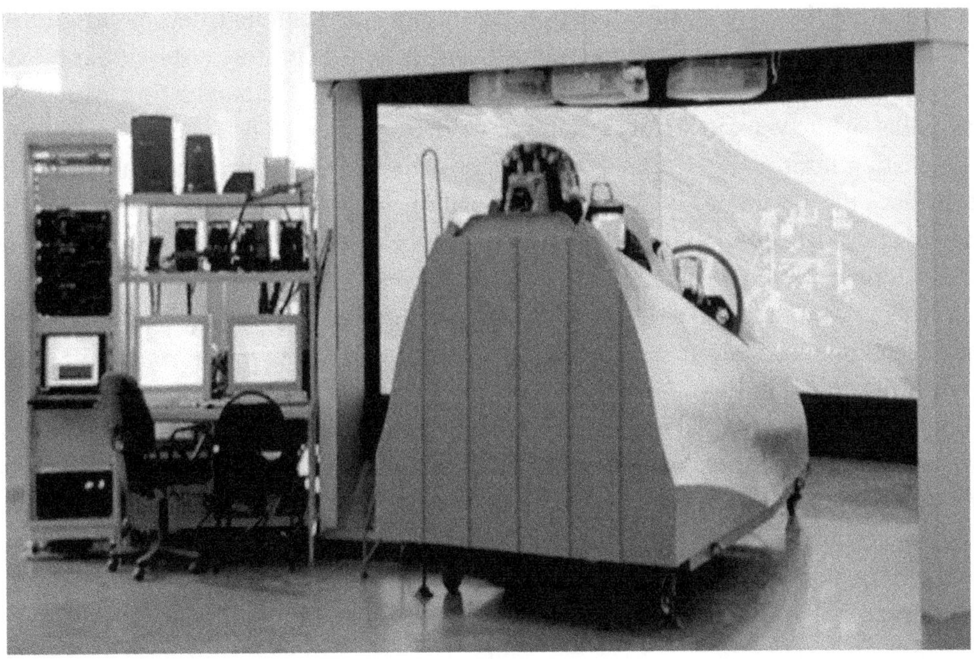

where z_k is the vector of state measuring, $\boldsymbol{H}_k$ is the measuring matrix, v_k is the discrete matrix of Gaussian white noise with zero mean and covariance matrix $\boldsymbol{R}_k$.

Results of the Study of Estimation Algorithms

As for estimation algorithms, the most applied algorithm might be the Kalman filter. The equations of nonlinear Kalman filter are

$$\hat{x}_k = \hat{x}_{k/k-1} + \boldsymbol{K}_k\left(\hat{x}_{k-1}\right)\left[z_k - \boldsymbol{H}_k \hat{x}_{k/k-1}\right]$$

$$\hat{x}_{k/k-1} = \boldsymbol{\Phi}_k\left(\hat{x}_{k-1}\right)$$

$$\boldsymbol{K}_k\left(\hat{x}_{k-1}\right) = \boldsymbol{P}_{k/k-1}\boldsymbol{H}_k^T\left[\boldsymbol{H}_k\boldsymbol{P}_{k/k-1}\boldsymbol{H}_k^T + \boldsymbol{R}_k\right]^{-1} \quad (3)$$

$$\boldsymbol{P}_{k/k-1} = \frac{\partial \boldsymbol{\Phi}_k\left(\hat{x}_{k-1}\right)}{\partial x_{k-1}^T}\boldsymbol{P}_{k-1}\left[\frac{\partial \boldsymbol{\Phi}_k\left(\hat{x}_{k-1}\right)}{\partial x_{k-1}^T}\right]^T + \boldsymbol{Q}_k$$

$$\boldsymbol{P}_k = \left[\boldsymbol{I} - \boldsymbol{K}_k\left(\hat{x}_{k-1}\right)\boldsymbol{H}_k\right]\boldsymbol{P}_{k/k-1}$$

where $\boldsymbol{K}_k$ – optimal Kalman gain, $\boldsymbol{P}_k$ – covariance matrix of estimation errors, $\boldsymbol{I}$ – identity matrix.

In order to analyze the efficacy of estimation algorithms, i.e. Kalman filter, modified Kalman filter with self-organization algorithms and genetic algorithms, the real laboratory tests were applied to evaluate the achievable estimation accuracy under a half-real environment. The navigation errors model serves as a system model, while velocity errors function as a measurement. The tested complex measurement system of navigation information CompaNav-2 can provide position coordinates and attitude angles of an aircraft. Besides these parameters above, the measurement system can also output accelerations and angular rates of aircraft-body motion with respect to the local-level frame. By utilizing various estimation algorithms for processing navigation information, Kalman Filter, Self-organization Algorithms and Genetic Algorithms were subsequently tested corresponding to structure blocks of novel measurement complex of navigation information.

The result is illustrated in Figure 3.

According to the results of laboratory experiments above, we can make conclusions as follows: In terms of navigation information processing, self-organization algorithms and genetic algorithms work better than Kalman filter; Kalman filter can reliably process navigation information, and thus the estimated values with Kalman filter might be function as a reference for evaluating the efficacy of navigation algorithms. The achievements of this research are beneficial for increasing the accuracy of navigation systems.

Figure 3. The estimated misalignment angles with different estimation algorithms
1 – estimated navigation errors with classical Kalman filter, 2 – estimated navigation errors with self-organization algorithms, 3 – estimated navigation errors with genetic algorithms, 4 – reference of navigation errors during laboratory experiments.

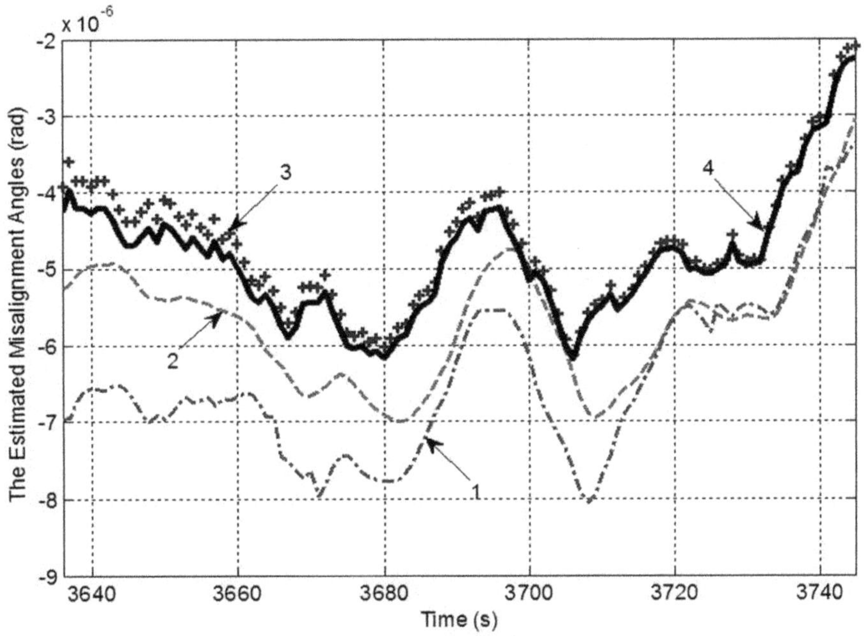

Results of the Study of the Degree of Observability

Generally, the criterion for the degree of observability (Ham, & Brown, 1983; Ma, Fang, Wang, & Li, 2014) has the following form as:

$$\lambda^i = \frac{E[(x^i)^2]R}{E[(y^i)^2]R^i} \tag{4}$$

where $E[(x^i)^2]$ is the variance of i-th state-variable, $E[(y^i)^2]$ is the variance of measured values.

The degree of observability (Yu, Cui, & Zhu, 2015; Wang, & Xia, 2015) can serve as an index to select better external aided navigation systems with time goes on. To clearly illustrate this function, we calculate the degree of observability with measurement from two sensors and show their relative relationship in Figure 4.

On the basis of the results illustrated above, it is easily to make the decision on which aided sensors should be selected during current time period and how to change the structure of integrated navigation systems to adapt to the changes of internal state and external environment.

Figure 4. The degree of observability of angular error with different sensors

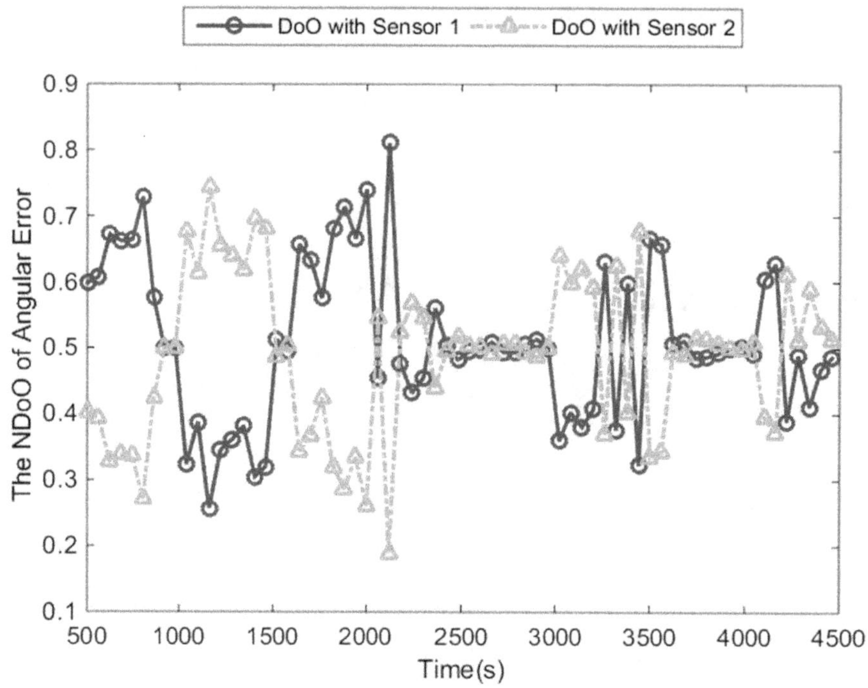

SUMMARY

The Avionics Engineering Center as a joint project was briefly introduced, and main research and educational directions of this center were also presented. Functioning as a scientific and engineering school of world level, the main tasks of Avionics Engineering Center were illustrated. Then, methods on implementation of perspective scientific-research and educational programs were developed for innovative development in the Center. The symbiosis of new developed methods and programs allows training and getting a set of better quality specialists and innovative technologies in the defense and aerospace industry.

FUTURE RESEARCH DIRECTIONS

Nowadays the establishment and successful application of engineering center allows to solve one of the most important and urgent tasks at the current stage, that is the creation of a new generation of young scientists capable of creating breakthrough technologies in industry.

One of the methodological foundations of innovative developments in the Avionics Engineering Center is the use of a large amount of "Russian method of engineering training". Within the framework of the advanced training system of "Russian method of engineering training", training programs for aviation specialists are implemented. At the Avionics Engineering Center, dynamic stands and aircraft cabin simulators are used to consolidate the acquired theoretical knowledge and develop the basic scientific knowledge.

In the process of preparation in Scientific and Educational Center (SEC), students get acquainted with the methods of work at specific enterprises, join scientific collectives, adopt the accumulated scientific and production experience, concentrated in the enterprises.

Students who have been trained in the system of advanced training in the SEC complex have priority in finding a job in specialized enterprises. Thus, the mechanism of adaptation of students and young specialists can be realized at the most difficult and fateful stage of individual professional trajectory. The annual summarizing of the student's activity allows him to correct individual professional trajectory.

The implementation of an individual professional trajectory in SEC provides for the self-realization of a student in various research directions. For example, a student has the opportunity to gain basic knowledge and try himself in the tasks of developing intelligent systems for general use in the engineering centers, and then try himself at the SEC "Avionics" when solving highly specialized tasks.

The implementation of the presented adaptation mechanism is carried out within the framework of advanced training system, the consumer organizations assess the level of graduates of the university. For example, the company "Mail.ru" carries out the integration of departments of "promising technologies" and "cloud computing" by 80% from students of BMSTU.

Geomification is a promising educational technology, which allows you to adjust the educational process so that the learner learns from his mistakes. For example, a student trained in cabin simulator of SEC "Avionics" systematically corrects the deviation when imitating the launch of a missile at a target. It is not possible to correct the situation by correcting the navigation process. There is an interest to correct the situation, to understand the technology of missile guidance, to understand the algorithms and models of guidance, and then improve it. The cabin-simulator of the SEC "Avionics" is a research cabin, that is, there is a possibility of access to the modeling complex.

A feature of gaming used in the advanced training system is the realization of the learner's ability to learn not only on his own mistakes, but also on other people's mistakes, working in small mixed creative groups. Competitions for teamwork of the two crews motivate the student to search for new tactical solutions when imitating the real situation, etc., instill team interaction skills, and strengthen the motivation to learn new things.

The peculiarity of the advanced training system applies for students who have already been working as part of small mixed creative groups. There is competition within a small creative group and between groups, which increases the effectiveness of the learning process.

Geomification calls for constant feedback from the teachers (the leaders of a small creative group, tutor) who performs:

- Constant correction of the educational process through gaming technologies;
- Setting new goals, increasing goals and maintaining motivation for the realization of goals;
- Combination of individual competition and teamwork;
- Monitor over the student's educational potential's enhancement and the accumulation of knowledge and skills, which enable the transition of specialist from the "scarlet ocean" to the "blue ocean".

In contrast to the current gaming format adopted in the advanced training system, simulators are auxiliary elements in the process of realizing gaming. Thus, through the implementation of advanced training system of "Russian method of engineering training", a personnel selection – selection of future employees for enterprises is carried out in the preliminary stage. The world view of the students are formed involved in the process of training and scientific research activities.

REFERENCES

Brown, R. G. (1973). Integrated navigation systems and Kalman filtering: A perspective. *Navigation, 19*(4), 355–362. doi:10.1002/j.2161-4296.1972.tb01706.x

Carvalho, H., Moral, P. D., & Monin, A. (1997). Optimal nonlinear filtering in GPS/INS integration. *IEEE Transactions on Aerospace and Electronic Systems, 33*(3), 835–850. doi:10.1109/7.599254

Ham, F. M., & Brown, R. G. (1983). Observability, eigenvalues, and Kalman filtering. *IEEE Transactions on Aerospace and Electronic Systems, 19*(2), 269–273. doi:10.1109/TAES.1983.309446

Holland, H. J. (1975). *Adaptation in natural and artificial systems*. Ann Arbor, MI: University of Michigan Press.

Ivachnenko, A. G. (1971). Polynomial theory of complex systems. *IEEE Transactions on Systems, Man, and Cybernetics, 1*(4), 364–378. doi:10.1109/TSMC.1971.4308320

Kalman, R. E. (1961). A new approach to linear filtering and prediction problems. *Transactions of the ASME –. Journal of Basic Engineering, 82*(1), 35–45. doi:10.1115/1.3662552

Knyaginin, V. N., Meshkov, N. A., & Utolin, K. V. (2016). Advanced education in the information society. *International Review of Management and Marketing, 6*(3), 89–99.

Ma, Y., Fang, J., Wang, W., & Li, J. (2014). Decoupled observability analyses of error states in INS/GPS integration. *Journal of Navigation, 67*(03), 473–494. doi:10.1017/S0373463313000829

McCulloch, W. S., & Pitts, W. (1943). A logical calculus of the ideas immanent in nervous activity. *The Bulletin of Mathematical Biophysics, 5*(4), 115–133. doi:10.1007/BF02478259

Neusypin, K. A., Proletarsky, A. V., Shen, K., Liu, R., & Guo, R. (2014). Aircraft self-organization algorithm with redundant trend. *Journal of Nanjing University of Science and Technology, 5*, 602–607.

Proletarsky, A. V., Shen, K., & Neusypin, K. A. (2015). Intelligent control systems: Contemporary problems in theory and implementation in practice. In *Proceedings of the 2015 5th International Workshop on Computer Science and Engineering: Information Processing and Control Engineering*. Moscow: The Science and Engineering Institute.

Rosenblatt, F. (1957). *The Perceptron: a perceiving and recognizing automaton*. New York: Cornell Aeronautical Laboratory.

Stepanov, O. A., & Toropov, A. B. (2010). A comparison of linear and nonlinear optimal estimators in nonlinear navigation problem. *Gyroscopy and Navigation*, *1*(3), 183–190. doi:10.1134/S2075108710030053

Stepanov, O. A., & Vasiliev, V. A. (2010). Comparison of the Bayesian and neural network algorithms in nonlinear navigation estimation problems. In *Proceedings of the 2010 International Joint Conference on Neural Networks*. Barcelona: IEEE Computational Intelligence Society. 10.1109/IJCNN.2010.5596834

Wang, L., & Xia, Y. (2015). Observability analysis of Mars entry integrated navigation. *Advances in Space Research*, *56*(5), 952–963. doi:10.1016/j.asr.2015.05.016

Yu, Z., Cui, P.-y., & Zhu, S.-y. (2015). Observability-based beacon configuration optimisation for mars entry navigation. *Journal of Guidance, Control, and Dynamics*, *38*(4), 643–650. doi:10.2514/1.G000014

KEY TERMS AND DEFINITIONS

Avionics: The electronic systems used on aircraft, artificial satellites, and spacecraft. The term avionics is a portmanteau of the words aviation and electronics.

Genetic Algorithm: A metaheuristic inspired by the process of natural selection that belongs to the larger class of evolutionary algorithms.

Global Navigation Satellite System: A system that uses satellites to provide autonomous geo-spatial positioning.

Group Method of Data Handling: A family of inductive algorithms for computer-based mathematical modeling of multi-parametric datasets that features fully automatic structural and parametric optimization of models.

Inertial Navigation System: A navigation aid that uses a computer, motion sensors (accelerometers) and rotation sensors (gyroscopes) to continuously calculate via dead reckoning the position, orientation, and velocity of a moving object without the need for external references.

Intelligent System: A system with artificial intelligence.

Kalman Filter: An algorithm that uses a series of measurements observed over time, containing statistical noise and other inaccuracies, and produces estimates of unknown variables that tend to be more accurate than those based on a single measurement alone, by using Bayesian inference and estimating a joint probability distribution over the variables for each timeframe. The filter is named after Rudolf E. Kálmán, one of the primary developers of its theory.

Neural Network: A computational model used in computer science and other research disciplines, which is based on a large collection of simple neural units (artificial neurons), loosely analogous to the observed behavior of a biological brain's axons.

APPENDIX

A student of BMSTU M. Selezneva has been successfully trained in the advanced training system of "Russian method of engineering training": laureate of the program "Step into the Future", student of the Department of Automatic Control and Systems Engineering, PhD student (thesis prepared in 2 years), the algorithm developer of the aiming and navigation complex of aircrafts (the results of the works were used in the Ramenskoye Design Company).

Graduates of the BMSTU, trained at Avionics Engineering Center, have the experience of solving real scientific and production problems, own modern engineering methods and software skills, published scientific works, skills and abilities to set prospective tasks and develop their original solutions, are included in the workflow of enterprises even as a student. The advanced training system of "Russian method of engineering training" can not only help the enterprises attract prospective specialists capable of meeting the challenges in the future, but also at the present stage obtain reliable personnel, ensure continuity and not lose the invaluable experience of generations. Young specialists from the school of SEC complex get the opportunity to realize the individual professional trajectory more effectively.

Figure 5. A student of the BMSTU with the cabin-simulator at Avionics Engineering Center

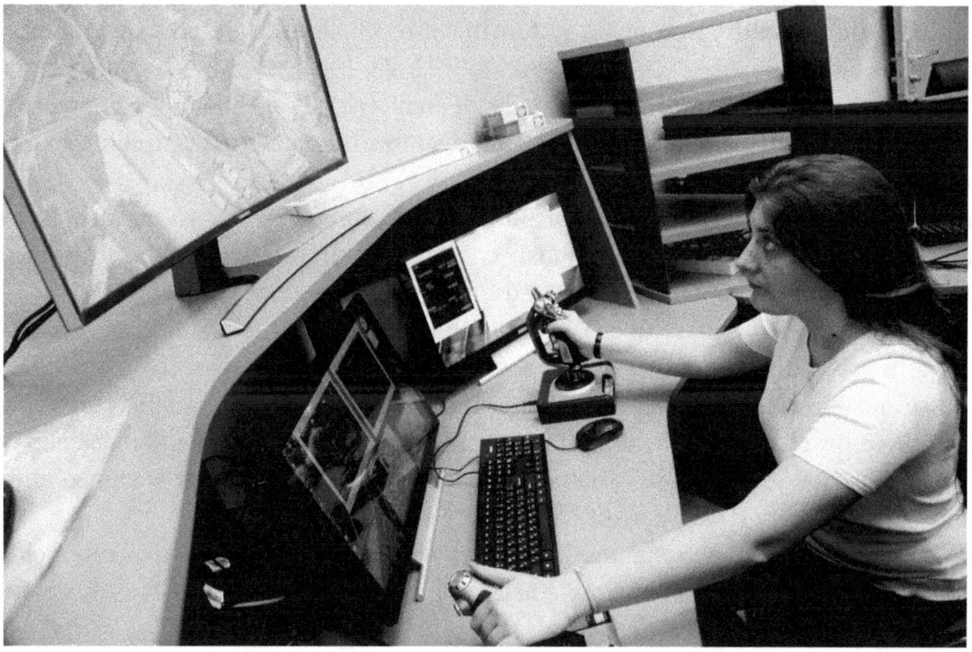

Figure 6. Exercise of group interaction in the cabin-simulator at Avionics Engineering Center

Chapter 20
Practice-Oriented Approach to the Study of Economics to Students of Engineer-Geological Specialties:
Using the Example of Solving a Task Concerning the Processing of Technogenic Mineral Resources

Vadim Vitalievich Ponkratov
Financial University Under the Government of the Russian Federation, Russia

Tatiana Alekseevna Bloshenko
Financial University Under the Government of the Russian Federation, Russia

Andrey Sergeevich Pozdnyaev
Bauman Moscow State Technical University, Russia

Alena Fedorovna Kireyeva
Belarus State Economic University, Belarus

ABSTRACT

Practice-oriented models are essential when teaching economics to engineering students. This chapter will discuss how to set and solve the applied scientific task of processing technogenic mineral reserves. Tools will be offered relating to engineering geological, economic, and mathematical sciences, as well as to form a group of students with various specialties. Experiments will aim to find solutions to these tasks with a generalized gradient method. This chapter will use evolutionary algorithms to calculate ad valorem MET rates. Technogenic raw materials are of economic interest to extract valuable components and produce finished goods. Often, the content of valuable components in technogenic deposits (TD) exceeds the content in natural fields. While secondary mineral resources harm the ecosystem, it is impossible to prevent environmental risks due to the lack of subsoil use. Differentiated rates will be selected based on maximum MET capacity on all valuable components extracted from deposits provided that each deposit is considered an investment project for the stated problem.

DOI: 10.4018/978-1-5225-3395-5.ch020

INTRODUCTION

Training students of engineering specialties is growing due to modern conditions. Graduates can apply their fundamental knowledge of basic economics and experience in making calculations in the following areas:

- Innovative entrepreneurship and technological start-up;
- Collaborative initiatives with economic agencies within large corporations;
- Integration into modern hi-tech businesses;
- High-level research and engineer studies, as well as their subsequent commercialization.

Aside from traditional workshops and lectures, practice-oriented teaching methods (i.e., a case study method, brainstorming sessions, etc.) are required. It is effective to conduct classes in mixed groups depending on their field specialization (i.e., merging engineers, technologists, economists, etc.). As Russian universities abolish the engineer-economist specialty, this approach will be valuable.

The following section will consider the extended example of interrelated application of engineering geological, economic, and mathematical tools to solve the applied scientific task concerning the processing of technogenic mineral reserves.

Statement of the Praktiko-Focused Task

The U.S. Geological Survey (USGS) collects information about the quantity and quality of all mineral resources. In 1976, the USGS and the U.S. Bureau of Mines developed a common classification and nomenclature, which was published as USGS Bulletin 1450-A—"Principles of the Mineral Resource Classification System of the U.S. Bureau of Mines and U.S. Geological Survey." Experience with this resource classification system showed that some changes were necessary in order to make it more workable in practice and more useful in long-term planning. Therefore, representatives of the USGS and the U.S. Bureau of Mines collaborated to revise Bulletin 1450-A. Their work was published in 1980 as USGS Circular 831— "Principles of a Resource/Reserve Classification for Minerals" (USGS Bulletin, 1976).

Domestic science gives several definitions of technogenic fields (TF). For example, K. Troubetzkoy and V. Umanets consider TF as the technogenic formations containing minerals, which are suitable for effective use in material production at the moment in quantity and quality (Troubetzkoy & Umanets, 1998). It should be noted that the research of K. Trubeckoy, V. Umanets and A. Tolumbaeva is very significant in science for the theoretical development of TF. According to their opinion, the comparative characteristics of technogenic objects should be evaluated using the criterion of the benefit maximization. From the standpoint of V. Chainikov and E. Goldman, TF is the accumulation of waste tonnage of mineral raw materials, which provides the economic effect by using (Chainikov & Goldman, 2000).

According to Pruss (2013), it was reasonable that the MET rate for mineral resources extracted from TDs should not be greater than 2%. Yet, no calculations or feasibility analysis of the indicated MET rate (Pruss, 2013) have been carried out. Other studies recommend to conduct MET differentiation in view of mining and geological, economic, and geographical peculiarities and the depleted status of deposits (Hung & Quyen, 2009; Lund, 2009; Pittel & Bretschger, 2010; Ponkrratov & Pozdnyaev, 2016). Bloshenko, Pozdnyaev, and Pozdnyaev (2016) and Ponkratov (2014) theoretically justified a method to identify the differentiated MET rates in relation to TDs.

D. Lund considers that at the same time, this tax is neutral in relation to companies that are well diversified. Although the information needed to implement an exactly optimal tax rate may be difficult to obtain (Lund, 2009).

Pittel K. and Bretschger L. (2010) analyze an economy in which sectors are heterogeneous with respect to the intensity of natural resource use. Long-term dynamics are driven by resource prices, sectoral composition, and directed technical change. We study the balanced growth path and determine stability conditions. Technical change is found to be biased towards the resource-intensive sector. Resource taxes have no impact on dynamics except when the tax rate varies over time. Constant research subsidies raise the growth rate while increasing subsidies have the opposite effect. We also find that supporting sectors by providing them with productivity enhancing public goods can raise the growth rate of the economy and additionally provide an effective tool for structural policy.

To engage the secondary mineral resources for processing, it is supposed to implement the investment project. For this the data on TF in Russia are necessary: their composition, volume, recycling, etc. Selection criterion to choose tax rates on mining is the maximum flow of MET in the budget for all TF (Bloshenko, 2013).

Currently, there is no formalized approach to define MET rates relating the resources extracted from TDs. To define differentiated MET rates, this article will show a balance of state's interest (determination of tax potential) and a subsoil user's interest (profitability of a project for developing new deposits).

To develop methods of defining ad valorem MET rates, this article will identify factors of rent that arise during mineral resources extraction and comprehensive processing of minerals. The contents of the valuable component and its amount in the base product is a fundamental requirement for economically feasible level of exemption from taxation at the stages of production of mineral resources and of procession of mineral resources. Common types of mineral resources are not considered when searching the solution of this problem as they may be taken for processing losses and by-products.

Research Methods

Within the proposed methods of identifying MET differentiated rates, it is expected that involvement of TDs into processing will be implemented with an investment project. To implement this approach, data are required on TD of Russia enshrined in a TD registry. This would contain:

- Mining and geological figures, including the amounts of valuable components in initial raw materials
- Amounts of valuable components extracted into the base product
- Prices, expenses, volumes of secondary mineral resources processing, development cost, etc.

In proposed methods, MET rate selection criterion is the maximum MET potential for the valuable components extracted from TD. The following symbols build a model:

- M: Number of considered TD
- r_m: Minimum rate of return at which the m-deposit development project is implemented (the m value may vary depending on region and deposit)
- K: Amounts of valuable components in considered TDs
- t_m: Duration of m-deposit development investment project

- c_{mt}: Cost of the m-investment project per t-year
- q_{mtk}: Finished product made up of chemically pure metal in which the value is a product of extracted minerals mass multiplied by the amount of valuable component (g/t) in this mass and throughout recovery of k-component at the m-deposit per t-year in accordance with the established procedure
- p_{tk}: Anticipated exchange price of a k-component per t-year
- η_k: Required MET rate for k-valuable component (%)
- d_t: Discount rate for t-year (defined as zero-component rate of return for state bonds or G-curve)

The following indicator function is introduced:

$$F(x) = \begin{cases} 0, x < 0 \\ 1, x \geq 0 \end{cases}; \tag{1}$$

It is assumed that the m-deposit development project will be implemented if (1) its internal rate of return (IRR rate) is not lower than the known minimum rate of return (r_m) or (2) if the project NVP calculated by the r_m rate is greater than or equal to zero, which is essentially the same. In case MET is calculated regarding the total component sales income, the condition of implementation of the m-deposit development project would be the following:

$$NPV_m\left(r_m\right) = \sum_{t=1}^{t_m} \frac{\sum_{k=1}^{K}\left(\left(1-\eta_k\right)q_{mtk}p_{tk}\right) - c_{mt}}{\left(1+r_m\right)^t} \geq 0; \tag{2}$$

Looking at the right-hand side of Equation 2, the product of $q_{mtk}p_{tk}$ proceeds from the sale of k-valuable component extracted from m-deposit during the t-year. Accordingly, $\left(1-\eta_k\right)q_{mtk}p_{tk}$ stands for the economic benefits on k-valuable component that arise after MET taxation. The term of a fraction of the Equation 2 is the surplus on m-deposit development after MET on the whole valuable component per t-year has been calculated and paid.

Given values are discounted at the r_m rate of return. This would provide the implementation of the m-deposit development investment project. Under given TD parameters, the fulfillment of Equation 2 condition depends on established MET rates.

Tax potential for MET shall be considered as the present value of revenues received from MET payments:

$$NPV_{TM}\left(\eta_k, k=1,\dots K\right) = \sum_{t=1}^{\max(t_m, m=1,\dots M)} \frac{\sum_{m=1}^{M} F\left(NPV_m\left(r_m\right)\right)\sum_{k=1}^{K}\eta_k q_{mtk}p_{tk}}{\left(1+d_t\right)^t}; \tag{3}$$

In this equation, $F\left(NPV_m\left(r_m\right)\right)=1$, if m – TD developed and $F\left(NPV_m\left(r_m\right)\right)=0$ in the contrary case. The production of $\eta_k q_{mtk} p_{tk}$ equals MET revenue to budgets of the budgetary system of the Russian Federation per year^t from a TD m upon its development. Therefore, the term of the fraction at the right-hand side of Equation 3 stands for MET tax revenues to budgets of the budgetary system of the Russian Federation per year (calendar tax period)^t from all the TDs^m under development.

To calculate present value of future MET payments, it is necessary to conduct discounting at rate of return of state bonds represented, as already stated, in the form of the G-curve.

To calculate the optimum differentiated MET rates, if the tax base is determined by the exchange prices for finished products (chemically pure metal) providing the development of cost-effective deposits and maximization of tax potential for MET, it is required to solve the optimization problem in Equation 4:

$$\max_{\eta_k, k=1,\dots K} \sum_{t=1}^{\max(t_m, m=1,\dots M)} \frac{\sum_{m=1}^{M} F\left(NPV_m\left(r_m\right)\right) \sum_{k=1}^{K} \eta_k q_{mtk} p_{tk}}{\left(1+d_t\right)^t}; \tag{4}$$

$$\eta_k \geq \eta_k^*, k=1,\dots K$$

In this equation, $\eta_k^*, k=1,\dots K$ is the minimum value of MET rates on the relevant components.

In a particular case, any or all of the values of $\eta_k^*, k=1,\dots,K$ can be negative. When Equation 4 is solved, the corresponding MET rates $\eta_k, k=1,\dots K$ can be also negative. Negative MET rates on certain valuable components extracted from the TD would mean subsidies payments. These are the financial payments made by the state (i.e., the Russian Federation Government Decree #1339 dated September 12, 2014, with amendments and additions), as well as other benefits stipulated by the existing legislation of the Russian Federation.

In some cases, negative MET rates promote the development of deposits that would not be developed without subsidies. As a result, tax potential will increase for MET calculated by using Equation 3. If the subsidy is not permitted, it will be required to determine $\eta_k^* \geq 0, k=1,\dots K$.

The following method for identifying differentiated MET rates consists of calculating the estimated value of the extracted minerals resources as a basis for the determination of the taxable base for MET. In this case, c_{mtk} is a part of the costs that relate to the k valuable component:

$$c_{mtk} = c_{mt} \frac{q_{mtk}}{\sum_{k=1}^{K} q_{mtk}}; \tag{5}$$

The condition of implementation of the investment project is:

$$NPV_m\left(r_m\right)=\sum_{t=1}^{t_m}\frac{\sum_{k=1}^{K}(q_{mtk}p_{tk}-(1+\eta_k)c_{mtk})}{\left(1+r_m\right)^t}\geq 0; \tag{6}$$

The term of a fraction stands for the benefits after MET on a specific deposit per t-calendar period has been calculated and paid. As in Equation 2, payment discounting is carried out at the r_m rate.

Given cost of MET revenue (tax potential) from the development of TD amounts to:

$$NPV_{TM}\left(\eta_k,k=1,\dots K\right)=\sum_{t=1}^{\max(t_m,m=1,\dots M)}\frac{\sum_{m=1}^{M}F\left(NPV_m\left(r_m\right)\right)\sum_{k=1}^{K}\eta_k c_{mtk}}{\left(1+d_t\right)^t}; \tag{7}$$

The optimization problem in identifying differentiated MET rates takes the following form:

$$\max_{\eta_k,k=1,\dots K}\sum_{t=1}^{\max(t_m,m=1,\dots M)}\frac{\sum_{m=1}^{M}F\left(NPV_m\left(r_m\right)\right)\sum_{k=1}^{K}\eta_k c_{mtk}}{\left(1+d_t\right)^t}; \tag{8}$$

$$\eta_k\geq\eta_k^*,k=1,\dots K$$

If the subsidies or/and tax benefits are not granted, the values of the $\eta_k^*\geq 0,k=1,\dots K$ parameters should be used.

Equations 1.4 and 1.8 are complicated optimization problems due to $F(x)$ constant function (Equation 1) in the functions to be optimized.

RESULTS

The following are consequences of built optimization models for identifying differentiated MET rates when developing TD:

- Clarify the number of discussed TDs, as well as invest in a development project for each TD
- Clarify the minimum internal rate of return for investment projects providing the implementation of the project and its profitability (the rates vary depending on region and deposit);
- Set the duration terms of the investment project according to the business plan
- Determine the percentage of valuable components contained in the deposit in 1 ton of extracted minerals according to the business plan;

- Determine the direct and indirect costs per 1 ton of each valuable component according to the business plan (in presentation currency such as USD);
- Set the value of products manufactured from the specific type of valuable component (without VAT), in USD per 1 ton, for the entire implementation phase;
- Write the expression for the net present value (NPV) depending on MET rates at given rate of return;
- Write the expression for calculating MET tax revenues to budget that arise from development of deposits, to determine the optimal value for the ad valorem MET rates for each component that is contained in TD, by solving the appropriate optimization problem;
- Solve the optimization problem using the MS Excel Solver add-in.

As an example of solving complex problems (Equations 1.4 and 1.8), three deposits containing three valuable components were considered: nickel, copper, and zinc. Prices for the specified period relating to the implementation of the investment projects were determined by Bloomberg (n.d.).

The notional value of project implementation has been calculated with allowance for the gradual increase in costs caused by inflation. It was expected that the extraction of each component would be conducted on a continuous basis. This does not consider the setting and the final stages of project implementation because the deposit is not natural. Therefore, there is no provision for enhancing the production capacity, extraction volumes, and extension of mining claim borders.

Results have been received by solving the optimization problem (Equation 4). The first calculation was conducted by considering the subsidy constraints. The second calculation allowed for a negative MET rate value up to -20%.

Identification of differentiated tax rates on the estimated value of extracted mineral resources contains the result of the (Equation 8) problem solution. The same project parameters were used.

Provided that exchange prices for the finished products were used when defining the tax base for MET, then subsidy of up to 20% of proceeds received for a valuable component would make it possible to initiate the process of developing two out of three considered TDs. As a result, the tax potential grows from US$9,012,293 to US$14,186,767. This is more than 57% higher.

Using the estimated cost of the mineral resources as a base for MET in given conditions has made it impossible to develop more than one deposit. In case of negative rates, the development of the first field remains profitable. However, the tax potential slightly increases due to changes in financial flows considered when using Equation 7.

To determine the tax base for MET with regard to the exchange prices of finished product, the differentiated MET rates have been determined for the following natural deposits:

- **Kamchatskaya Copper Company, LLC:** This is an investment project titled "Construction of Maletoyvayamsky Area Ore Mining and Processing Enterprise." Its value is 534 mln roubles with VAT (to calculate MET potential and to determine the optimal MET rate, the value given is without VAT). The project duration is estimated to be 20 years with projected capacity of 140 tons.
- **Bystrinskaya Mining Company, CJSC:** This is an investment project titled "Construction of 'Kumroch' Ore Mining and Processing Enterprise." Its value is 10,590 mln rubles with VAT (to calculate MET potential and determine the optimal MET rates, the value given is without VAT). The project duration is estimated to be 15 years with 30 tons of reserves and 170 tons of resources.

Industry classification of projects—extraction of precious metals ores (gold, silver, platinum, and palladium). Additional mechanisms for implementing the projects include receiving the status of the regional investment project in accordance with Section 3.3 Part 1 of Tax Code of the Russian Federation (n.d., 2002, 2016).

At present, taxation of precious metals is conducted in accordance with Section 5, Article 340 of the Tax Code of the Russian Federation. This defines that the estimated value of precious metals extracted from the primary, placer, and TDs is calculated based on taxpayer's sale prices put on the chemically pure metals without VAT, reduced by the taxpayer's expenses for refining and delivery, for the specific taxable period (in the absence thereof, for the nearest preceding taxable period). The methodology used to determine the tax base on the basis of the estimated value for the specific deposits cannot be applied.

Regarding the fact that the format of the specified investment projects does not contain information on the content of the valuable component and the volume of its extraction to the finished products, the following values were used to solve the problem: content of gold (4 g/t); silver (8 g/t); platinum (5 g/t); palladium (6 g/t); and average coefficient of extraction (0.75) considering the ore dilution factor. To define MET potential, the resources and the reserves are aggregated.

As an assumption for calculating the price on palladium, the value of US$670 per ounce was increased by 10% on an annual basis. This was in accordance with terms of the project implementation and inflation.

When conducting calculations using the method suggested in this section (see Equation 4), the optimal ad valorem MET rate will be: gold (10%); silver (8.6%); platinum (3.2%); and palladium (0%). Calculations are conducted with regard to non-negativity constraint.

In accordance with the Tax Code of the Russian Federation, current MET rates compared to those received show that MET rate for gold comprise 6% and MET rate for multicomponent raw materials comprises 8% (n.d., 2002, 2016). According to this example, the tax potential for MET comprises US$731,536. Both deposits are under development.

CONCLUSION

The authors suggest to determine the database of investment projects (mining of solid minerals), and based on this database to calculate differentiated MET rate at which the profitability of investment projects is ensured. MET rates are determined by solving optimization problems, the basic conditions of which are stock prices for finished products, key performance indicators of the investment project (NPV, IRR), and G- curve. The economic implications of the suggested method are as follows: with positive MET rates the field should be developed, and with negative ones the field mining needs the subsidies from the state. Methods for determination of optimal MET-rates calculate income to the budget with secondary mineral resources recycling in TD. Involvement of secondary resources in recycling is directed to the full withdrawal of all useful components from the mining wastes. Rational use of resources is aimed at reducing the negative impact on the environment.

An alternative idea is to use not the best severance tax rate for the taxation of TM. Then taxes will be lower. This algorithm is implemented in a standard Excel add-in. Here it is necessary to specify in the article. A description of the algorithm wo uld require a separate article (maybe not the same). The article makes it clear that there are some variables, what is the criterion of optimization and constraint.

Methods for determination of optimal MET - may be used for Russian Federation, because Tax Code of the Russian Federation hadn't the Differentiated MET Rates Relating to the Recovery of Solid Minerals the Differentiated MET Rates Relating to the Recovery of Solid Minerals. If we wont to use these Methods for determination of optimal MET for anything Countries we mast know Tax Code of this Countries.

FUTURE RESEARCH DIRECTIONS

The countries that use the JORC system to calculate reserves and resources must identify the procedure for forming the cadastre of secondary mineral resources containing useful components. The conducted work will require applying the models presented above.

To adapt these models to foreign countries, it is advisable to consider the discount rates and the possibility to substitute the Heaviside step function with Monte Carlo methods and to impose restrictions depending on the amount of extraction and processing waste and the scope of state support, namely to establish the linkage between tax and environment policies of a country.

Foreign countries preform complex processing of minerals, so on the basis of the foreign experience it is possible to make a conclusion that complex processing of minerals implies zero waste. The tax assessment of by-products acquired in integrated technological process in various taxation systems of China, Kazakhstan, the USA, the UK, Germany and France is of greatest interest.

In India, for example, differentiated royalty rates are defined by the content of useful components and the processing level. High rate level means higher content of useful components in raw materials. Calculation of differentiated rates requires identifying whether the component is a staple or a by-product. Ad-valorem rate is higher for metals produced during the integrated technological process known as by-production.

International fiscal systems provide various tools to perform mine rent withdrawal, so the international experience will provide basis to adapt models aimed at defining tax rates and resource tax potential.

REFERENCES

Bloomberg Market. (n.d.). Retrieved December 18, 2016, from http://www.bloomberg.com/markets

Bloshenko, T., Ponkratov, V., & Pozdnjaev, A. (2016). Development of methods for determining differentiated rates of mineral extraction tax in recovery of solid commercial components from technogenic and natural fields. *Indian Journal of Science and Technology*, *9*(27). doi:10.17485/ijst/2016/v9i27/97683

Hung, N. M., & Quyen, N. V. (2009). Sales tax: Specific or ad valorem tax for a non-renewable resource? *Economics Letters*, *102*(2), 132–134. doi:10.1016/j.econlet.2008.12.001

Lund, D. (2009). Rent taxation for nonrenewable resources. *Annual Review of Resource Economics*, *1*(1), 287–308. doi:10.1146/annurev.resource.050708.144216

Pittel, K., & Bretschger, L. (2008). *Sectoral heterogeneity, resource depletion, and directed technical change: Theory and policy* (Working paper 08/96). CER-ETH – Center of Economic Research at ETH Zurich, Switzerland. Retrieved from https://www.ethz.ch/content/dam/ethz/special-interest/mtec/cer-eth/cer-eth-dam/documents/working-papers/wp_08_96.pdf

Ponkratov, V. V. (2014). Tax maneuver in Russian oil production industry. *Neftyanoe khozyaystvo – Oil Industry, 9*, 58-61.

Ponkratov, V. V., & Pozdnyaev, A. S. (2016). The oil production taxation in Russia – consequences of tax maneuver. *Neftyanoe khozyaystvo – Oil Industry, 3*, 24-27.

Pruss, Y. V. (2013, February). To the problem of exploration of technogenic complex of Kolyma. *Gornyi Zhurnal, 2*, 38–40.

Tax Code of the Russian Federation. (2017). Russian Federation.

ADDITIONAL READING

Anisimov, V. N., Adamchuk, A. M., Uvarov, V. M., & Scherbakov, A. Y. (2016). Environmental safety in mining and processing of mineral raw materials under the conditions of rational nature management as exemplified by the Kursk magnetic anomaly region. *Indian Journal of Science and Technology, 46*, 74–86.

Bloshenko, T. A. (2014). *Methods for Determination of the Optimal Mining tax Rates when Secondary mineral resources are involved. Social Educational Project of Improving Knowledge in Economics: Journal L'Association 1901 "SEPIKE". – Poitiers* (pp. 102–104). Los Angeles: Osthofen.

Bloshenko, T.A. (2014). Taxation Of Mineral Products In Russian Federation. *Review of European Studies / Canadian Center of Science and Education, 6-4,* 91-99.

Lipina, S. A., Zaikov, K. S., & Lipina, A. V. E. (2017). Introduction of innovation technology as a factor in environmental modernization in Russian arctic. *Economic and Social Changes: Facts, Trends. Forecast, 2*, 164–180.

Ponkratov, V. (2015). Oil production taxation in Russia and the impact of the tax maneuver. *Journal of Tax Reform, 1*(1), 100–112. doi:10.15826/jtr.2015.1.1.007

Rylnikova, M. V. (2011). Combined geo-technology is a way to increase of fullness and complex using of mineral resources. *Twentieth International Symposium on Mine Planning and Equipment Selection MPES*, 267-279.

Saliyeva, R.N. & Fatkudinov Z.M. (2011). Some points concerning legislation regulating relations in the sphere of technogenic deposits. *Concise compilation of articles from Representative power – 21 st century: legislation, commentary, problems, 1-4,* 6-15.

Samarina, V., Skufina, T., Samarin, A., & Baranov, S. (2016). Some System Problems of Russian Mining Enterprises of Ferrous Metallurgy. *International Review of Management and Marketing. Special Issue for «Socio-Economic and Humanity-Philosophical Problems of Modern Sciences», 6,* 90-94. URL: http://econjournals.com/index.php/irmm/article/view/1882/pdf

KEY TERMS AND DEFINITIONS

Cadastre of Technogenic Deposit: Should be supplemented by the information: 1) capital and operating project costs on the development of each TD; 2) annual extraction of each useful component in the case of realization the investment project when NPV is more than zero; 3) price forecast for all components in the raw materials for the term of investment project.

Categories of Reserves: A, B, C1, C2. Reserves of categories A and B are geologically the most detail investigated deposits of operating mines where the current or prior operational prospecting by underground drilling has been fulfilled. Reserves of category C1 are the basic category created by detail prospecting which are used together with A and B for compilation of mine projects, estimation of profitability of ores refinement. Reserves of category C2 are amassed on the flanks of deposits of rich cupriferous ores and deep mines. Besides balance and off-balance reserves in deposits and ore regions epy the forecasted recourses of categories P1, P2 and P3 are estimated.

IRR: Internal rate of return.

MET: Severance tax. It is paid when developing fields proceeding from quantity or cost of the extracted mineral raw materials.

NPV: Net present value, the sum of the discounted values of the flow of payments given to today.

Tax Base: Cost, physical or other characteristic of an object of the taxation.

Tax Rate: The size of tax charges on a unit of measure of tax base.

Technogenic Deposit (TD)/Technogenic Fields (TF): Mining mineral wastes. Technical difficulties to reproduce them are usually absent.

Chapter 21
The Use of Active Learning in Biotechnical Engineering Education

Sergey I. Suyatinov
Bauman Moscow State Technical University, Russia

ABSTRACT

This chapter presents information-computing complex of modular type to perform interdisciplinary laboratory work. The object of the study is a complex biosystem – the human body. Feature of information-computing complex is the developed hardware and software for identification and study of systems and processes. Unique biosignal sensors allow to record electrocardiograms and sphygmograms in the process of laboratory work, to realize various algorithms of digital processing of signals and to use them in the process of structural and parametrical identification of cardiovascular system. Other sensors estimate an individual's psychophysiological state in different conditions. Thus, the student becomes the object of the research. This, undoubtedly, increases their motivation to assimilate new knowledge.

INTRODUCTION

Creating and delivering meaningful learning opportunities require a thoughtful and deliberate approach. Often, learning opportunities use innovative technology in course design to motivate and entertain students (Moats, 2015).

In this case, it is necessary to consider the characteristics and content of innovative education. Innovative education involves the purposeful formation of certain knowledge, skills, and methodological culture. This imposes special demands on the educational program content and teaching methodology. The following principles of an organization's educational process in innovative education system contains the following:

DOI: 10.4018/978-1-5225-3395-5.ch021

- Problem-oriented interdisciplinary approaches to the study of natural and technical disciplines
- Active methods of contextual learning, learning from experience, and learning through research
- Case studies of methods (based on practice)

This chapter represents a conceptual model for implementing the noted principles of an innovative education system. Their implementation is illustrated through an example of a methodology for studying a complex biological system (Anishchenko et al., 2009). This chapter presents information-computing complex for the study of methods and algorithms to assess the functional state of human health based on the registration and processing of biosignals (Buldakova & Suyatinov, 2002).

Problem Orientation and the Research Object

The problem orientation of innovative education is determined by the dominant strategy of scientific and technological development. A new technological paradigm based on the synthesis of nano-, bio-, information, and cognitive technologies (NBIC) is dominant (Roco & Bainbridge, 2003). This primarily applies to studies of natural processes for technologization of knowledge and the creation of new anthropomorphic systems. By their efficiency and profitability, these would be like living systems. Successful implementation of this strategy will provide essential competitive advantages and create conditions for economic and social domination (Kamensky, 2015).

NBIC-systems are complex and poorly formalized. Therefore, the key moments in the development of NBIC-technologies are the application of interdisciplinary approaches and convergence of sciences and technologies in the study of complex systems characterized by synergistic behavior (Fedorov, Norenkov, & Korshunov, 2006). Poorly formalized complex systems represent real objects in the fields of industry, medicine, economics, and ecology. They are the object of close theoretical study because the successful solution of the problems of diagnostics and forecasting depends on the formal description of the complex system and the construction of its model.

The main difficulties in studying complex systems are their abstract representation and complexity of their experimental study. There is a limited choice of complex systems convenient for experimental research under laboratory conditions.

The human body is a complex system of natural origin. In modern technical, physical, economic, and other sciences, such complex systems are considered to behave like biological systems. Therefore, using the example of identification and research of the human body, it is possible to study fundamental principles of the functioning of complex systems of various nature.

From the standpoint of synergetics, a complex system of natural origin has many degrees of freedom. However, in the process of natural evolution, several degrees are distinguished as order parameters (the others are adjusted accordingly). The dynamics of a limited number of these parameters reflect basic properties of the entire complex system. This chapter will use this to base the principle of the model representation of complex systems of different physical nature.

The essence is that it is a priori assumed that, firstly, there are characteristic types of motion inherent in systems of different physical nature. Secondly, the entire physical variety can be presented in the form of enough simple model equations.

Implementation of Active Learning Methods

The methodology of innovative education includes active methods of contextual learning, learning from experience, and learning through research.

It is shown that learning opportunities have innovative technology integrated into the course design to engage and entertain the learners (Moats, 2015). One example is the use of virtual reality simulations to train police officers and firefighters on proper response to high-risk situations. While these appear to be full-size simulations, the interactive video games provide learners with stressful experiences and kinesthetic activities without the risk of serious or grave injury. Another example is the use of simulated disaster response props to teach team building and decision making to students in the master of business administration (MBA) program. However, these examples are more student entertainment than contributions to the assimilation of knowledge.

Experience shows that laboratory and practical works are of great interest to students. The organization of laboratory works depends on (or is defined by) the level of technological development. The content is determined by the dominant strategy of scientific and technological progress. The paradigm of technological development considered by NBIC has its own peculiarities. The complexity of NBIC systems is not determined by the number of known elements with deterministic constraints. It is determined by the impossibility of decomposition.

The leading features of such systems are evolutionary development, synergy of action, and complex interconnections between subsystems. Such systems are available only from the results of indirect measurements and observations. Therefore, it is not possible to use known identification methods.

This circumstance imposes the special requirement to a training system in higher education, particularly the organization of laboratory occupations.

There is an increase in student training on systems analysis and information processing. These disciplines are the cornerstone of the comprehensive analysis of various natural processes and complex systems.

Natural systems are classified as complex. Their research methods use a study course without adequate reflection. If, for example, concepts like research modeling, multi-scale models, basic models, or entropy assessment of the state are discussed at lectures, then their practical demonstration in the laboratory is not possible.

Substantially it is caused by the absence of a suitable object of research. At the same time, the cardiovascular system represents a unique object for research, as well as approbation of new methods and approaches to the study of complex systems (Lantsberg, Troitzch, & Buldakova, 2011; Lia, Fua, & Dong, 2008).

Practical lessons on the study of complex systems and methods of data processing can be conducted with biotechnical complex. Biotechnical complex, which registers biosignals, allows for the use of various methods of information processing. The student becomes the object of research. This, in turn, increases the student's motivation. At laboratory sessions, the student studies the methods of identification and learns the state of his/her organism. The initial information for identification is presented in the form of recorded biosignals.

Organization of the Biotechnical Laboratory

The current organization of laboratory work in a technical university is characterized by the following features:

- Various courses
- Unique equipment, particularly specific for a course object
- Educational (yet, not research nature of laboratory work)

The dominant feature in technological development adds requirements to the practical training of specialists. This includes:

- Development of laboratory work using interdisciplinary integration
- Research-focused laboratory work
- Research of certain laws rather than private manifestations (fundamental principles of the organization and functioning of living systems)

In addition, there are limited resources for the purchase of expensive equipment and fixed space under laboratories. This adds requirements of aggregation, flexibility, extensibility, and interchangeability of equipment.

Until recently, the study of the human state was possible in special laboratories with the use of bulky and expensive hardware. Improvements to microelectronic devices—primarily sensory and computer technology—allows for the creation of safe compact information-computing systems (Paradiso, Loriga, & Taccini, 2005; Winters & Wang, 2003). These systems are used for studying and testing methods of complex systems on the human body.

The generalized structure of the complex is shown in Figure 1. The structure is an implemented modular principle of design and organization of mathematical software providing easy reconfiguration. Reconfiguration allows for the adaption of the complex for various laboratory works.

The presented biotechnical complex can carry out laboratory works on various disciplines, including:

- Programming and debugging of microprocessor technology
- Digital signal processing
- Methods and algorithms of identification
- Methods of diagnosing systems
- Research of nonstationary systems
- Identification and methods for assessing the state of complex systems

A feature of a complex is the use of measuring instruments made by the "lab-on-a-chip" technology. With several sensors, the measuring instrument turns on when the microprocessor controls the measuring process and data preprocessing. The lab-on-a-chip expands the functionality of the complex, which makes it suitable for laboratory work in related disciplines (i.e., microprocessor-based systems, telecommunication systems, digital signals processing, etc.). Producer companies do not charge for the delivery of debugging tools.

Figure 1. Generalized structure of the complex

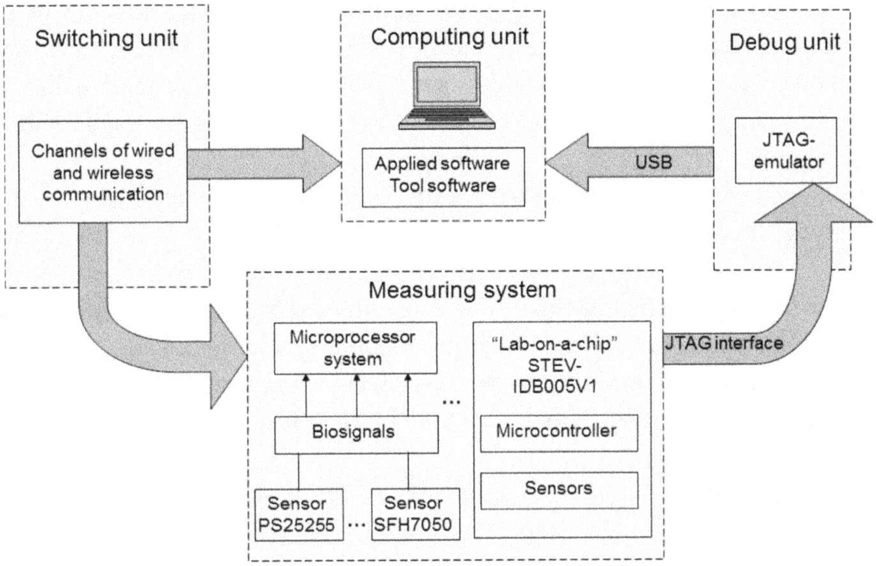

The complex uses different sensors to satisfy the ease of use requirement. For example, in the presented version of the complex, biosignals of the cardiac cycle are registered by PS25201B and SFH 7050 sensors.

PS25201B is an ultra-high impedance solid state electrocardiograph (ECG) sensor. It can be used as a dry contact ECG sensor without the need for potentially dangerous low impedance circuits across the heart. The resolution is good as conventional wet electrodes. The device uses active feedback techniques to lower the effective input capacitance of the sensing element and boost the input resistance. These techniques realize a sensor with a frequency response suitable for both diagnostic and monitoring of ECG applications.

The new SFH 7050 is the first integrated optical sensor from OSRAM Opto Semiconductors for automatic fitness tracking. The compact sensor includes three light-emitting diodes of different wave lengths. These are based on high-efficiency chip technology to ensure low energy consumption and high signal quality for extremely reliable simultaneous measurements. A built-in photo detector receives the reflected optical signals, which are separated from the emitters by an opaque barrier.

Methods of Nonlinear Dynamics in the Identification of Complex Systems

The presented biotechnical complex investigates methods of nonlinear dynamics in the identification of complex systems. Recently, methods of identification or reconstruction of equations of dynamic systems on the registered signals developed within nonlinear dynamics. The solution of the identification problem allows us to answer the following question: What are the parameters of the system that generated this signal? These parameters can identify (or recognize) the system (i.e., distinguish it from others).

A similar problem arises in the decision of application-oriented tasks. For example, in tasks of medical diagnostics, it is necessary to distinguish a norm from various pathologies by using measurement characteristics (i.e., heart rate, external breathing functions, pulse signal, etc.). The organization of biological systems is extremely complex and interdependent. These systems belong to the class of nonlinear dissipative systems functioning in the conditions of self-organization. The study of such systems is called synergetics.

Synergetics relies on the fundamental laws of nature. Therefore, it operates with abstract mathematical relationships satisfying these laws. These relationships can be referred to as a "model" in the sense that they are applicable to a variety of systems (i.e., physical, biological, economic, etc.).

An absolute majority of these model equations were obtained before the emergence of synergetics in the research of concrete physical processes. A typical example is the model equation of the Van der Pol oscillator. In comparison with biological systems, physical systems described by model equations are relatively simple and available for experimental research. Therefore, it seems reasonable to use them as a basis for identifying biosystems.

Model approach is perspective in medical applications. The most important biomedical information about the state and dynamics of the subsystem is not contained in the amplitude-frequency spectrum and structure of the recorded biosignals (i.e., ECG, sphygmogram, pneumotachogram, phonogram, etc.). It is contained in models of the studied subsystems and the nature of their interrelations.

A big class of complex systems of a natural origin, including the human body, represents various ensembles of interacting subsystems functioning in the oscillatory mode. For example, a person's heart and vascular systems can be presented in the form of the coupled dynamic systems. The description of the interacting oscillators in the cardiovascular system with use of the basic models used for example in the theory of nonlinear fluctuations is advisable. These relationships can be called "model" in the sense that they are applicable to a variety of systems (i.e., physical, biological, economic, etc.).

Biosignals information is available surrounding an organism's activity. Each signal has its own specificity caused by the concrete physiological nature of its generative organ. Together, they reflect collective and purposeful interaction of the main subsystems of an organism. Their joint analysis judges both the state of the concrete organ and the functional state of the organism.

Among recorded biosignals, a special role is played by the signal group generated by elements of the heart-vessels-lungs system. This group includes ECG, phonocardiogram, sphygmogram of the carotid artery, pneumotachogram, and kinetocardiogram. It is proposed to use the mathematical model of the heart-vessels-lungs biosystem as the base object for determining the functional state of the human body. This biosystem is the basis of the body's oxygen supply. Dynamics of the system elements' (i.e., heart, vessels, lungs) mechanical movements obey the laws of mechanics.

Circumstance makes it possible to construct the components of the structural information models of these elements using the laws of mechanics. Hidden or unknown functioning mechanisms of considered elements, which are available in the form of observable biosignals, are proposed to be described in the form of neural network information models. Thus, the task of structural identification of the model of the heart-vessel-lung biosystem is solved. Parametric identification is carried out on the basis of registered biosignals.

Knowledge of the functional state of one's organism is a motivating factor in the learning process. This is an advantage of the proposed approach to the organization of laboratory and practical works.

In summary, the subjects of possible laboratory work in shown in Figure 2 (with participation of students as objects of complex systems).

Figure 2. Subjects of laboratory work

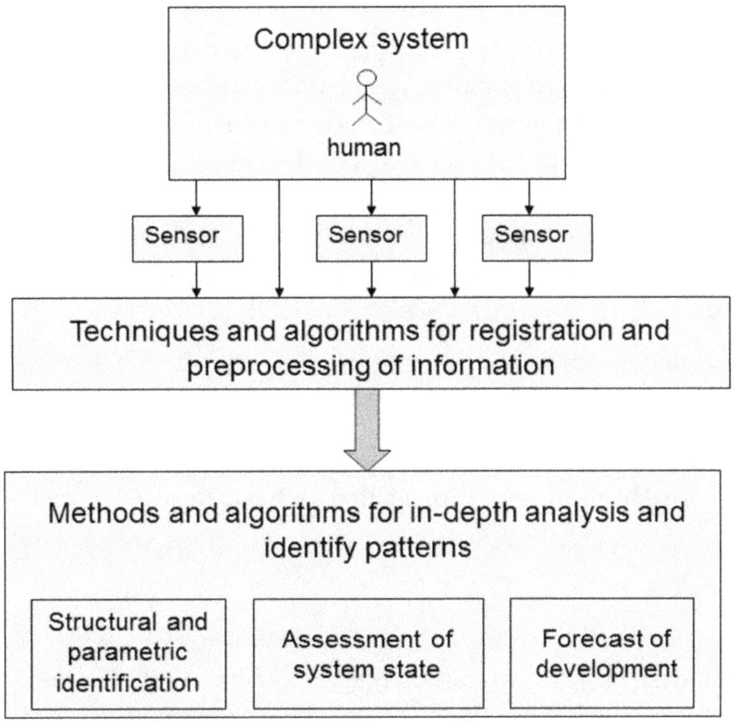

As an example, consider the solution of a problem of structural and parametrical identification of the human cardiovascular system (Buldakova & Suyatinov, 2014). The problem is solved with research modeling and basic model approaches.

Electrocardiogram signal inputs and sphygmogram signal outputs define parameters of the neural network model describing mechanics of the heart and parameters of the Van der Pol-Rayleigh equation as a model of vessel wall dynamics. At the same time, the state of the system is determined by calculating the approximated value of entropy (Yentes et al., 2013).

Optimum (the maximizing of the functional capabilities) implementation of the complex is achieved on the national instrument technical platform and LabVIEW system design software.

FUTURE RESEARCH DIRECTIONS

NBIC technological paradigm is based on the fundamental laws inherent in living systems. These laws, in particular, are manifested in the cardiovascular system of man. The presented software and hardware complex allows the student to study the manifestations of these laws on the example of own organism.

On the other hand, the described the software and hardware complex allows to combine the learning and research processes. This will improve the presented method of active learning, allowing the student to acquire the skills of the researcher. Such skills are very important characteristic for the modern engineer. Future research can be aimed at developing an active learning methodology that combines learning and research processes in a unified process.

CONCLUSION

This chapter presents an approach to the implementation of active learning methods involving the use of a biotechnical complex in the study of complex systems. The information-computing biotechnical complex intends to study methods and algorithms for the assessment of the functional state of human health on the registration and processing of various biosignals. It is noted that student motivation increases as they become the object of research.

Biotechnical complex can be used during laboratory work in disciplines like system analysis and identification, methods and information processing tools, Internet technologies, and automated information-control systems.

ACKNOWLEDGMENT

This research was supported by the Russian Foundation for Basic Research [grant number 16-07-00878].

REFERENCES

Anishchenko, V. S., Buldakova, T. I., Dovgalevskij, P. Y., Lifshic, V. B., Gridnev, V. I., & Suyatinov, S. I. (2009). Konceptual'naya model' virtual'nogo centra ohrany zdorov'ya naseleniya [Conceptual model of the virtual center of public health protection]. *Informacionnye tekhnologii - Information Technology, 15*(12), 59-64.

Buldakova, T., & Suyatinov, S. I. (2014). Reconstruction method for data protection in telemedicine systems. *Progress in Biomedical Optics and Imaging - Proceedings of the Society for Photo-Instrumentation Engineers, 94481U*, 1–6.

Buldakova, T. I., & Suyatinov, S. I. (2002). Registration and identification of pulse signal for medical diagnostic. *Proceedings of SPIE-The International Society for Optical Engineering, 4707*, 343–350.

Fedorov, I. B., Norenkov, I. P., & Korshunov, S. V. (2006). Organizacionnaya podgotovka specialistov v oblasti komp'yuternyh nauk, tekhniki i tekhnologij [Organizational training of specialists in the field of computer science, technics and technology]. *Informacionnye tekhnologii - Information Technology, 12*(9), 73-77.

Kamensky, E. G. (2015). Context of NBIC-technologies development: Institutions, ideology and social myths. *Mediterranean Journal of Social Sciences, 6*(6), 181–185.

Lantsberg, A. V., Troitzch, K. G., & Buldakova, T. I. (2011). Development of the electronic service system of a municipal clinic (based on the analysis of foreign Web resources). *Automatic Documentation and Mathematical Linguistics, 45*(2), 74–80. doi:10.3103/S0005105511020075

Lia, B. N., Fua, B. B., & Dong, M. C. (2008). Development of a mobile pulsewaveform analyzer for cardiovascular health. *Computers in Biology and Medicine*, *38*(4), 438–445. doi:10.1016/j.compbiomed.2008.01.008 PMID:18328471

Moats, J. (2015). Influences on the acceptance of innovative technologies used in learning opportunities: A theoretical perspective. In F. M. Nafukho & B. J. Irby (Eds.), *Handbook of Research on Innovative Technology Integration in Higher Education* (pp. 262–281). Hershey, PA: IGI Global. doi:10.4018/978-1-4666-8170-5.ch013

Paradiso, R., Loriga, G., & Taccini, N. (2005). A wearable health care system based on knitted integrated sensors. *IEEE Transactions on Information Technology in Biomedicine*, *9*(3), 337–344. doi:10.1109/TITB.2005.854512 PMID:16167687

Roco, M. C., & Bainbridge, W. S. (Eds.). (2003). *Converging technologies for improving human performance: Nanotechnology, biotechnology, information technology and cognitive science*. Kluwer Academic Publishers. doi:10.1007/978-94-017-0359-8

Winters, J., & Wang, Y. (2003). Wearable sensors and telerehabilitation. *IEEE Engineering in Medicine and Biology Magazine*, *22*(3), 56–65. doi:10.1109/MEMB.2003.1213627 PMID:12845820

Yentes, J. M., Hunt, N., Schmid, K. K., Kaipust, J. P., McGrath, D., & Stergiou, N. (2013). The appropriate use of approximate entropy and sample entropy with short data sets. *Annals of Biomedical Engineering*, *41*(2), 349–365. doi:10.100710439-012-0668-3 PMID:23064819

ADDITIONAL READING

Bolshakov, A., Glaskov, V., Egorov, I., Lobanov, V., Perova, L., & Pchelintseva, S. (2014). Methods and tools for software development to improve the effectiveness of engineering education in the direction of "mechatronics" using grid-computing technologies. In Communications in Computer and Information Science (466 CCIS, pp. 123-133). doi:10.1007/978-3-319-11854-3_12

Buldakova, T. I., & Dzalolov, A. S. (2012). Analysis of data processes and choices of data-processing and security technologies in situation centers. *Scientific and Technical Information Processing*, *39*(2), 127–132. doi:10.3103/S0147688212020116

Chistyakova, T. B., & Novozhilova, I. V. (2016). Intelligence computer simulators for elearning of specialists of innovative industrial enterprises. In *Proceedings of the 19th International Conference on Soft Computing and Measurements* (SCM 2016, pp. 329-332). 10.1109/SCM.2016.7519772

Chistyakova, T. B., Novozhilova, I. V., & Zelezinsky, A. L. (2016). Electronic information and education environment as instrument of forming and quality evaluation of professional competences of the international industrial enterprises specialists.In *IEEE 5th Forum Strategic Partnership of Universities and Enterprises of Hi-Tech Branches, Science. Education. Innovations* (pp. 12-14). 10.1109/IVForum.2016.7835839

Gavrilina, E., Zakharov, M., Karpenko, A., Smirnova, E., & Sokolov, A. (2016). Model of integral assessment quality of training graduates of higher engeneering education. In CEUR Workshop Proceedings (vol. 12, no. 3-2, pp. 11-16).

Kamensky, E. G. (2015). Context of NBIC-technologies development: Institutions, ideology and social myths. *Mediterranean Journal of Social Sciences*, *6*(6), 181–185.

Kilpatrick, W. H. (1918). The project method. *Teachers College Record*, *19*, 319–335.

Ryneveld, L. V. (2016). Introducing educational technology into the higher education environment: A professional development framework. In K. Dikilitaş (Ed.), *Innovative Professional Development Methods and Strategies for STEM Education* (pp. 126–136). Hershey, PA: IGI Global. doi:10.4018/978-1-4666-9471-2.ch008

Shpak, M. A., Smirnova, E. V., Karpenko, A. P., & Proletarsky, A. V. (2016). Mathematical models of learning materials estimation based on subject ontology. *Advances in Intelligent Systems and Computing*, *450*, 271–276. doi:10.1007/978-3-319-33609-1_24

KEY TERMS AND DEFINITIONS

Anthropomorphic Systems: The humanoid system.

Biosignals: Signals recorded by sensors on the human body (i.e., ECG, sphygmogram, pneumotachogram, phonogram, etc.).

Cardiovascular System: The circulatory system, including the heart and blood vessels.

Information-Computing Complex: Sensors connected to a computer to convert recorded signals.

Laboratory Work: Training sessions to carry out experiments.

Chapter 22
The Significance of Interdisciplinary Projects in Becoming a Research Engineer

Tatyana I. Buldakova
Bauman Moscow State Technical University, Russia

Sergey I. Suyatinov
Bauman Moscow State Technical University, Russia

ABSTRACT

The chapter presents the importance of interdisciplinary projects for modern engineer education. The components of the educational process influencing the quality of engineer education are allocated. The importance of the "learning through research" principle is noted. Examples of projects centered on the creation of information-analytical systems in various subject domains, and also, professional tasks which were solved by students are given. The conclusion is drawn that experience in the development of systems in one area can effectively be used in the development of systems in another area. The features of the project centered on the creation of a remote monitoring system of the human state, and the protection of the transferred physiological data are considered.

INTRODUCTION

The quality of engineer learning depends on many components directly related to the organization of the educational process (Ryneveld, 2016). The most important ones among them are: the use of modern forms and methods of learning based on information technology; high-quality software and information support of the educational process; integration of the scientific and educational activities of students; widespread introduction of the results of scientific research into the educational process; and practical orientation of the graduation projects (Sinicyna, Kurdyukova, & Abramova, 2013).

DOI: 10.4018/978-1-5225-3395-5.ch022

These components are most effective when using active forms of learning; for example, ones based on the project approach (Knoll, 1997). The application of the project method of teaching has a long history. It appeared in the first 20-30 years of the 19th century in the USA and was based on the principle of "Learning-by-doing" (Dewey, 2009) (Kilpatrick, 1918). At present, the methods of project-based learning are the subject of many publications. For engineering education, the fundamental ideas (Hall, 1962) that were described over half-a-century ago have remained relevant.

The project approach is also widely applied in the industry when developing new equipment and techniques. Therefore, it is logical to acquaint students with the main ideas of the project approach at the implementation of not only educational, but also real-world, projects.

The Internet and modern information technologies provide quick access to a huge amount of diverse information (Proletarskij & Neusypin, 2014). In Aleksandrov et al.'s work (Aleksandrov, Proletarskij, Neusypin, & Sharkov, 2012), features of innovative information technologies are highlighted for application in education. In particular, it becomes possible to set simple but real-world project tasks and to find their solutions in a short time. Although the obtained solutions may not always be correct, the process of obtaining them – in terms of learning – is effective.

In modern engineering education, the combination of the project approach with innovative information technologies is most effective when implementing the "Learning-through-Research" principle (Fedorov, Norenkov, & Korshunov, 2006). It involves engaging students in scientific research and development work. At the same time, researches of interdisciplinary character are of great importance. Moreover, in terms of modern technological development, interdisciplinarity is the key factor.

It should be noted that in this case, unlike in the classical project method, students not only use the knowledge they have gained, but they also acquire new competencies—which increases their qualifications. In the process of implementing interdisciplinary projects the students acquire skills of joint collective work on projects and competencies related to searching the solutions of professional interdisciplinary tasks.

The proposed approach complements the traditional forms of education. First of all, it is aimed at students who wish to acquire additional competencies. Similar situations arise with the target preparation of specialists, early employment, etc.

In this chapter, the examples of projects involving the creation of information-analytical systems for various purposes, and the problems solved by students during the implementation of those systems, are given. The professional knowledge gained during the work on such projects helps future engineers to design, develop, and implement information-analytical systems; to perform their operations and upgrade; to conduct experimental research of information processes; and to make informed management decisions.

The Generalized Process of Data Collection and Data Processing in Information-Analytical Systems: Identification, Diagnosis, and Management

In information-analytical systems (as realized in various subject areas), the data transformation process is performed according to the same scheme, including: data collection, identification of the functional state, diagnosis, and decision-making.

Although the descriptions of researched objects are different, the methods and algorithms of data processing and decision-making have a common theoretical basis. Therefore, the structure of the information-analytical (expert) system has a universal type (Figure 1).

Figure 1. Generalized structure of the system

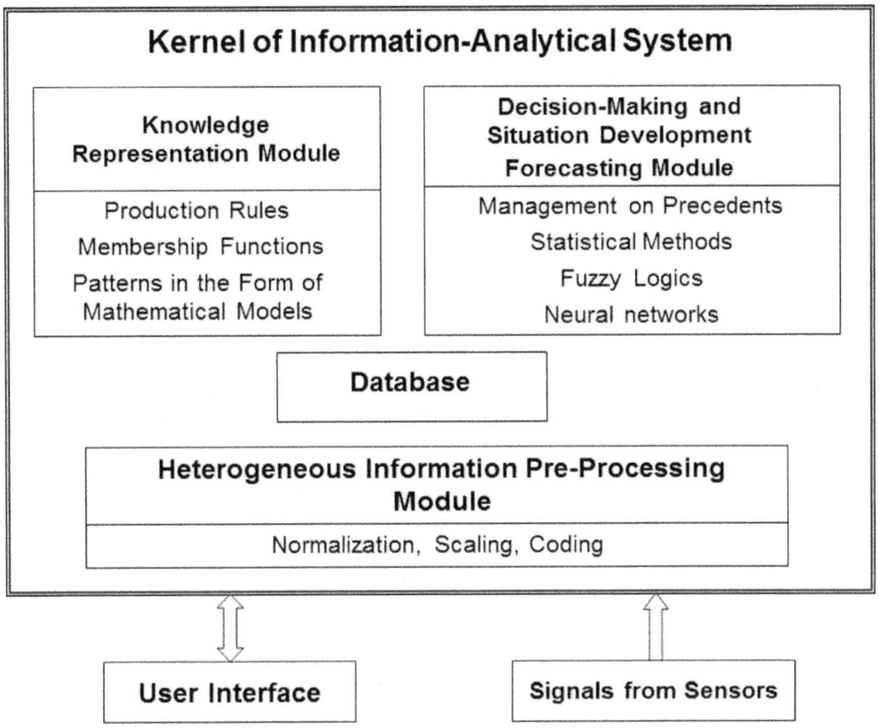

Considering the unified theoretical basis of methods and algorithms, and considering the universal structure of information-analytical systems and the technology of databases and knowledge bases, it is possible to claim that the experience of developing systems in one area can effectively be used in the development of systems in another area.

Tasks typical to the development of decision support systems are allocated as follows (Buldakova & Dzalolov, Analysis of Data Processes and Choices of Data-processing and Security Technologies in Situation Centers, 2012):

1. Formalization of the subject domain, and the creation of a conceptual representation (model) of the information-analytical system – including definitions of the types of information sources (the major entities) and the relations between them; object models of data (type of data, nature of data, and use of data); and solvable problems;
2. The development and analysis of the information models of business processes (inputs and outputs, objectives and types of processing, and existing relationships). The creation of a hierarchy of IDEF0-diagrams;
3. The formation of state classes and diagnostic feature space (based on developed models, and taking the opinions of experts into account);
4. Coding and generalization of heterogeneous information (qualitative, quantitative, verbal, and interval) for the development of integrated identification indicators;
5. Data structuring and knowledge model creation. Development of decision rules;
6. Choice of software and hardware; and also, physical implementation of the system;

7. Testing of components and the system as a whole.

The technology presented here – the creation of information-analytical systems – has been realized with students' participation in the following projects:

- The development of a new technological means of intelligent information processing in management systems for the supply and production of technological equipment and tools (Buldakova & Suyatinov, Informacionno-analiticheskaya sistema upravleniya snabzheniem i proizvodstvom instrumenta [Information-analytical system of supply and production management], 2002);
- The development of a new technological means of state identification, failure prediction, and control of grinding equipment; and its implementation on the SWaAGL-50 grinding machine (Buldakova & Suyatinov, Informacionno-analiticheskaya sistema upravleniya snabzheniem i proizvodstvom instrumenta [Information-analytical system of supply and production management], 2002);
- The development of an information-analytical medical complex for monitoring the cardiovascular system and decision-making support (Buldakova, Koblov, & Suyatinov, Informacionno-izmeritel'nyj kompleks sovmestnoj registracii i obrabotki biosignalov [Informational and measuring complex of the joint registration and processing of biosignals], 2008).

Currently, the RFBR (Russian Foundation for Basic Research) project is being implemented, which relates both to the creation of a remote monitoring system of the human state of health and to the protection of the transmitted physiological data. Students and postgraduates participate in this project, which makes it possible to implement the "Learning-through-Research" principle.

Features of Remote Monitoring System

The process of modernization of the health care system (Lantsberg, Troitzch, & Buldakova, 2011) is followed by the active introduction of information and communication technologies and the creation of telemedicine systems by which doctors in medical centers can remotely provide high-quality assistance to patients in remote areas (Prado, Roa, & Reina-Tosina, 2002).

The modern telemedicine complex integrates powerful computers which easily interface with various medical equipment by means of (both area network and satellite network) wireless communication; videoconferencing tools; and IP broadcasting. A most promising direction of development of remote monitoring of the human state is the integration of sensors into clothes and various accessories including mobile phones (Paradiso, Loriga, & Taccini, 2005).

A built-in sensor system allows you to control a variety of physiological parameters including heart rate, respiration rate, electrocardiogram (ECG) readings, body temperature, blood oxygen saturation level, etc. (Winters & Wang, 2003).

In biomonitoring systems, assessment of the human state of health can be carried out in different ways (Anishchenko, et al., 2009): under the supervision of an attending physician; based on the analysis of monitored parameters (ranging from normal to pathological); or automatically via computer model of the patient (virtual physiology) (Prado, Roa, & Reina-Tosina, 2002).

Regardless of the assessment method, the sources of primary information are various sensors allowing registration of the person's biosignals. Therefore, they are the most important part of the monitoring system, and their characteristics determine the effectiveness of the entire system. Pulse, ECG, and respiration rate sensors are the most widespread such sensors.

A modern pulse sensor usually consists of a steel sensor and a smartphone. The principle of pulsator operation is based on the measurement of a potential difference at the time of reduction of a cardiac muscle. Further, the pulsator transmits pulse signals to the receiving device (smartphone), which processes these signals by means of an installed application. If necessary, data can be transferred to a medical center for a more in-depth analysis by the physician.

Mobile ECG sensors are capable of recording the electric activity of the heart and transferring that data to the smartphone for a preliminary assessment of the human state. For in-depth analysis, the signals must also be sent to a specialist.

A respiration rate sensor is designed to study the process of human respiration. It is a belt within which is embedded a hollow rubber chamber with a gas pressure sensor. Further results of the sensor's measurements may be transformed into analyses of the frequency and period of human respiratory movements.

Thus, the technological level which has been reached allows for the creation of compact mobile terminals which can perform primary analysis functions and transmit physiological data to medical centers for more insightful analysis (Lia, Fua, & Dong, 2008).

The inclusion of such mobile measuring systems in the common information space will allow interested parties (for example, insurance companies, local clinics, research hospitals, privately-owned companies) to carry out continuous monitoring of the human state, irrespective of a person's geographically location (Figure 2).

Figure 2. Remote monitoring, based on the common information space

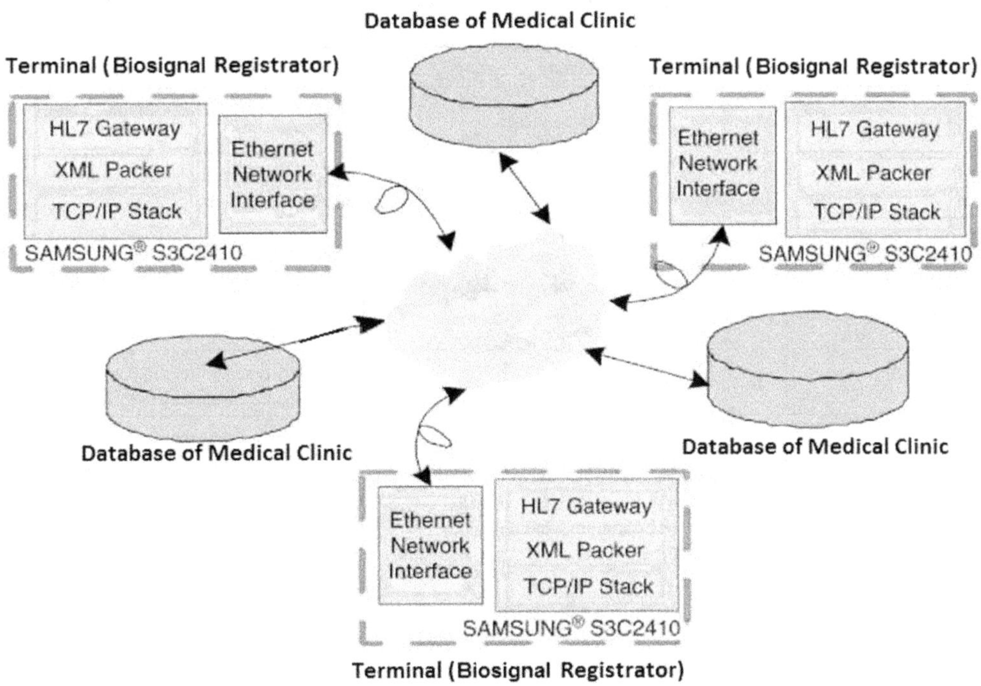

At the same time, however, there is a problem with ensuring the integrity, confidentiality, and availability of the transferred physiological data. Moreover, violation of the integrity and confidentiality of information, and the theft of personal medical data, would lead not only to financial losses, but also to undesirable social consequences, causing moral damage to the patient.

Results of the analysis of possible threats of information security – applicable to all monitoring system components – are given in Table 1.

Analysis of the threat model showed that there is a problem with the information security of patient data transmitted from the sensor to the repository. At the same time, the protection of personal medical data –transmitted through a communication channel from the sensor to the cloud-based medical database – is of crucial importance. In this regard, for the protection of personal information transmitted, it is necessary to select a method of cryptography data protection.

Despite the growing flow of research in the field of information security (Appari & Johnson, 2010), very few research efforts are directed towards the studying of information security risks in the health sector – a sector which is substantially regulated and which uses business models, rather different from the models of other industries (Malhotra, Gardner, & Patz, 2007). The analysis of various approaches to distribution of cryptographic keys is made in work (Buldakova & Suyatinov, 2014).

All traditional approaches to the safety of health systems are based on asymmetric cryptosystems. Asymmetric encryption uses two different keys: one for encryption (also called a "public" key) and another for decryption (called a "private" or "secret" key). Such an approach is rather reliable for ensuring the confidentiality and integrity of transmitted data; but due to the large key lengths, a regular data exchange in a real-time system is expensive as it demands large expenditures of resources and time. Therefore, it is inexpedient to use asymmetric enciphering in remote monitoring systems where the data is being processed in real time.

Table 1. Possible information security threats

Components	Threats	Comment
Sensors	Malefactor access to the sensor	It is necessary to use reliable sensors; limiting access
Communication	Malefactors can overhear all types of talk, and also distort signals	In-system communication is unreliable; therefore, it is necessary to encrypt signals
Smartphone	Malefactor can affect the performance of the smartphone	Protection of applications on the smartphone
Data storage in the cloud	Possible access to the data in the cloud	Only after successful authorization can the physician access a patient's information
Medical staff	Transmission of information to malefactor	It is assumed that medical staff will not open access to information under the influence of malefactors
Patient	Transmission of information to malefactor	It is assumed that the patient will not open access to information under the influence of malefactors
Patient's body	Malefactor can have physical contact with the patient (for example, shake his hand), therefore patient's biological signals can be distorted by the malefactor's signals	Reliable sensors don't allow the malefactor to distort signals. Besides, all information on the past history of a patient's health is unknown to the malefactor

An alternative approach is to protect the transmitted data by creating a symmetric key pair for the sensor and the receiver. In a symmetric cryptosystem, the same cryptographic key is applied both for encryption and for decryption. As a result, algorithms with the closed key work three orders faster than do algorithms with an open key—which is very important for real-time telemedicine systems. However, a problem with the lack of symmetric codes is the impossibility of their use for confirmation of authorship, as the key is known to each party.

In a number of works, it is suggested that biosignals registered by sensors be used in the construction of cryptographic keys. Such biosignals reflect physiological features of the patient and can be used for the concealment of information. For example, in Banerjee et al.'s work (Banerjee, Gupta, & Venkatasubramanian, 2013), some morphological features of biosignals, which are unique for each person and which change a little eventually, are marked out. It is suggested that these features be used in the construction of cryptographic keys and for receiving a model of an individual "physiological" signature. However, in this method it is necessary to apply matching of functional dependences by the form of the registered biosignals that it is not really effective.

In Buldakova & Suyatinov's work (Buldakova & Suyatinov, 2014) it is suggested that the mathematical model of a biosignal generator (in the form of a system of differential equations) be used as morphological features. This approach is more universal than that offered in the work by Banerjee et al. (Banerjee, Gupta, & Venkatasubramanian, 2013).

The problems to be performed in this project require interdisciplinary knowledge in the fields of biomedicine, system engineering, signal processing, web technologies, modeling, information security, and other fields. At the same time, it is easy to find – online – the solution prototypes for most of the technical issues considered in this project.

FUTURE RESEARCH DIRECTIONS

The problem of improving the quality of engineering education necessitates a thoughtful and deliberate approach to the organization of educational process and the use of innovative technologies to motivate students. Currently, active learning forms (for example, based on the project approach) are widely used for this purpose. The combination of the project approach with innovative information technologies is most effective when implementing the "Learning-through-Research" principle. Inclusion of elements of research activity in educational process is a distinctive (characteristic) tendency of the development of engineering education.

An important idea that comes out of this chapter is follow: experience of development of information-analytical systems in one area can effectively be used by development of systems in other area. However, the considered examples are local in nature and relevant for the specific university community.

At present, there is no generalized conceptual model illustrating the embedding of elements of research activity in traditional educational forms of education. Future researches in this area are warranted.

CONCLUSION

In the context of the chapter, such educational aspects of active learning as project approach, learning through research, and interdisciplinarity are presented. The emphasis is on expanding the capabilities of the project approach in engineering education based on modern information technologies, and on implementing the "Learning-through-Research" principle.

Modern information technologies allow for improving the classical methods of the project approach in education. The Internet, and the availability of information, provides promptness in the formulation and solution of real-world project problems. Each such problem is unique. Its solution includes the research stage and is based on an interdisciplinary approach.

The practical orientation of the educational-scientific process allows future engineers to quickly master new information technologies in solving interdisciplinary problems, and to acquire the necessary professional competencies.

The article contains examples of interdisciplinary projects which have already been implemented, along with an example of a project currently being implemented.

Problems solvable upon the implementation of projects centered on the creation of information-analytical systems in various subject domains, are given.

Features of remote monitoring systems of the human state of health are delineated; and possible ways of protecting transmitted data are analyzed.

The solution of the problems under consideration contributes, in the authors' opinion, to improving the quality of learning for engineers.

ACKNOWLEDGMENT

This research was supported by the Russian Foundation for Basic Research [grant number 16-07-00878].

REFERENCES

Aleksandrov, A. A., Proletarskij, A. V., Neusypin, K. A., & Sharkov, A. A. (2012). Koncepcija kompleksnogo nepreryvnogo obuchenija s ispol'zovaniem innovacionnyh informacionnyh tehnologij [The concept of integrated continuous learning using innovative information technologies]. *Nauchnoe obozrenie: gumanitarnye issledovanija - Scientific Review: Humanitarian researches, 4*, 38-42.

Anishchenko, V. S., Buldakova, T. I., Dovgalevskij, P. Y., Lifshic, V. B., Gridnev, V. I., & Suyatinov, S. I. (2009). Konceptual'naya model' virtual'nogo centra ohrany zdorov'ya naseleniya [Conceptual model of the virtual center of public health protection]. *Informacionnye tekhnologii - Information Technology, 15*(12), 59-64.

Appari, A., & Johnson, M. E. (2010). Information security and privacy in healthcare: Current state of research. *International Journal of Internet and Enterprise Management*, *6*(4), 279–314. doi:10.1504/IJIEM.2010.035624

Banerjee, A., Gupta, S. S., & Venkatasubramanian, K. K. (2013). PEES: Physiology-based end-to-end security for mHealth. *Proceedings of the 4th Conference on Wireless Health*, 2.

Buldakova, T., & Suyatinov, S. I. (2014). Reconstruction method for data protection in telemedicine systems. *Progress in Biomedical Optics and Imaging - Proceedings of the Society for Photo-Instrumentation Engineers*, *9448*(94481U), 1–6.

Buldakova, T. I., & Dzalolov, A. S. (2012). Analysis of data processes and choices of data-processing and security technologies in situation centers. *Scientific and Technical Information Processing*, *39*(2), 127–132. doi:10.3103/S0147688212020116

Buldakova, T. I., Koblov, A. V., & Suyatinov, S. I. (2008). Informacionno-izmeritel'nyj kompleks sovmestnoj registracii i obrabotki biosignalov [Informational and measuring complex of joint registration and processing of biosignals]. *Pribory i sistemy. Upravlenie, kontrol', diagnostika - Devices and systems. Management, control, diagnostics*, *6*, 41-46.

Buldakova, T. I., & Suyatinov, S. I. (2002). Informacionno-analiticheskaya sistema upravleniya snabzheniem i proizvodstvom instrumenta [Information-analytical system of supply and production management]. *Informacionnye tekhnologii - Information Technology*, *8*(11), 28-33.

Dewey, J. (2009). *Democracy and education: An introduction to the philosophy of education*. New York: WLC Books. (Original work published 1916)

Fedorov, I. B., Norenkov, I. P., & Korshunov, S. V. (2006). Organizacionnaya podgotovka specialistov v oblasti komp'yuternyh nauki, tekhniki i tekhnologij, otrazhennaya v rossijskih i zarubezhnyh dokumentah [Organizational training of specialists in the field of computer science, technics and technology; as reflected in Russian and foreign documents]. *Informacionnye tekhnologii - Information Technology*, *12*(9), 73-77.

Hall, A. D. (1962). *A Methodology for Systems Engineering*. Princeton, NJ: Van Nostrand.

Kilpatrick, W. H. (1918). The project method. *Teachers College Record*, *19*, 319–335.

Knoll, M. (1997). The project method: Its vocational education origin and international development. *Journal of Industrial Teacher Education*, *34*, 59–80.

Lantsberg, A. V., Troitzch, K. G., & Buldakova, T. I. (2011). Development of the electronic service system of a municipal clinic (based on the analysis of foreign web resources). *Automatic Documentation and Mathematical Linguistics*, *45*(2), 74–80. doi:10.3103/S0005105511020075

Lia, B. N., Fua, B. B., & Dong, M. C. (2008). Development of a mobile pulsewaveform analyzer for cardiovascular health. *Computers in Biology and Medicine*, *38*(4), 438–445. doi:10.1016/j.compbiomed.2008.01.008 PMID:18328471

Malhotra, K., Gardner, S., & Patz, R. (2007). Implementation of elliptic-curve cryptography on mobile healthcare devices. *IEEE International Conference on Networking, Sensing and Control*, 239-244. 10.1109/ICNSC.2007.372784

Paradiso, R., Loriga, G., & Taccini, N. (2005). A wearable health care system based on knitted integrated sensors. *IEEE Transactions on Information Technology in Biomedicine*, *9*(3), 337–344. doi:10.1109/TITB.2005.854512 PMID:16167687

Prado, M., Roa, L., Reina-Tosina, J., Palma, A., & Milan, J. A. (2002). Virtual center for renal support: Technological approach to patient physiological image. *IEEE Transactions on Biomedical Engineering*, *49*(12), 1420–1430. doi:10.1109/TBME.2002.805454 PMID:12542237

Proletarskij, A. V., & Neusypin, K. A. (2014). Osobennosti ispol'zovanija sovremennyh informacionnyh tehnologij v obrazovanii [Features of using modern information technologies in education]. *European Social Science Journal, 1-1*(40), 63-65.

Ryneveld, L. V. (2016). Introducing educational technology into the higher education environment: A professional development framework. In K. Dikilitaş (Ed.), *Innovative Professional Development Methods and Strategies for STEM Education* (pp. 126–136). Hershey, PA: IGI Global. doi:10.4018/978-1-4666-9471-2.ch008

Sinicyna, E. Y., Kurdyukova, G. N., & Abramova, E. Y. (2013). Razvitie paradigmy kreativnogo obrazovaniya posredstvom obucheniya cherez issledovaniya [Creative education paradigm development via 'Learning through Research']. *Vestnik MJEI - MPEI Vestnik, 6*, 217-219.

Winters, J., & Wang, Y. (2003). Wearable sensors and telerehabilitation. *IEEE Engineering in Medicine and Biology Magazine*, *22*(3), 56–65. doi:10.1109/MEMB.2003.1213627 PMID:12845820

ADDITIONAL READING

Bolshakov, A., Glaskov, V., Egorov, I., Lobanov, V., Perova, L., & Pchelintseva, S. (2014). Methods and tools for software development to improve the effectiveness of engineering education in the direction of "mechatronics" using grid-computing technologies. In Communications in Computer and Information Science (466 CCIS, pp. 123-133). doi:10.1007/978-3-319-11854-3_12

Buldakova, T. I., & Suyatinov, S. I. (2002). Registration and identification of pulse signal for medical diagnostic. *Proceedings of SPIE-The International Society for Optical Engineering*, *4707*, 343–350.

Chistyakova, T. B., & Novozhilova, I. V. (2016). Intelligence computer simulators for elearning of specialists of innovative industrial enterprises. In *Proceedings of the 19th International Conference on Soft Computing and Measurements* (SCM 2016, pp. 329-332). 10.1109/SCM.2016.7519772

Chistyakova, T. B., Novozhilova, I. V., & Zelezinsky, A. L. (2016). Electronic information and education environment as instrument of forming and quality evaluation of professional competences of the international industrial enterprises specialists.In *IEEE 5th Forum Strategic Partnership of Universities and Enterprises of Hi-Tech Branches, Science. Education. Innovations* (pp. 12-14). 10.1109/IVForum.2016.7835839

Gavrilina, E., Zakharov, M., Karpenko, A., Smirnova, E., & Sokolov, A. (2016). Model of integral assessment quality of training graduates of higher engeneering education. In CEUR Workshop Proceedings (vol. 12, no. 3-2, pp. 11-16).

Kamensky, E. G. (2015). Context of NBIC-technologies development: Institutions, ideology and social myths. *Mediterranean Journal of Social Sciences*, *6*(6), 181–185.

Moats, J. (2015). Influences on the acceptance of innovative technologies used in learning opportunities: A theoretical perspective. In F. M. Nafukho & B. J. Irby (Eds.), *Handbook of Research on Innovative Technology Integration in Higher Education* (pp. 262–281). Hershey, PA: IGI Global. doi:10.4018/978-1-4666-8170-5.ch013

Roco, M. C., & Bainbridge, W. S. (Eds.). (2003). *Converging technologies for improving human performance: Nanotechnology, biotechnology, information technology and cognitive science*. The Netherlands: Kluwer Academic Publishers. doi:10.1007/978-94-017-0359-8

Shpak, M. A., Smirnova, E. V., Karpenko, A. P., & Proletarsky, A. V. (2016). Mathematical models of learning materials estimation based on subject ontology. *Advances in Intelligent Systems and Computing*, *450*, 271–276. doi:10.1007/978-3-319-33609-1_24

KEY TERMS AND DEFINITIONS

Active Forms of Learning: The methods that stimulate cognitive activity of students in the process of mastering the educational material.

Biosignals: Signals recorded by sensors located on the human body (ECG, sphygmogram, pneumotachogram, phonogram, etc.).

Information Security: The process of ensuring the confidentiality, integrity, and accessibility of information.

Information-Analytical Systems: Automated systems which carry out the storage, processing, analysis, and provision of information in a user-friendly form.

Interdisciplinarity: The combining of two or more academic disciplines into one activity (e.g., a research project).

Learning Through Research: The principle of learning in which the basic knowledge is acquired by students themselves during research activities.

Remote Monitoring: The assessment of an object's state at a distance.

Scientific Research: The systematic collection and analysis of data to obtain new knowledge.

Chapter 23
Spectral Algorithms for Signal Generation as Learning-Methodical Tool for Engineer Preparation

Vladimir V. Syuzev
Bauman Moscow State Technical University, Russia

Elena Smirnova
Bauman Moscow State Technical University, Russia

Kirill Kucherov
Bauman Moscow State Technical University, Russia

Vladimir Gurenko
Bauman Moscow State Technical University, Russia

Gurgen Khachatrian
American University of Armenia, Armenia

ABSTRACT

In this chapter, a spectral method of deterministic signals simulation is proposed. The target is to solve the teaching methodological problem of individual tasks creation for students: future engineers in the field of real-time control systems' development and research. The tasks are related to the correlation theory. The method of simulation algorithms tuning with the help of specified spectrally correlation of signal characteristics and their parameters in the harmonic Fourier and Hartley basis is presented. To expand the set of task's variants for the independent student self-preparation the task of signals simulation has been formulated mathematically and solved in arbitrary complex and real systems of orthogonal basic functions. The requirements have been underlined on how teacher can teach basic functions for the practical applications. During such kind of decision task, the student will be able to counts the multiplicity, precision, and computational complexity of the resulting spectral simulation algorithms.

DOI: 10.4018/978-1-5225-3395-5.ch023

RELEVANCE

Practical classes including course design, homework assignments and research works are important components of the teaching process in engineering field. The teaching efficiency depends on individual tasks for the student's independent work. These individual tasks should be prepared for the large number of students, should be multivariate and informative, as well as should cover a large number of competencies, which are specified in the curriculum.

Disciplines, which are related to the design of computer systems and real-time control systems, need training and methodological tools for student's independent work. This paper gives an example of such kind of methodological tool using signal's simulation methods and algorithms. The methodological tool set could be applied to the tasks of the subject on "Information's storing, processing and transformation" teaching.

INTRODUCTION

This paper proposes Science-Based Approach in Engineer's Preparation which is a continuation of Project-Based Approach. Students took part in a research project which been supported by Ministry of Education and Science of Russian Federation (Project Contract ID 2.7782.2017/BC). The project is concern to simulation algorithms, which use spectral representation of signal in different orthogonal bases. Such an approach allows students to produce simulation algorithms, which are differ in accuracy and computation complexity due to the proper basis function systems choice. The approach provides the multivariate tasks for the student's individual work. Moreover, the independent research on the variety of spectral representations will expand the scope of learner's knowledge and his skills in analog and digital signal spectral processing - theoretically and in practice. The results of the students work give statistical material for further ideas of the scientists.

From the large number of known base systems the most attractive ones to meet different authors' goals are parametrical bases containing changing parameters in the structure of their functions. They affect their properties. Examples of such bases systems are complex exponential functions, Vilenkin – Chrestenson functions, the generalized Hartley functions. The control over their properties is achieved by using variation of different bases.

In this paper the deterministic signals are under consideration only in the framework of the correlation wide sphere of practical applications because they are useful components of input signal processing in different control and management systems. The variation of such signal's frequency characteristics provides an extension for many options in simulation algorithms, which are helpful for student's individual task's creation from the side of teacher too.

The authors of the chapter develop an enrolled course named "Digital Signal Processing" (DSP) supported by industry enterprises working in the field of automation.

First part of the chapter is about how teacher could create variants of tasks for the individual student's work. The authors paid attention to the mathematical description of tasks assigned to students (see the Appendix) for performing laboratory work with the aim that teacher who carries on similar DSP course could use this mathematical device to create their own versions of tasks. The main emphasis in the paper is made precisely on the fact that this is a methodical material for conducting new form laboratory works, while each student or each group of students investigates new mathematical calculations and creates new solutions.

Second part of the chapter is about software environment which student use for his/her individual research work. An example of student's research work is given. The explanation is given about how student's engagement ensured and promoted.

Third part concerns to students' experiences feedbacks in their research of the fast algorithms of signal's transformation, which confirms the interactive and motivated science based approach to engineer's preparation.

Fourth part contains some students' feedbacks.

The Conclusion is dedicated to further development of the individual task's variants using different theoretical bases.

VARIANTS OF TASKS FOR INDIVIDUAL STUDENT'S WORK

Variants of Tasks for Individual Student's Work

Background for signal simulation algorithms may be found in the Appendix.

The common part for the student's homework was as follows:

Using the theoretically grounded connection of the signal spectrum and the Power Spectral Density Function (PSDF), as well as the initial data of the personal variant of the task (see Table 1), write the program using programming language on your choice to perform algorithm's modeling. Construct and show the autocorrelation functions of the original and simulated signals and after that compute absolute modelling error for 20 realizations (present it in table form) and estimate it by computing its average value.

The Table 1 presents the variants of individual tasks for students' homework based on theoretical knowledge listed above for the complex exponential basis.

SOFTWARE FOR THE STUDENT'S RESEARCH

The students were advised to use programming language Python 3.5 to obtain results faster. This programming language has tools for visual modeling and for errors' estimation as well as for vector operations.

At the beginning of laboratory work every student gets Python virtual environment with packages that are necessary to perform individual study: "matplotlib" for plotting (A. Devert, 2014) "numpy" for efficient calculations and vector operations and "scipy" for advanced usage (E. Bressert, 2012).

The structure of individual virtual environment's working directory is shown at the Figure 1.

Using the Python virtual environment student writes his own script to perform simulation algorithm due to individual task variant (Table 1). The script can be started using local environment Python interpreter installed at Windows (cmd command) or Linux operation system (console command).

Initially the script plays role of testing one – it allows student to check whether Python virtual environment works properly.

Table 1. The variants of individual tasks for students' homework

Variant's Number	Signal Type	Time Interval	Spectral Density Maximum/Dispersion	Cut-Off Frequency
NºNº	$x(t)$	$T[s]$	$S_0 / \sigma^2 [J]$	$\omega_2 \left[\frac{rad}{s}\right]$
1	White noise	100	5	3π
2		60	10	20π
3		160	15	5π
4	Triangular spectral density signal	150	3	3π
5		180	9	20π
6		80	8	25π
7	Exponential spectral density signal	50	6	4π
8		120	4	10π
9		240	10	2π

Figure 1. Structure of Python virtual environment

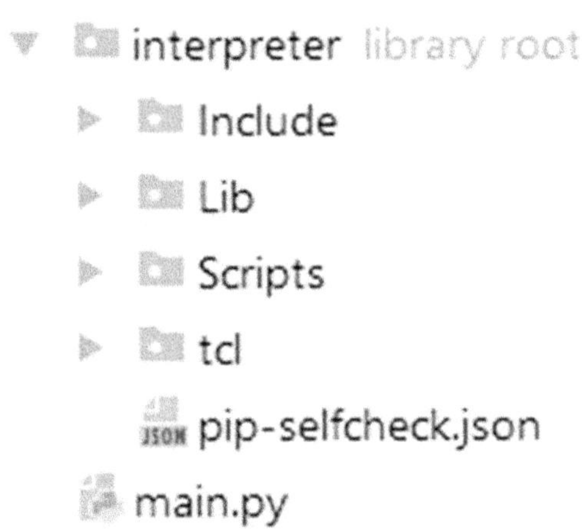

AN EXAMPLE OF STUDENT'S EXPERIENCE

Initial data:

Type of signal is white noise;

The spectral power density (SD) function has the form $S(\omega) = \begin{cases} S_{0,} |\omega| \le \omega_2 \\ 0, |\omega| > \omega_2 \end{cases}$, $S_0 = 5[J]$;

Cut-off frequency ;

The duration of the signal reception interval is T = 50 [s],

The interval is two-sided, that is, $[-25s, 25s]$;

Initial conditions

The simulation algorithm takes as an input data the following set of parameters:

1. The interval of the signal T;
2. Spectral density signal, specified in the form of an analytical expression, a table of values or some other way. Within the framework of this version, an approximate approach to analytical formulations is considered (a predetermined number of SDs). For white noise, the SD has the form, where are taken from the initial conditions;
3. Cut-off frequency of the SD ;
4. The time sampling interval $\Delta t\left[c\right] \leq \Delta t_k$, where $\Delta t_k = \frac{\pi}{\omega_\varepsilon}$ is the sampling step taken by the Kotel'nikov theorem;
5. The frequency sampling interval
6. The number of discrete signal samples $N = \frac{T}{\Delta t}$.

The particular variant of student task is shown at the Table 2.

The number of discrete signal samples

$$N = \frac{T}{\Delta t} = 150,$$

however, in order to obtain a performance gain when converting spectra, the condition $N_{id} = 2^n$, where $n = \left]\log_2\left(N\right)\right[$ (operation $]x[$ here is rounding x to the nearest larger integer).

Hence $N = N_{id} = 2^8 = 256$

Table 2. The particular variant of student task.

Parameter	Meaning
T	$50\left[s\right]$
ω_2	$3\pi\left[\frac{rad}{s}\right]$
S_0	$5\left[J\right]$
Δt	$0.33\left[s\right]$
$\Delta\omega$	$0{,}1257\left[\frac{rad}{s}\right]$

The simulated signal is shown at the Figure 2 for the exponential basis.

The accuracy of signal reception can be estimated by comparing the ideal autocorrelation function (ACF) with the experimental one (obtained from the simulated signal). The graphs of the theoretical (from above) and the experimental ACF are shown at the Figure 3.

Obviously the forms of the graphs do not coincide. This is due to the fact that the ideal form of the ACF graph can be obtained only for $N \to \infty$, which could not be achieved in practice. The larger N, the stronger the shape of the graph of the experimental ACF approaches the theoretical one.

To estimate the accuracy, the absolute error at points N, calculated as

$$E = \max\left[\left|X - Y\right|\right],$$

where $-$ is the operation of two vectors term difference. Here **X** is a vector of the theoretical ACF samples and **Y** is the sample vector of the experimental ACF.

The Table 3 gives estimates of the maximum error for 20 realizations at N = 256.

THE STUDENTS' FEEDBACKS

After the students passed the enrolled course, they were asked to give some feedback. The questions were concerned to:

1. Whether they find course applicable in their working experience (see student's feedback about practical applicability of course at Table 4);

Figure 2. Signal in the exponential basis

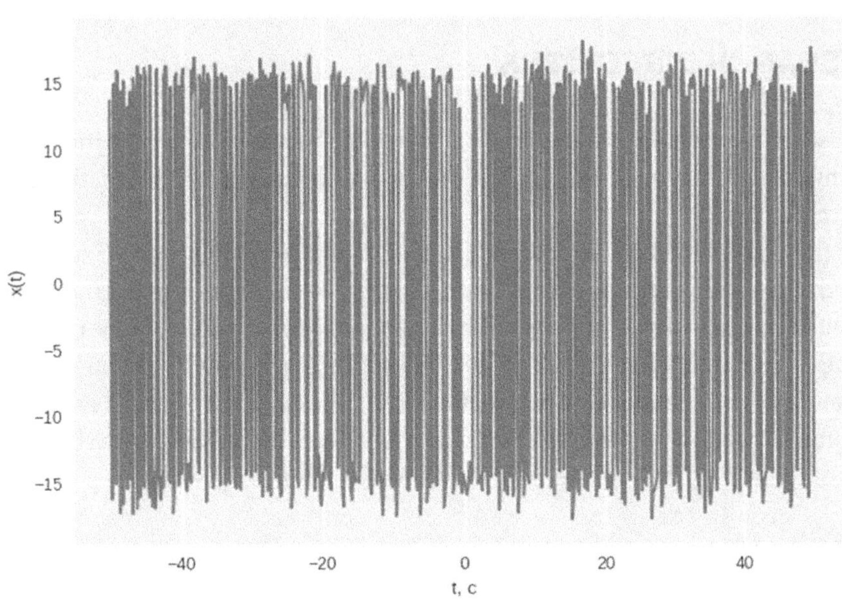

Figure 3. The graphs of the theoretical (above) and the experimental (below) autocorrelation function

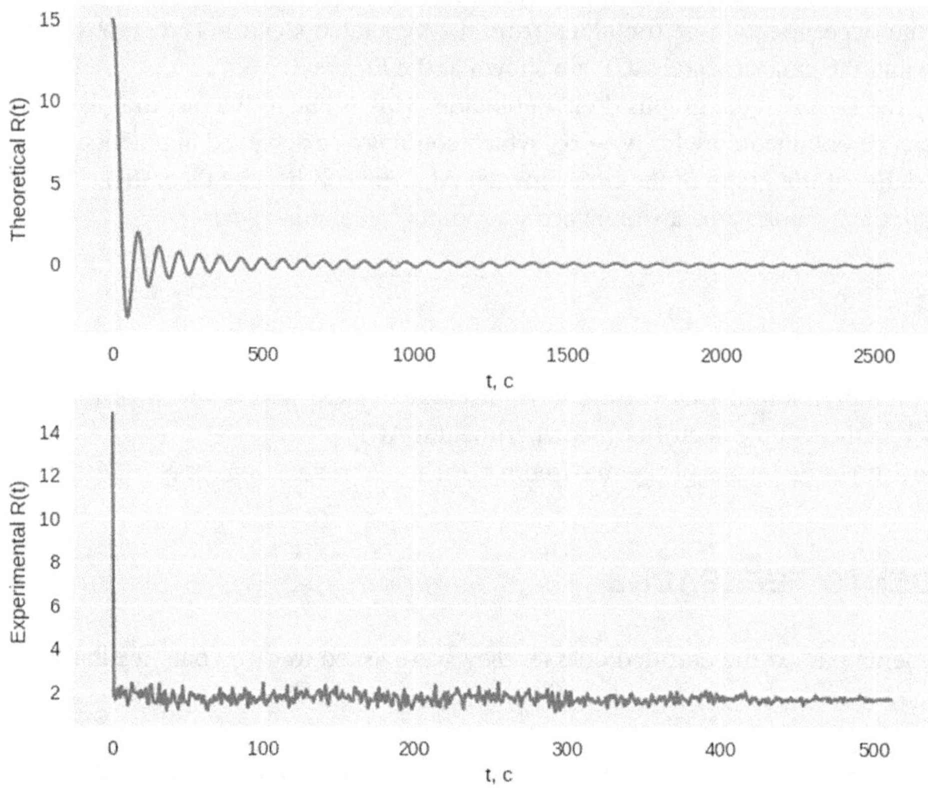

2. How they consider complexity of course (see Table 5);
3. Whether they believe the course should be improved (see Table 6).

FUTURE RESEARCH DIRECTIONS

The main emphasis in the chapter is a methodical material for conducting new innovative form of laboratory works, while each student or each group of students investigates new mathematical calculations and creates new solutions.

It can be concluded that the task of deterministic signals simulation in the framework of correlation theory, and spectral algorithms for its solution are an effective methodological tool for the formation of wide variety of customized tasks on all types of independent work of students (homeworks, course work, projects in scientific research work in the subject area related to the study of real-time control systems).

As well the authors paid attention to the mathematical description of tasks assigned to students for performing laboratory work with the aim that teacher who carries on similar DSP course could use this mathematical tool to create their own versions of tasks for laboratory works.

Table 3. The estimates of errors

Number of Experiment	Maximum Error, %
1	10.540251200154879
2	10.049834351918072
3	11.022740100197371
4	7.8523315419693045
5	8.18375876274209
6	10.8202909669208
7	11.067793977583067
8	10.143977032452643
9	10.285152165281206
10	9.184353083065337
11	10.981953587902993
12	8.397526266566993
13	10.276303918586438
14	9.93328327131869
15	7.681415524479086
16	8.798219826675322
17	10.505741873344858
18	8.798184061014213
19	9.520308068680409
20	9.6916559124436
The average error value	9.686753774664869

Table 4. Students' feedback about practical applicability of course

I have found the course...	Surely Applicable	Partially Applicable	Neutral	Not Applicable at All
	5	12	3	0

Table 5. Students' feedback about complexity of course

I Have Found the Following Component of Course...	Very Clear for Understanding	Understandably	Clear, but Needs Hard Work to Understand	Not Clear at All
Theory	0	7	13	0
Special calculations	0	12	8	0
Laboratory works	3	5	12	0

Table 6. Students' feedback about improvement of course

I Have Found the Following Component of Course Should Be Improved...	Deeply	Some Minor Improvements	Some Aspects	Doesn't Need Improvement
Theory	0	0	4	16
Special calculations	0	3	7	10
Laboratory works	0	0	12	8

Variable parameters providing a multiplicity of the proposed algorithms are function characteristics on the power spectral density, the interval definition signal, the step of sampling time, accuracy requirements and the complexity of the simulation and, especially, chosen system of bases functions.

By changing the radix and choosing different ways to sort the functions in the systems, the teacher can obtain a large family of ortho-normalized basis systems with fast algorithms of spectral transformations. Special cases of these systems are been widely used in signal processing with the basic system of the Walsh, Paley, Hartmut, exponential and Haar for the methodological educational materials' development (Syuzev V.V., 2017). All mentioned systems will be used to start the student's engagement into research project.

CONCLUSION

The authors of the chapter gave a description of innovative enrolled course named "Digital Signal Processing" (DSP) where the student's laboratory work results are the raw material for the scientific research. Such Science Based Approach in Engineering Education is been supported by Russian industrial enterprises working in the field of automation. Authors paid a lot of attention to the theoretical bases of fast algorithms of signal's transformation because they guess that other teachers could create their own variants of tasks for the individual student's work.

To sum up the student's feedbacks, it might be said that they understood material on a good level. Authors plan to work on complicated theoretical aspects, as well as to work on mathematical background to make students better understand special calculations. Beside the questions the students also noted that it would be desirable to use the toolkit of unit tests in Python and the version control system to automate testing procedures. This note will be taken into account during course improvement.

ACKNOWLEDGMENT

The authors would like to thank students at Computer Systems and Network Department of the Bauman State Technical University who willing their insights while they carried on research homework.

The part of this chapter concerned to new bases for *fast algorithms of spectral transformation* is done under support of Ministry of Education and Science of the Russian Federation (Project Contract ID 2.7782.2017/BC).

REFERENCES

Ahmed, N., & Rao, K. (1975). *Orthogonal Transforms for Digital Signal Processing*. Berlin: Springer. doi:10.1007/978-3-642-45450-9

Bracewell, R. (1990). *The Hartley Transform*. Oxford, UK: Oxford University Press.

Bressert, E. (2012). *SciPy and NumPy: An Overview for Developers*. Sebastopol, CA: O'Reilly Media.

Devert, A. (2014). *A matplotlib Plotting Cookbook*. Birmingham, UK: Packt Publishing.

Harmuth, H. (1977). *Sequence Theory: foundations and applications*. New York: Academic Press.

Kotel'nikov, V. (1959). *The Theory of Optimum Noise Immunity*. New York: McGraw-Hill Book Co.

Oppenhein, A., & Schafer, R. (2012). *Digital Signal Processing*. Englewood Cliffs, NJ: Prentice-Hall.

Smirnova, E., Suzev, V., Proletarsky, A., & Gurenko, V. (2017) Signal's Simulation Methodologies used in Scientific and Educational tasks of real-time Information System's Modelling. In *Procceedings of 11th International Technology, Education and Development Conference INTED2017* (vol. 11, pp. 9140 – 9143). Valencia, Spain: IATED Academy.

Syuzev, V. V. (2015). Fast generalized Hartley transform in single-base number systems. [in Russian]. *Bauman University Journal of Machinary Devices*, *6*, 63–81.

Syuzev, V.V. (2014). *Introduction to digital signal processing*. Moscow: RTSoft. (in Russian)

KEY TERMS AND DEFINITIONS

Fast Transformation (FT) Algorithms: A computational procedure using fast spectral transformations in order to save computational resources, for example, fast Fourier transform (FFT).

Independent Study: Is a form of education offered by many high schools, colleges, and other educational institutions. It is sometimes referred to as directed study, and is an educational activity undertaken by an individual with little to no supervision.

Science-Based Approach in Engineer Preparation: New trend in future engineer preparation when students do their homework as a practical research on of theoretical problem given by teacher.

Spectral Method for Engineering and Scientific Computing Applications: Actual modern course in engineer preparation curricula.

APPENDIX

SIGNAL SIMULATION THEORY

Algorithms for Simulation in Complex Exponential Bases

While students solve a signal simulation task in the framework of correlation theory they need to use a spectral field of the complex exponential bases. The power spectral density function (PSDF) has been presented by (Ahmed N., Rao K., 1975; Bracewell R., 1990; Oppenhein A. and Schafer R., 2012) and others. The equation for the spectral field of the complex exponential bases see at (1):

$$S(\omega) = \lim_{T\to\infty} \frac{X(\omega)X^*(\omega)}{T} = \lim_{T\to\infty} \frac{|X(\omega)|^2}{T}, \quad (1)$$

where ω is a cyclic frequency;

$X(\omega)$ is an integral Fourier transform of the signal $x(t)$. An integral Fourier transform of the signal could be represented by equation (2):

$$X(\omega) = \int_{-\infty}^{+\infty} x(t)\exp(-j\omega t)\,dt. \quad (2)$$

An integral Fourier transform of the signal $x(t)$, is determined at the double-side infinite intervals $(j = \sqrt{-1})$. The spectral function $X^*(\omega)$ in the expression (1) is the complex conjugate spectrum $X(\omega)$ and it could be described by expression:

$$|X(\omega)| = \sqrt{\mathrm{Re}[X(\omega)]^2 + \mathrm{Im}[X(\omega)]^2}.$$

The equation (3) describes the amplitude spectrum, where the notations *Re* and и *Im* mean here the real and imaginary parts of the complex quantity.

If the signal $x(t)$ is an integrable function with a square function, it has final average power and it is determined at the final symmetrical time interval with duration *T*, then infinite interval in spectrum (2) becomes finally equal to 1 (one), and the passage to the limit in the expression (1) could be removed. Thus in this case

$$S(\omega) = S_T(\omega) = \frac{1}{T}|X_T(\omega)|^2 = \frac{1}{T}\{\mathrm{Re}[X_T(\omega)]^2 + \mathrm{Im}[X_T(\omega)]^2\},$$

where

$$X_T(\omega) = \int_{-T/2}^{T/2} x(t)\exp(-j\omega t)\,dt.$$

As such the teacher needs to consider a power spectral density function and an integral Fourier's spectrum in selected points of the frequency axis, which is taken with the constant sampling interval $\Delta\omega = 2\pi / T$. The functions in these points will be accordingly:

$$S_T(k\Delta\omega) = \frac{1}{T}\left|X_T(k\Delta\omega)\right|^2, k = 0,1,2,\ldots; \quad (3)$$

$$X_T(k\Delta\omega) = \int_{-T/2}^{T/2} x(t)\exp\left(-j\frac{2\pi}{T}kt\right)dt, k = 0,1,2,\ldots \quad (4)$$

Teacher then should ask students to compare the Fourier spectrum (4) with the spectral coefficients of the direct Fourier transformation of the signal $x(t)$ limited in time in the complex exponential Fourier functions $\left\{\exp\left(j\frac{2\pi}{T}kt\right)\right\}$ which have been described in books (Ahmed N., Rao K., 1975; Bracewell R., 1990; Oppenhein A. and Schafer R., 2012).

The expression to obtain k's Fourier transform spectral coefficient is shown here:

$$X_F(k) = \frac{1}{T}\int_{-T/2}^{T/2} x(t)\exp\left(-j\frac{2\pi}{T}kt\right)dt,$$

Teacher should draw the student's attention to the following expression (5):

$$X_T(k\Delta\omega) = TX_F(k). \quad (5)$$

Then from the expressions (5) and (3) the following dependence (6) for the the specified function of power spectral density can be obtained:

$$S_T(k\Delta\omega) = T\left|X_F(k)\right|^2 = T\left\{\mathrm{Re}\left[X_F(k)\right]^2 + \mathrm{Im}\left[X_F(k)\right]^2\right\}. \quad (6)$$

The expression (6) could be used for the definition of spectral coefficients $X_F(k)$ using the specified function of power spectral density $S_T(k\Delta\omega)$:

$$X_F(k) = \mathrm{Re}\left[X_F(k)\right] - j\,\mathrm{Im}\left[X_F(k)\right]. \quad (7)$$

It should be taken into account that an expression (6) alone does not allow to find two unknown components of the spectrum (7). It is due to the fact that power spectral density could consider amplitude component of the complex spectrum only $X_T(\omega)$, and does not consider his phase component as other researchers described already (Bracewell R., 1990; Oppenhein A. and Schafer R., 2012). Therefore function $S_T(\omega)$ defines not only one proper signal but some cortex of signals with different phase components. If the phase density is as the following expression

$$\psi(\omega) = arctg\left\{\mathrm{Im}[X_T(\omega)] / \mathrm{Re}[X_T(\omega)]\right\},$$

and if the phase density has already determined, then by doing quantization with the same frequency interval $\Delta\omega$, the expression (8) will be obtained:

$$tg\psi(k\Delta\omega) = \mathrm{Im}[X_T(k\Delta\omega)] / \mathrm{Re}[X_T(k\Delta\omega)] = \mathrm{Im}\left[X_F(k)\right] / \mathrm{Re}\left[X_F(k)\right] = \lambda_k, k = 0,1,2,\ldots \quad (8)$$

The dependence (8) for the phase coefficients λ_k gives the second expression for Fourier spectrum definition (7).

The dependence (8) along with the expression (6) determines a tuning procedure of the signal simulation algorithm in the complex exponential Fourier bases. The algorithm itself could be shown by the inverse Fourier transform in the same bases, as such it has the following view:

$$x(t) = \sum_{k=0}^{\infty} X_F(k) \exp\left(j\frac{2\pi}{T}kt\right), -\frac{T}{2} \le t < \frac{T}{2}. \quad (9)$$

If the phase component does not set, then phase dependence (8) for simulation needs to be provided. The simplest case is when all coefficients λ_k are equal to 1.

The tuning algorithm for a given amplitude and phase spectrum (6) and (8) could be written more conveniently for the practical usage. To do this, students should know that the real component of the complex Fourier functions is an even function $\left(\cos\frac{2\pi}{T}kt\right)$, but the imaginary component is an odd function $\left(\sin\frac{2\pi}{T}kt\right)$. Therefore the Fourier spectrum $X_F(k)$ could be presented by even $X_{Fe}(k)$ and odd $X_{Fo}(k)$ components as follows:

$$X_F(k) = X_{Fe}(k) - jX_{Fo}(k), \quad (10)$$

where

$$X_{Fe}\left(k\right) = \mathrm{Re}\left[X_{F}\left(k\right)\right], X_{Fo}\left(k\right) = \mathrm{Im}\left[X_{F}\left(k\right)\right]. \tag{11}$$

Thus, the expressions (6) and (8) in the tuning procedure can be rewritten as:

$$\left.\begin{array}{l} S_{T}\left(0\right) = TX_{Fe}^{2}\left(0\right); S_{T}\left(k\Delta\omega\right) = T\left[X_{Fe}^{2}\left(k\right) + X_{Fo}^{2}\left(k\right)\right], \\ X_{Fo}\left(k\right) = \lambda_{k} X_{Fe}\left(k\right). \end{array}\right\} \tag{12}$$

The solution of the expression system (12) lead to more compact expressions (13) for $X_{Fe}\left(k\right)$ and $X_{Fo}\left(k\right)$:

$$\left.\begin{array}{l} X_{Fe}\left(0\right) = \sqrt{\dfrac{S_{T}\left(0\right)}{T}}, X_{Fe}\left(k\right) = \sqrt{\dfrac{S_{T}\left(k\Delta\omega\right)}{T\left(1+\lambda_{k}^{2}\right)}}, \\ X_{Fo}\left(k\right) = \lambda_{k}\sqrt{\dfrac{S_{T}\left(k\Delta\omega\right)}{T\left(1+\lambda_{k}^{2}\right)}}, k = 1, 2, \ldots. \end{array}\right\} \tag{13}$$

Simulation algorithm (9) is a continuous Fourier series in the complex exponential bases. The series with infinite number of terms is reduced to the signal $x\left(t\right)$. The existence of the negative time values in the series does not have a principal importance for simulation. In practice the realization of the algorithm (9) of Fourier series is limited by *N* members, and the signal is calculated on the basis of *M* discrete time points $i \in \left[-M/2, M/2\right)$ in total, where $i = t / ” t$, a $M = T / ” t$. The magnitude of the sampling interval at time $” t$ is selected under consideration of desired fidelity of the signal $x\left(t\right)$, and the number of set terms *N* – spectral-correlation properties of the signal.

The algorithm (9) is interesting, because his Fourier series, truncated to *N* terms, allows to playback of signal $x\left(t\right)$ with the power spectral density function matching with a given function $S_{T}\left(\omega\right)$ exactly at points $k\Delta\omega$. In the interim frequency values theoretical and practical densities will be different from each other. With the increase of number *N* of matched points, the error of the representation $S_{T}\left(\omega\right)$ in the interim will decrease accordingly. In general it is not necessary for the series (9) that values *N* and *M* match.

The discussed simulation algorithm could be used also for the discrete signals representation $x\left(i\right)$, defined on the interval $\left[-N/2, N/2\right)$. In this case the simulation algorithm uses a discrete Fourier series in the lattice complex exponential basis $\left\{\exp\left(j\dfrac{2\pi ki}{N}\right)\right\}$, the expression (14) for the simulated signal is shown:

$$x(i) = \sum_{k=0}^{N-1} X_F(k) \exp\left(j\frac{2\pi ki}{N}\right), -\frac{N}{2} \le i < \frac{N}{2}, \tag{14}$$

where the spectrum $X_F(k)$ consists of even $X_{Fe}(k)$ and odd $X_{Fo}(k)$ components (see (10)), which are calculated on the ratio of (13).

In discrete series (14) the values *N* and *M* match, and the sampling interval in time " t must be confirmed by Nyquist theorem. In Russia there is the same theorem and Russian author (Kotel'nikov V., 1959). This theorem proofs the upper frequency (the cutoff frequency) ω_2 of the frequency spectrum of the continuous signal. The value *N* is selected from the conditions required fidelity of signal representation $S_T(\omega)$ between points $S_T(k\Delta\omega)$.

In this described signal simulation algorithm the nature of the transformation of the spectral density function is simple: it is sampled with a frequency of $\Delta\omega$ in the interval of its definition. For its detection the autocorrelation function of a time limited signal $x(t)$ is as:

$$R_T(\tau) = \frac{1}{T}\int_{-T/2}^{T/2} x(t)x(t+\tau)dt$$

Substitute it by the expression of the signal through the double-sided Fourier spectrum in the complex exponential basis.

Autocorrelation function is shown here

$$R_T(\tau) = \frac{1}{T}\int_{-T/2}^{T/2}\left[\sum_{k=-\infty}^{\infty} X_F(k)\exp(jk\Delta\omega t)\right]x(t+\tau)dt$$

By multiplying the integrand in this expression to additional expression

$$\exp(jk\Delta\omega\tau)\cdot\exp(-jk\Delta\omega\tau) = 1$$

obtain:

$$R_T(\tau) = \sum_{k=-\infty}^{\infty} X_F(k)\left[\frac{1}{T}\int_{-T/2}^{T/2} x(t+\tau)\exp\left(jk\Delta\omega(t+\tau)\right)dt\right]\exp(-jk\Delta\omega\tau).$$

Bracketed part of this expression defines complex-conjugate spectrum $X_F^*(k)$ of the original signal. Thus

$$R_T(\tau) = \sum_{k=-\infty}^{\infty} X_F(k) X_F^*(k) \exp(-jk\Delta\omega\tau) = \sum_{k=-\infty}^{\infty} \left|X_F(k)\right|^2 \exp(-jk\Delta\omega\tau).$$

Taking into account the parity summable functions on the index k, this series could be simplified and the autocorrelation function will have a the following view:

$$R_T(\tau) = \left|X_F(0)\right|^2 + 2\sum_{k=1}^{\infty} \left|X_F(k)\right|^2 \cos(k\Delta\omega\tau) \tag{15}$$

The relationship between the communication spectra $X_F(k)$ and $X_T(k)$ (expression (5)) as well as with the function of power spectral density (expression (3)) allows to represent $R_T(\tau)$ as it shown at the expression (16):-

$$R_T(\tau) = \frac{\left|X_T^2(0)\right|}{T^2} + \frac{2}{T}\sum_{k=1}^{\infty} \frac{\left|X_T(k\Delta\omega)\right|^2}{T} \cos(k\Delta\omega\tau) = \frac{S_T(0)}{T} + \frac{2}{T}\sum_{k=1}^{\infty} S_T(k\Delta\omega)\cos(k\Delta\omega\tau). \tag{16}$$

The expressions (15) and (16) could be used for practical study of the autocorrelation function of the signal $x(t)$.

Algorithms for Signal Simulation in Specific Bases

There exist simulation algorithms in specific bases, namely: for complex exponential bases; for trigonometric bases, for Fourier and for Hartley bases. All four type of algorithms could be used by teachers to create variants for student's independent work. The list of Russian chapters describing these simulation algorithms in detail can be found in a chapter, which authors published in the Russian journal in Russian language (Syuzev V.V., Gurenko V.V., Smirnova E.V., 2016).

All above mentioned simulation spectral algorithms use the same analytical relation between simulation series spectrum and given spectral-correlation characteristics of the signal. The same is true for all bases systems which use the same harmonic even and odd functions. In this sense, the algorithms of the simulation are related, reproduce the same signal and differ only in their *computational complexity*.

However note that in the complex Fourier basis and the real Harley basis, there are so called *fast algorithms of spectral transformation* for students learning (Ahmed N., Rao K., 1975; Harmuth H., 1977; SyuzevV.V., 2014), and for the trigonometric bases there are special recursive algorithms for computing the trigonometric functions (SyuzevV.V., 2014). Their application allows to create fast algorithms' modifications in harmonic bases.

Simulation algorithms in harmonic bases are core simulation algorithms. Imitation Algorithms in Harmonic bases can be used also as base imitation algorithms in an arbitrary orthogonal bases.

Algorithms for Signal Simulation in an Arbitrary Orthogonal Bases

If the simulation signal spectrum is defined in harmonic bases, then by using mathematical apparatus of generalized spectrum analysis (SyuzevV.V., 2014) it is possible to determine the spectra of the same signal in any other orthogonal basis. As such the signal itself could be restored by using an inverse Fourier transform in this basis. The teacher chooses a proper basis while he/she creates tasks for independent homework, otherwise student could choose himself the proper basis considering optionally conditions of computational complexity and accuracy of needed simulation algorithms.

In the case if the chosen basis will have a complex character, it is advisable to take the simulation algorithm in complex exponential Fourier bases. If the real bases was chosen, then the simulation algorithm should be created using a real simulation algorithms in trigonometric bases or in Hartley bases.

Let's obtain the tung algorithms for both basis choose cases. Let's start from the complex base $\{\varphi(m,t)\}$ with the power P_m, defined at the interval $[-T/2, T/2)$. The signal spectrum (9) in these bases is:

$$X_\varphi(m) = \frac{1}{TP_m}\int_{-T/2}^{T/2} x(t)\varphi^*(m,t)\,dt = \sum_{k=0}^{\infty} X_F(k)\left[\frac{1}{TP_m}\int_{-T/2}^{T/2} \exp\left(j\frac{2\pi}{T}kt\right)\varphi^*(m,t)\,dt\right].$$

The obtained expression describes an operator for Fourier spectra conversion into the в spectra of arbitrary basis $\{\varphi(m,t)\}$ for the same signal $x(t)$. An expression in square brackets gives a Fourier core $\mathrm{F}(m,k)$ for spectra conversion:

$$\mathrm{F}(m,k) = \frac{1}{TP_m}\int_{-T/2}^{T/2} \exp\left(j\frac{2\pi}{T}kt\right)\varphi^*(m,t)\,dt. \tag{17}$$

Mathematically it is in a range of basic exponential functions $\exp\left(j\frac{2\pi}{T}kt\right)$ in basic function $\{\varphi(m,t)\}$. The function $\varphi^*(m,t)$ here is a complex conjugate function $\varphi(m,t)$.

Taking into account (17) the desired spectrum $X_\varphi(m)$ will be equal:

$$X_\varphi(m) = \sum_{k=0}^{\infty} X_F(k)F(m,k), m = 0,1,2,\ldots. \tag{18}$$

The spectrum $X_F(k)$. here will be determined by expression (13) as well. For normalized basis $\{\varphi(m,t)\}$ all $P_m = 1$, and the formula for core calculation will be simplified.

The signal simulation algorithm in basis $\{\varphi(m,t)\}$ will take a view of the following Fourier series:

$$x(t)=\sum_{m=0}^{\infty}X_{\varphi}(m)\varphi(m,t), t\in[-T/2,T/2). \tag{19}$$

s tuning to set the function of the power spectral density is performed according to formulas (13), (17) and (18). The tuning procedure becomes complicated in general case, but however it has no principal importance, since the procedure could be executed before simulation and with use of computer.

During a practical realization the series (19) is truncated just as it was done in the harmonic bases. The spectrum transformation core series (18) is also truncated.

In discrete version the simulation and tuning algorithms have the following view:

$$x(i)=\sum_{m=0}^{N-1}X_{\varphi}(m)\varphi(m,i), i\in[-N/2,N/2),$$

$$X_{\varphi}(m)=\sum_{k=0}^{N-1}X_{F}(k)F(m,k), m=0,1,2,\ldots,N-1,$$

$$F(m,k)=\frac{1}{NP_{k}}\sum_{i=-N/2}^{N/2-1}\exp\left(j\frac{2\pi}{N}ki\right)\varphi^{*}(m,i).$$

Note, that the basis $\{\varphi(m,t)\}$ is a system of valid basis functions. To create the simulation algorithm, it is advisable to use the algorithms in trigonometric bases or in Hartley bases. In the case of Hartley bases the spectrum $X_{\varphi}(m)$ is determined by the signal $x(t)$ and will be equal:

$$X_{\varphi}(m)=\frac{1}{TP_{m}}\int_{-T/2}^{T/2}x(t)\varphi(m,t)\,dt=$$

From this expression it follows that:

$$X_{\varphi}(m)=\sum_{k=0}^{\infty}\left[X_{Fe}(k)+X_{Fo}(k)\right]F(m,k), m=0,1,2,\ldots, \tag{20}$$

$$F(m,k)=\frac{1}{TP_{m}}\int_{-T/2}^{T/2}cas\left(\frac{2\pi}{T}kt\right)\varphi(m,t)\,dt, \tag{21}$$

and

$$x(t)=\sum_{m=0}^{\infty}X_{\varphi}(m)\varphi(m,t),t\in[-T/2,T/2). \tag{22}$$

The last formula (22) for the Fourier series matches with expression (19). Components $X_{Fe}(k)$ and $X_{Fo}(k)$ of the Fourier spectrum in expression (20) could be obtained from expression (13) and will have a view:

$$x(t)=\sqrt{\frac{S_T(0)}{T}}+\sum_{k=1}^{N-1}\sqrt{\frac{S_T(k\Delta\omega)(1+\lambda_k)^2}{T(1+\lambda_k^2)}}cas\left(\frac{2\pi}{T}kt\right),$$

where $t=i\Delta t, i=-M/2,-M/2+1,\ldots,M/2-1$ and values *N*, *M* and Δt are chosen, according to the same requirements as for the bases described above.

In discrete version

$$x(i)=\sum_{m=0}^{N-1}X_{\varphi}(m)\varphi(m,i),i\in[-N/2,N/2);$$

$$X_{\varphi}(m)=\sum_{k=0}^{N-1}\sqrt{\frac{S_T(k\Delta\omega)(1+\lambda_k)^2}{T(1+\lambda_k^2)}}F(m,k),$$

$$F(m,k)=\frac{1}{NP_k}\sum_{i=-N/2}^{N/2-1}cas\left(\frac{2\pi}{N}ki\right)\varphi(m,i).$$

Here $\Delta\omega=2\pi/(N\Delta t)$, and $\lambda_0=0$.

Compilation of References

«Svetotekhnika» vnedryayet printsipy berezhlivogo proizvodstva. (2014, July 7). ["Svetotechnika" is implementing the principles of careful manufacturing]. Retrieved from http://www.up-pro.ru/library/production_management/lean/svetotehnika-lean.html

Abuatiq, A., & Davis, C. (2014). *Integration of Virtual Clinical Excursions in Medical Surgical Nursing Course* [Course]. San Bernardino, CA: California State University. Retrieved June 2, 2017, from https://contentbuilder.merlot.org/toolkit/html/snapshot.php?id=26184150973506

Adams, M., & Brown, S. (2011). Reigniting the Voice of Disabled People in Higher Education. In D. Moore, A. Gorra, M. Adams, J. Reaney, & H. Smith (Eds.), *Disabled Students in Education: Technology, Transition, and Inclusivity* (pp. 305–315). IGI Global. doi:10.4018/978-1-61350-183-2.ch016

Agamba, J. (2015). Optimizing Blended Teaching and Learning in Brick-and-Mortar Institutions. In J. Keengwe & J. Agamba (Eds.), *Models for Improving and Optimizing Online and Blended Learning in Higher Education* (pp. 1–11). Hershey, PA: IGI Global. doi:10.4018/978-1-4666-6280-3.ch001

Ageeva, T. I., Baldin, A. V., & Baryshnikov, V. A. (2009). Informacionnaja upravljajushhaja sistema MGTU im. N. Je. Baumana "Jelektronnyj universitet": koncepcija i realizacija. In I. B. Fedorov (Ed.), *Information management system of the Bauman MSTU "Electronic University": the concept and the realization of* (p. 376). Moscow: BMSTU.

Aggarwal, C. C., & Subbian, K. (2012, April). Event detection in social streams. In *Proceedings of the 2012 SIAM International Conference on Data Mining* (pp. 624-635). Philadelphia, PA: Society for Industrial and Applied Mathematics. 10.1137/1.9781611972825.54

Aggarwal, C. C., & Yu, P. S. (2010). On clustering massive text and categorical data streams. *Knowledge and Information Systems*, *24*(2), 171–196. doi:10.100710115-009-0241-z

Aghvami, S. A., & Kolomeitseva, M. B. (2011). Sintez adaptivnogo neuro-regulatora dlya upravleniya nelineinogo MIMO ob'ekta [Adaptive neural network controller synthesis for non-linear MIMO object]. *Vestnik MEI*, *6*, 209–215.

Ahluwalia, G. (2010). Language Learning with Internet-based Projects: a Student-centered Approach for Engineering Students. *ESP World*, *9*(27). Retrieved June 5, 2016, from http://www.esp-world.info/Articles_27/lang%20learn.pdf

Ahmed, A., Ho, Q., Eisenstein, J., Xing, E., Smola, A. J., & Teo, C. H. (2011). Unified analysis of streaming news. In *Proceedings of the 20th International Conference Companion on World Wide Web* (pp. 267-276). New York, NY: ACM. DOI: 10.1145/1963405.1963445

Ahmed, N., & Rao, K. (1975). *Orthogonal Transforms for Digital Signal Processing*. Berlin: Springer. doi:10.1007/978-3-642-45450-9

AIRR. (2017). *Rating of innovation processes in 2016.* Retrieved 06/07/2017 from http://i-regions.org/images/files/presentations/AIRR_26.12.pdf

Alam, F., Hadgraft, R. G., & Alam, Q. (2014). eLearning: Challenges and opportunities. In F. Alam (Ed.), Using technology tools to innovate assessment, reporting, and teaching practices in engineering education (pp.217-226). Hershey, PA: IGI Global.

Alcock, S., & Nelson, R. (2012). *Libprotoident: Traffic classification using lightweight packet inspection.* Technical report, University of Waikato. Retrieved from https://wand.net.nz/sites/default/files/lpi.pdf

Aleksandrov, A. A., Fang, K., Proletarsky, A. V., & Neusypin, K. A. (2012). Conception of complex continuous education with innovative information technologies. In *Proceedings of the 2ndInternational Conference on Education and Education Management. S. "Advances in Education Research"*(pp. 374-378). Academic Press.

Aleksandrov, A. A., Proletarskij, A. V., Neusypin, K. A., & Sharkov, A. A. (2012). Koncepcija kompleksnogo nepreryvnogo obuchenija s ispol'zovaniem innovacionnyh informacionnyh tehnologij [The concept of integrated continuous learning using innovative information technologies]. *Nauchnoe obozrenie: gumanitarnye issledovanija - Scientific Review: Humanitarian researches, 4*, 38-42.

Aleksandrov, A. A., Proletarsky, A. V., Ke, F., & Neusypin, K. A. (2012). Conception of complex continuous education with innovation technologies. In *Proceedings of the 2012 2nd international Conference on Education and Education Management.* Hong Kong: Information Engineering Research Institute.

Alimov, A., & Moffat, D. C. (2015). Adaptive Model of Multiobjective Agent Behavior in Real-Time Systems. In A. Bikakis & X. Zheng (Eds.), Lecture Notes in Computer Science: Vol. 9426. *Multi-disciplinary Trends in Artificial Intelligence* (pp. 51–60). Cham: Springer; doi:10.1007/978-3-319-26181-2_5

Allen, J. F., & Ferguson, G. (1994). Actions and events in interval temporal logic. *Journal of Logic and Computation, 4*(5), 531–579. doi:10.1093/logcom/4.5.531

Allen, J., & Van der Velden, R. (Eds.). (2011). *The flexible professional in the knowledge society.* Dordrecht, The Netherlands: Springer; doi:10.1007/978-94-007-1353-6

Allied Market Research. (2015). *Enterprise application market.* Allied Market Research. Retrieved July 18, 2017 from https://www.alliedmarketresearch.com/enterprise-application-market

Altbach, P. G., & Salmi, J. (Eds.). (2011). *The road to academic excellence: the making of world-class research universities.* Washington, DC: The International Bank for Reconstruction and Development / The World Bank. doi:10.1596/978-0-8213-8805-1

Alvesson, M., & Sandberg, J. (2011). Generating research questions through problematization. *Academy of Management Review, 36*(2), 247–271.

Alvesson, M., & Sandberg, J. (2013). *Constructing Research Questions: Doing Interesting Research.* Los Angeles, CA: SAGE Publications. doi:10.4135/9781446270035

Amazon. (2017). *Amazon Web Services.* Retrieved July 18, 2017 from https://aws.amazon.com/ products/?nc1=h_ls

Ancona, D., Malone, T. W., Orlikowski, W. J., & Senge, P. M. (2007, February). In praise of the incomplete leader. *Harvard Business Review.* Retrieved from hbr.org/2007/02/in-praise-of-the-incomplete-leader PMID:17345683

Andrews, J., Clark, R., & Higson, H. (2012). *Enhancing Employability: Making the Most of University.* Report for National HE STEM Programme.

Andrews, J., & Clark, R. (2015). *BME SDG Final Report.* York, UK: Higher Education Academy.

Anishchenko, V. S., Buldakova, T. I., Dovgalevskij, P. Y., Lifshic, V. B., Gridnev, V. I., & Suyatinov, S. I. (2009). Konceptual'naya model' virtual'nogo centra ohrany zdorov'ya naseleniya [Conceptual model of the virtual center of public health protection]. *Informacionnye tekhnologii - Information Technology, 15*(12), 59-64.

Anisimov, A. M. (2009). *Rabota v sisteme distancionnogo obuchenija Moodle.* Kharkov: HNAGH.

Antsupova, G. N. (2005). *MGTU – glazami istorika* [BMSTU – eyes of the historian]. Bauman Moscow State Technical University Publishers.

Apache. (2017a). *Spark overview.* Retrieved July 18, 2017 from http://spark.apache.org/docs/ latest/

Apache. (2017b). *Apache Flink documentation.* Retrieved July 18, 2017 from https://ci.apache. org/projects/flink/flink-docs-release-1.3/

Apache. (2017c). *Apache Apex.* Retrieved July 18, 2017 from http://apex.apache.org/docs/apex/

Appari, A., & Johnson, M. E. (2010). Information security and privacy in healthcare: Current state of research. *International Journal of Internet and Enterprise Management, 6*(4), 279–314. doi:10.1504/IJIEM.2010.035624

Arakcheeva, O.E. (2012). Propedevticheskaya podgotovka uchitelya k formirovaniyu sotsial'noy kompetentnosti uchashchikhsya [Propaedutic teacher's preparation for the formation of students' social competence]. *Obrazovanie i Samorazvitie, 3*, 43-46.

Archan, S. (2005). Modularisierung: ein Weg zur Steigerung der Attraktivität der Lehre in Österreich [Modularisation: A Pathway to increase the attraction of apprenticeship training in Austria]. *BWP Berufsbildung in Wissenschaft und Praxis, 4*, 23–25.

Arnove, R. F., Torres, C. A., & Franz, S. (2012). *Comparative education: The dialectic of the global and the local.* Lanham, MD: Rowman & Littlefield Publishers. Retrieved from https://rowman.com/Page/RLContact

Arutyunova, N. D. (1999). *Yazyk i mir cheloveka* [Human Language and World]. Moscow, Russia: Yazyki russkoi kul'tury.

ASEC. (2017). *Aston STEM Education Centre.* Retrieved 11/8/17 from http://www.aston.ac.uk/eas/research/groups/asec/

ASI. (2016, January 27). *National Technology Initiative. EnergyNet Market Development Road Map.* Prepared for discussion at the Government Expert Council meeting. Moscow, Russia: Agency for Strategic Initiatives, RVC.

ASI. (2017). *Natsionalnaya tehnologicheskaya initsiativa* [National Technology Initiative]. Retrieved from: http://asi.ru/nti/

ASI. (2017). *The national rating of the investment climate in Russian regions.* Retrieved 06/07/2017 from https://asi. ru/investclimate/rating/

Attfield, I., & Vu, B. T. (2013). A rising tide of primary school standards: The role of data systems in improving equitable access for all to quality education in Vietnam. *International Journal of Educational Development, 33*(1), 74–87. doi:10.1016/j.ijedudev.2012.02.003

Avachev, A. P., Vishnyakov, N. V., Gololobov, G. P., & Mitrofanov, K. V. (2012). *Methodical manuals to laboratory work in the field of "Nanomaterials".* Ryazan: Ryazan State Radio Engineering University.

Avdeeva, Z. K., Kovriga, S. V., Makarenko, D. I., & Maksimov, V. I. (2007). Cognitive approach in control. Control, 2007(3), 2-8.

Avraamova, E. M., Karavay, A. V., Klyachko, T. L., & Loginov, D. M. (2016). *Monitoring dopolnitel'nogo professional'nogo obrazovaniya v Rossii* [Monitoring of additional professional education in Russia. Moscow: Publishing]. Moscow: Delo, Russian Presidential Academy of National Economy and Public Administration.

Avraamova, E. M., Klyachko, T. L., & Loginov, D. M. (2015). *Monitoring nepreryvnogo profes-sionalnogo obrazovaniya: pozitsii rabotodateley i rabotnikov* [Monitoring of Professional Education: Positions of Employers and Employees]. Moscow: Delo, Russian Presidential Academy of National Economy and Public Administration.

Avtomaticheskaja sistema upravlenija Universitet. (n.d.). [Automatic control system university]. Retrieved from http://www.akvadra.ru/ catalog/7.html

Awad, M., & Khanna, R. (2015). *Efficient learning machines: Theories, concepts, and applications for engineers and system designers*. New York, NY: Apress Media, LLC. doi:10.1007/978-1-4302-5990-9

Azoev, G. L., Afanasiev, V. Y., & Balyakin, A. A. (2012). *Innovatsionnyie klasteryi nanoindustrii* [Innovative clusters for nanoindustry]. Russia, Moscow: BINOM. Laboratoriya znaniy.

Baillie, C. (2002). Enhancing creativity in engineering students. *Engineering Science and Education Journal, 11*(5), 185–192. doi:10.1049/esej:20020503

Balakrishnan, B., & Woods, P. (2009). Virtual laboratories in engineering education: The simulation lab and remote lab. *Computer Applications in Engineering Education, 17*(1), 108–118. doi:10.1002/cae.20186

Ball, Ph., Kelly, K., & Clegg, Jh. (2015). *Putting CLIL into practice*. Oxford, UK: Oxford University Press.

Banerjee, A., Gupta, S. S., & Venkatasubramanian, K. K. (2013). PEES: Physiology-based end-to-end security for mHealth. *Proceedings of the 4th Conference on Wireless Health*, 2.

Banik, G. (2016). *Strategies and techniques for new tenure-track faculty to become successful in Academia*. Oklahoma State University. doi:10.18260/p.25886

Barnes, E. (2016). *The Minority Body: A Theory of Disability (Studies in Feminist Philosophy Series)*. Oxford, UK: OUP. doi:10.1093/acprof:oso/9780198732587.001.0001

Barrett, T. (2010). The problem-based learning process as finding and being in flow. *Innovations in Education and Teaching International, 47*(2), 165–174. doi:10.1080/14703291003718901

Barr, R. B., & Tagg, J. (1995). From Teaching to Learning: A New Paradigm for Undergraduate Education. *Change, 27*(6), 12–25. doi:10.1080/00091383.1995.10544672

Baryshev, G. K., Berestov, A. V., & Konashenkova, N. A. (2016). The Application of Online Team Project Training in Nuclear Engineering Education. In A. Chugunov, R. Bolgov, Y. Kabanov, G. Kampis, & M. Wimmer (Eds.), *Digital Transformation and Global Society. DTGS 2016. Communications in Computer and Information Science* (Vol. 674, pp. 370–379). Cham: Springer; doi:10.1007/978-3-319-49700-6_35

Bass, L., Clements, P., & Kazman, R. (2012). *Software Architecture in Practice*. New York, NY: Addison-Wesley Professional.

Basturkmen, H. (2006). *Ideas and options in English for Specific Purposes*. Mahwah, NJ: Lawrence Erlbaum Associates.

Basturkmen, H. (2010). *Developing courses in English for Specific Purposes*. Basingstoke, UK: Palgrave Macmillan. doi:10.1057/9780230290518

Basu, A., & Blanning, R. (2007). *Metagraphs and Their Applications*. New York: Springer.

Batyshev, S. Ya. (1987). Reforma professional'noy shkoly: opyt, poisk, puti realizatsii [Reforming professional schooling: Experience, research, and implementation]. Moscow: Vysshaya Shkola Publishers.

Bauman Moscow State Technical University. (2016). Retrieved from http:// www.bmstu.ru:8093/mstu/admissions/specialities/

Bauman Moscow State Technical University. (n.d.). In *Wikipedia*. Retrieved April 20, 2017, from https://en.wikipedia.org/wiki/Bauman_Moscow_State_Technical_University

Bayes-Brown, G. (2015). *Best practices of universities of technology commercialization at an early stage commercialization of technologies at an early stage. The study of global practices: Universities, corporations and government.* Retrieved 06/15/2017 from https://www.rusventure.ru/ru/programm/analytics/ docs/ESR_RVC_2015_RU.pdf

Beal , V. (2008). *What is open source software?* Retrieved March 28, 2017, from http://www.webopedia.com/DidYouKnow/Computer_Science/open_source.asp

Bedny, A., Erushkina, L., & Kuzenkov, O. (2014). Modernising educational programmes in ICT based on the Tuning methodology. *Tuning Journal for Higher Education, 1*(2), 387–404. doi:10.18543/tjhe-1(2)-2014pp387-404

Belous, V. V., Bobrovsky, A. V., Dobrjkov, A. A., Karpenko, A. P., & Smirnova, E. V. (2012). Multicriterion integral alternatives' estimation: mentally-structured approach to education. *2nd International Conference on Education and Education Management, 3*, 215-224.

Belous, V. V., Bobrovsky, A. V., Dobrjkov, A. A., Karpenko, A. P., & Smirnova, E. V. (2012). Multi-criterion integral alternatives' estimation: Mentally-structured approach to education. In *Proceedings of the 2nd International Conference on Education and Education Management. S. "Advances in Education Research"* (pp. 215-224). Academic Press.

Benjaminsson, S., & Djurfeldt, M. (2008). *Artificial neural networks, advanced course, DD2433 Lab 1: Accelerated Back Propagation and Regularisation via Pruning*. Retrieved June 18, 2016, from http://www.csc.kth.se/utbildning/kth/kurser/DD2433/annfk08/lab1-mlp.pdf

Berestov, A.V., Baryshev, G.K., Birjukov, A.P., & Rodko, I.I. (2016). Izmenenie trebovanij k kompetencijam v oblasti inzhiniringa v obrazovatelnyh programmah issledovatelskogo universiteta. *Science and Education: Scientific Publication,* (7), 115-122.

Bevinakoppa, S., Ray, B., & Sabrina, F. (2016). Effectiveness of problem-based learning implementation. *International Journal of Quality Assurance in Engineering and Technology Education, 5*(3), 46–58. doi:10.4018/IJQAETE.2016070104

Biggeri, M., & Santi, M. (2012). The missing dimensions of children's well-being and well-becoming in education systems: Capabilities and philosophy for children. *Journal of Human Development and Capabilities: A Multi-Disciplinary Journal for People-Centered Development, 13*(3), 373-395.

Biggs, J. (1999). What the Student Does: Teaching for enhanced learning. *Higher Education Research & Development, 18*(1), 57–75. doi:10.1080/0729436990180105

Biggs, J. B. (2011). *Teaching for quality learning at university: What the student does.* McGraw-Hill Education.

Biktimirov, R.L., Valiev, R.A., Galiullin, L.A., & Zubkov, E.V. (2014). Automated test system of diesel engines based on fuzzy neural network. *Research Journal of Applied Sciences., 9*, 1059–1063.

Blackboard Learning System (Release 6). (n.d.). Retrieved from http://blackboardsupport.calpoly.edu/ content/about/Print/Bb6LearnWP.pdf

Blanck, P. (2016). eQuality: The Struggle for Web Accessibility by Persons with Cognitive Disabilities. Cambridge University Press.

Bloomberg Market. (n.d.). Retrieved December 18, 2016, from http://www.bloomberg.com/markets

Bloshenko, T., Ponkratov, V., & Pozdnjaev, A. (2016). Development of methods for determining differentiated rates of mineral extraction tax in recovery of solid commercial components from technogenic and natural fields. *Indian Journal of Science and Technology*, *9*(27). doi:10.17485/ijst/2016/v9i27/97683

Boev, O. (2014). *BEng Project: Outputs and Further Development.* Retrieved from MyBookLibrary Web site: http://www.mybooklibrary.com/beng-project-outputs-and-further-development.html

Bogdanova, N. N., & Semenova, E. L. (2009). Uchebnik nemeckogo yazyka dlya texnicheskix universitetov i vuzov (s interaktivnymi uprazhneniyami i tekstami na kompakt-diske) [Textbook in German dedicated for technical universities and institutions (with interactive exercises and texts on CD)]. Moscow, Russia: Bauman MSTU Press.

Bogoslovskiy, V. A., Karavaeva, E. V., Kovtun, E. N., Melehova, O. P., Rodionova, S. E., Tarlyikov, V.A., & Shehonin, A.A. (2007). Guidelines for the design of assessment tools for the implementation of multi-level educational programs HPE with competent approach. In Higher education in Russia (vol. 10, pp. 3-10). Moscow: Moscow Polytechnic University Press.

Boldyrev, N. N. (2001). *Kognitivnaya semantika: Kurs lektsii po angliiskoi filologii* [Cognitive Semantics: Lectures on English Philology]. Tambov, Russia: Tambov State University.

Bologna Process - European Higher Education Area. (2017). Retrieved April 12, 2017 from http://www.ehea.info

Booker, S. J. (2013). Are scientists with disabilities the forgotten underrepresented minority? *ASBMB Today*, *12*(3), 34.

Borisov, S. R., & Korobets, B. N. (2015). Engineering education, business and management of intellectual property. *Vysshee obrazovanie v rossii, 4,* 91-97.

Bozhko, Y. V., Maksimkin, A. I., Baryshev, G. K., Voronin, A. I., & Kondratyeva, A. S. (2016). Digital Transformation as the Key to Synthesis of Educational and Innovation Process in the Research University. In A. Chugunov, R. Bolgov, Y. Kabanov, G. Kampis, & M. Wimmer (Eds.), *Digital Transformation and Global Society. DTGS 2016. Communications in Computer and Information Science* (Vol. 674, pp. 386–391). Cham: Springer; doi:10.1007/978-3-319-49700-6_37

Bracewell, R. (1990). *The Hartley Transform.* Oxford, UK: Oxford University Press.

Brants, T., Chen, F., & Farahat, A. (2003, July). A system for new event detection. In *Proceedings of the 26th Annual International ACM SIGIR Conference on Research and Development in Information Retrieval* (pp. 330-337). New York, NY: ACM. DOI: 10.1145/860435.860495

Bressert, E. (2012). *SciPy and NumPy: An Overview for Developers.* Sebastopol, CA: O'Reilly Media.

Brice, A. (2009). *Is desktop software dead.* Successful software. Retrieved March 31, 2017, from http://successfulsoftware.net/2013/10/28/is-desktop-software-dead

Briguglio, C. (2014). Linguistic and Cultural Skills for Communication in Global Workplaces of the 21st Century. In Information Resources Management Association (Ed.), Cross-Cultural Interaction: Concepts, Methodologies, Tools, and Applications (pp. 832-848). Hershey, PA: IGI Global. Doi:10.4018/978-1-4666-4979-8.ch047

Brockmann, M., Clarke, L., Mehaut, P., & Winch, C. (2008). Competence-based vocational education and training (VET): The cases of England and France in a European perspective. *Vocations and Learning*, *1*(3), 227–244. doi:10.100712186-008-9013-2

Brodeur, D. (2012). Aligning Accreditation Criteria and Processes with the CDIO Approach. In *Proceedings of the 8th International CDIO Conference*. Queensland University of Technology.

Brown, G. (1999). The Quality of Learning of Learning and How to Assess it. *Lifelong Learning in Europe.*, *IV*(99), 47–54.

Brownlee, J. (2014). *Gentle introduction to Scikit-Learn: A Python machine learning library*. Retrieved April 21, 2017, from http://machinelearningmastery.com/a-gentle-introduction-to-scikit-learn-a-python-machine learning-library/

Brown, R. G. (1973). Integrated navigation systems and Kalman filtering: A perspective. *Navigation*, *19*(4), 355–362. doi:10.1002/j.2161-4296.1972.tb01706.x

Büchner, A. (2016). *Moodle 3 Administration*. Birmingham: Packt Publishing.

Bujlow, T., Carela-Españolb, V., & Barlet-Ros, P. (2014). *Extended independent comparison of popular deep packet inspection (DPI) tools for traffic classification*. Retrieved from http://people.ac.upc.edu/pbarlet/reports/extended_dpi_report.pdf

Buldakova, T. I., & Suyatinov, S. I. (2002). Informacionno-analiticheskaya sistema upravleniya snabzheniem i proizvodstvom instrumenta [Information-analytical system of supply and production management]. *Informacionnye tekhnologii - Information Technology, 8*(11), 28-33.

Buldakova, T. I., Koblov, A. V., & Suyatinov, S. I. (2008). Informacionno-izmeritel'nyj kompleks sovmestnoj registracii i obrabotki biosignalov [Informational and measuring complex of joint registration and processing of biosignals]. *Pribory i sistemy. Upravlenie, kontrol', diagnostika - Devices and systems. Management, control, diagnostics, 6*, 41-46.

Buldakova, T. I., & Dzalolov, A. S. (2012). Analysis of data processes and choices of data-processing and security technologies in situation centers. *Scientific and Technical Information Processing*, *39*(2), 127–132. doi:10.3103/S0147688212020116

Buldakova, T. I., & Suyatinov, S. I. (2002). Registration and identification of pulse signal for medical diagnostic. *Proceedings of SPIE-The International Society for Optical Engineering*, *4707*, 343–350.

Buldakova, T., & Suyatinov, S. I. (2014). Reconstruction method for data protection in telemedicine systems. *Progress in Biomedical Optics and Imaging - Proceedings of the Society for Photo-Instrumentation Engineers*, *94481U*, 1–6.

Bursztein, E. (2011). *Analyzing web application performance, elie*. Retrieved March 31, 2017, from http://www.elie.net/blog/web/analyzing-web-application-performance

Campbell, J. D., & Narduzzi, J. L. (2015). The More Things Change: Reflections on the State of Marketing in Continuing Higher Education. In Centennial Conversations: Essential Essays in Professional, Continuing, and Online Education (pp. 309-17). Washington, DC: UPCEA.

Cantillon, P., Irish, B., & Sales, D. (2004). Using computers for assessment in medicine. *BMJ (Clinical Research Ed.)*, *329*(7466), 606–609. doi:10.1136/bmj.329.7466.606 PMID:15361445

Carvalho, H., Moral, P. D., & Monin, A. (1997). Optimal nonlinear filtering in GPS/INS integration. *IEEE Transactions on Aerospace and Electronic Systems*, *33*(3), 835–850. doi:10.1109/7.599254

Caschera, M. C., D'Ulizia, F., Ferri, F., & Grifoni, P. (2014). Multiculturality and Multimodal Languages. In Information Resources Management Association (Ed.), Cross-Cultural Interaction: Concepts, Methodologies, Tools, and Applications (pp. 1027-1042). Hershey, PA: IGI Global. Doi:10.4018/978-1-4666-4979-8.ch058

CDIO. (2017). Retrieved from http://cdio.org/about

Ceibal, P. (2006). *What is Plan Ceibal?* Retrieved 18.12.2017 from https://www.ceibal.edu.uy/en/institucional

Ceibal, P. (2014). *Plan Ceibal over the years – 2014*. Retrieved 18.12.2017 from https://www.ceibal.edu.uy/en/institucional

Ceibal, P. (2015). *Plan Ceibal over the years – 2015*. Retrieved 18.12.2017 from https://www.ceibal.edu.uy/en/institucional

Ceibal, P. (2016). *Plan Ceibal over the years – 2016*. Retrieved 18.12.2017 from https://www.ceibal.edu.uy/en/institucional

Champandard, A. (2007). *Popular Approaches to Behavior Tree Design*. Retrieved February 13, 2017, from http://aigamedev.com/open/article/popular-behavior-tree-design/

Chang, Y., Atkinson, D., & Hirleman, E. D. (2009). International research and engineering education: impacts and best practices. *Online J. Global Eng. Educ.*, *4*(2), Article 1.

Chan, P., Cooper, R., & Tzortzopoulos, P. (2005). Organizational learning: Conceptual challenges from a project perspective. *Construction Management and Economics*, *23*(7), 747–756. doi:10.1080/01446190500127021

Chapman, C., Emmerich, W., Márquez, F. G., Clayman, S., & Galis, A. (2012). Software architecture definition for on-demand cloud provisioning. *Cluster Computing*, *15*(2), 79-100. Retrieved July 18, 2017 from https://www.ee.ucl.ac.uk/~sclayman/docs/HDPC2010.pdf

Chappell, L. (2014). Troubleshooting with Wireshark: Locate the source of performance problems. Reno, NV: Protocol Analysis Institute (dba "Chappell University").

Chevalier, A., & Jia, X. (2016). Subject-Specific League Tables and Students' Application Decisions. *Manchester School*, *84*(5), 600–620. doi:10.1111/manc.12124

Chistyakova, T. B. (2016). Elektronnaya obrazovatel'naya sreda dlya kompetentnostno-oriyentirovannogo obucheniya spetsialistov inzhenernogo profilya [Electronic educational environment for competence-oriented training of engineering specialists]. *Science and Education of Bauman MSTU*, *7*, 230–241.

Chistyakova, T. B., & Novozhilova, I. V. (2016). Intelligence computer simulators for elearning of specialists of innovative industrial enterprises. In *Proceedings of the 19th International Conference on Soft Computing and Measurements, SCM 2016* (pp. 329-332). St. Petersburg, Russia: IEEE. 10.1109/SCM.2016.7519772

Chistyakova, T. B., Novozhilova, I. V., & Zelezinsky, A. L. (2016). Electronic information and education environment as instrument of forming and quality evaluation of professional competences of the international industrial enterprises specialists. In *Proceedings of the IEEE 5th Forum Strategic Partnership of Universities and Enterprises of Hi-Tech Branches, Science. Education. Innovations 2016* (pp. 12-14). St. Petersburg, Russia: IEEE. 10.1109/IVForum.2016.7835839

Chomsky, N. (2005). *Karteziankaya lingvistika* [Cartesian Linguistics]. Moscow, Russia: KomKniga.

Choshanov, M. A. (1996). *Flexible technology of problem-modular training*. Public Education.

Chubik, P. S., & Zamyatina, O. M. (2013). Training elite specialists in engineering and technologies. In *Proceedings of the 9th International CDIO Conference*. Massachusetts Institute of Technology and Harvard University School of Engineering and Applied Sciences. Retrieved from www.cdio.org/files/document/file/T3C2_Chubik_061.pdf

Chubik, P., & Zamyatina, O. (2013). Training Elite Specialists in Engineering and Technologies. *Proceedings of the 9th International CDIO Conference.*

Chuchalin, A. (2013). TPU-SKOLKOVO Project: Modernization of BEng Programs in Russia. *Proceedings of the 9th International CDIO Conference.*

Chuchalin, A. I., Soloviev, M. A., Zamyatina, O. M., & Mozgaleva, P. I. (2013). *Elite Engineering Programme in Tomsk Polytechnic University - The way to attract talented students into engineering*. Paper presented at the IEEE Global Engineering Education Conference (EDUCON-2013), Berlin, Germany.

Chuchalin, A. (2012). RAEE Accreditation Criteria and CDIO Syllabus: Comparative Analysis. In *Proceedings of the 8th International CDIO Conference*. Queensland University of Technology.

Chuchalin, A., Daneikina, N., & Fortin, C. (2016). Application of CDIO Approach to Engineering BEng, MSc and PhD Programs Design and Implementation. *Proceedings of the 12th International CDIO Conference*.

Chuchalin, A., Petrovskaya, T., Kulyukina, E., & Tayurskaya, M. (2012). Benchmarking of TPU Academic Standard and CDIO Standards in Engineering Education. In *Proceedings of the 8th International CDIO Conference*. Queensland University of Technology.

Chuchalin, A., Tayurskaya, M., & Malmqvist, J. (2015). Professional Development of Russian HEIs' Management and Faculty in CDIO Standards Application. *European Journal of Engineering Education*, *40*(6), 426–437. doi:10.1080/03043797.2015.1085837

Chung, C. (2011). Changing engineering curriculum in the globalizing world. *New Horizons in Education*, *59*(3).

Clark, R., & Andrews, J. (2012a). *Innovation in the Academic / Vocational Interchange: Developing and Achieving Good Practice in Employer Engagement.* Report for National HE STEM Programme.

Clark, R., Andrews, J. & Gorman, P. (2013). Tackling Transition: The Value of Peer Mentoring. *Journal of Widening Participation and Lifelong Learning*, *14*, 57-75.

Clark, B. R. (2004). *Sustaining Change in Universities. Continuities in Case Studies and Concepts.* New York, NY: McGraw-Hill.

Clarke, J. (2016). *Key statistics on people with disabilities in employment.* Briefing paper, 7540(7).

Clarke, M. (2007). The Impact of Higher Education Rankings on Student Access, Choice and Opportunity. *Higher Education in Europe*, *32*(1), 59–70. doi:10.1080/03797720701618880

Clark, R., & Andrews, J. (2010). Researching Primary Engineering Education: An Exploratory Study from a UK Perspective. *European Journal of Engineering Education*, *35*(3), 585–595. doi:10.1080/03043797.2010.497551

Clark, R., & Andrews, J. (2012b). Engineering the Future: CDIO as a tool for combating retention difficulties. In M. Rasul (Ed.), *Developments in Engineering Education Standards: Advanced Curriculum Innovations* (pp. 143–155). IGI Global. doi:10.4018/978-1-4666-0951-8.ch008

Clark, R., & Andrews, J. (2014). Relationships, Variety & Synergy [RVS]: The Vital Ingredients for Scholarship in Engineering Education? A Case-Study. *European Journal of Engineering Education*, *39*(6), 585–600. doi:10.1080/03043797.2014.895707

Clark, R., & Andrews, J. (2016). A Community Based Participatory Research Study into Why Some Girls Don't 'Do' Engineering. *International Journal of Engineering Education*, *32*(6).

Clark, R., Glew, B., & Andrews, J. (2014). Developing the Engineering Talent Pipeline using Work Based Learning. *Engineering Leaders Conference.*

Clark, R., & Thomson, G. (2015). Exploring the Form and Impact of a CDIO Implementation – A Case Study. *European CDIO Meeting.*

Cloud News Daily. (2015). *Guide to Big Data Analytics: Platforms, software, companies['] tools, solutions and Hadoop.* Retrieved July 18, 2017 from http://cloudnewsdaily.com/ big-data-analytics/

Cole, J., & Foster, H. (2008). *Using Moodle* (2nd ed.). Sebastopol, CA: O'Reilly Media, Inc.

Collins, T. R., Berivudes, M. G., Youngblood, A. D., & Pazos, P. (2004). Professional Development Training for Engineering Managers. *Engineering Management Journal, 16*(3), 3–9. doi:10.1080/10429247.2004.11415251

Columbia University Center for Career Education. (2017). *Virtual Internship Program.* Retrieved June 2, 2017, from https://www.careereducation.columbia.edu/programs/virtual-internship-program-vip

Computer Managed Instruction Guidelines for Interoperability (CMI001) Version 3.5. (2001). Aviation Industry CBT Committee (AICC).

Conole, G., & Warburton, B. (2005). A review of computer-assisted assessment. *ALT-J, 13*(1), 17–31. doi:10.3402/rlt.v13i1.10970

Construction Guidelines. (2000). *MDS 80-12.2000: Metodicheskie rekomendacii po razrabotke uslovij (trebovanij) investora (zakazchika) pri podgotovke podryadnyh torgov* [Construction Guidelines 80-12.2000: Methodological Recommendations on Working Out of the Investor's (Customer's) Conditions (Requirements) during the Preparation for Contract Tendering. Gosstroy (Russian State Committee for Construction)]. Moscow, Russia: SUE Centre of Construction Design Products.

Cook, D. (2016). *Practical machine learning with H2O: Powerful, scalable techniques for deep learning and AI.* Sebastopol, CA: O'Reilly Media, Inc.

Cotič, M., & Zuljan, M. V. (2009). Problem-based instruction in mathematics and its impact on the cognitive results of the students and on affective-motivational aspects. *Educational Studies, 35*(3), 297–310. doi:10.1080/03055690802648085

Coyle, D., Hood, P., & Marsh, D. (2010). *CLIL: Content and Language Integrated Learning.* Cambridge, UK: Cambridge University Press.

Crawley, E. F. (2001). *The CDIO syllabus: a statement of goals for undergraduate engineering education.* Cambridge, MA: The Department of Aeronautics and Astronautics, Massachusetts Institute of Technology.

Crawley, E., Malmqvist, J., Ostlund, S., Brodeur, D., & Edström, K. (2014). *Rethinking Engineering Education, the CDIO Approach* (2nd ed.). New York, NY: Springer. doi:10.1007/978-3-319-05561-9

Cruz, B. A., & Eremeeva, E. V. (2013). Definition of meta-subject competences of younger school student. *Modern Problems of Science and Education, 2013*(6). Available at: http://www.science-education.ru/113-11014

Cypriot Council for the Academic Recognition of Study Titles. (2017). Retrieved April 14, 2017 from http://www.kysats.ac.cy

Dainotti, A., Pescape, A., & Claffy, K. (2012). Issues and future directions in traffic classification. *IEEE Network, 26*(1), 35–40. doi:10.1109/MNET.2012.6135854

Dalton-Puffer, Ch. (2011). Content-and-Language Integrated Learning: From Practice to Principles? *Annual Review of Applied Linguistics, 31*, 182–204. doi:10.1017/S0267190511000092

Databricks. (2016). *Apache Spark Survey 2016: Report highlights.* Retrieved July 18, 2017 from http://cdn2.hubspot.net/hubfs/438089/DataBricks_Surveys_-_Content/2016_ Spark_ Survey/2016_Spark_Infographic.pdf

De Amicis, R., Stojanovic, R., & Conti, G. (2009). *GeoSpatial Visual Analytics: Geographical Information Processing and Visual Analytics for Environmental Security.* Dordrecht, The Netherlands: Springer. doi:10.1007/978-90-481-2899-0

Delbanco, A. (2012). *College. What It Was, Is, and Should Be.* Princeton, NJ: Princeton University Press.

Delgado, J. C. (2015). Distributed Interoperability in Heterogeneous Cloud Systems. In S. Bagchi (Ed.), *Emerging Research in Cloud Distributed Computing Systems* (pp. 1–40). IGI Global; doi:10.4018/978-1-4666-8213-9.ch001

Deligirova, O. A. (2013). Informacionnye tehnologii v upravlenii personalom vuza [Information technologies in personnel management]. *Perspektivy razvitija informacionnyh tehnologij, 14*, 67-73.

DeLisle, M. (2016). *Mastering phpMyAdmin 2.8 for effective MySQL management* (3rd ed.). Birmingham: Packt Publishing.

Deri, L., Martinelli, M., Bujlow, T., & Cardigliano, A. (2014). nDPI: Open-source high-speed deep packet inspection. In *Proceeding of Wireless Communications and Mobile Computing Conference (IWCMC) 2014 International*. IEEE Xplore. 10.1109/IWCMC.2014.6906427

Devert, A. (2014). *A matplotlib Plotting Cookbook*. Birmingham, UK: Packt Publishing.

Dewey, J. (2009). *Democracy and education: An introduction to the philosophy of education*. New York: WLC Books. (Original work published 1916)

Dickinson, P., & Montgomery, J. L. (2016). The Role of Teacher Leadership for Promoting Professional Development Practices. In K. Dikilitaş (Ed.), *Innovative Professional Development Methods and Strategies for STEM Education* (pp. 91–114). Hershey, PA: IGI Global. doi:10.4018/978-1-4666-9471-2.ch006

Dikilitaş, K. (Ed.). (2015). *Innovative professional development methods and strategies for STEM education*. Gaziantep: Hasan Kalyoncu University.

Dinham, S. (2013). The quality teaching movement in Australia encounters difficult terrain: A personal perspective. *Australian Journal of Education, 57*(2), 91–106. doi:10.1177/0004944113485840

Dinsmore, T. W. (2017). *Spark is the future of analytics*. Retrieved July 18, 2017 from https://thomaswdinsmore.com/

Dionysiou, I., & Ktoridou, D. (2012). Enhancing dynamic-content courses with student-oriented learning strategies: The case of computer security course. *International Journal of Cyber Ethics in Education, 2*(2), 24–33. doi:10.4018/ijcee.2012040103

Dobrov, B. V., & Pavlov, A. M. (2010). Issledovanie kachestva bazovyh metodov klasterizatsii novostnogo potoka v sutochnom vremennom okne [Basic line for news clusterization methods evaluation]. In *Elektronnye biblioteki: perspektivnye metody i tekhnologii, elektronnye kollektsii: Trudy XII Vseros. nauch. konf. (RCDL'2010)* (pp. 287–295). Kazan, Russia: Kazan Federal University Publishing House. Retrieved from http://rcdl.ru/doc/2010/287-295.pdf

Dodge, B. (n.d.). *WebQuest.Org*. Retrieved June 5, 2016, from https:// www.webquest.org

Dohmen, G. (1999). Lifelong Learning for All – innovative perspectives of continuing education. *Lifelong Learning in Europe., 4*(99), 154–158.

Doing Business. (n.d.). Retrieved 06/13/2017 from http://russian.doingbusiness.org/

Doneckaya, S. (2014). Analysis of the competitiveness of Russian universities in world rankings. In Methodological issues of evaluation of the project 5-100 in the ratings of universities. Moscow: Center of sociological researches.

Donovan, M. S. (2013). Generating improvement through research and development in education systems. *Science, 340*(6130), 317–319. doi:10.1126cience.1236180 PMID:23599484

Duckett, T. (2005). *Artificial Neural Networks, Course homepage*. Retrieved June 18, 2016, from http://aass.oru.se/~tdt/ann/index-english.html#Labbar

Dudley-Evans, T. (1999). *Developments in English for Specific Purposes: A multi-disciplinary approach*. Cambridge, UK: Cambridge University Press.

Dvoretskii, D. S., Dvoretskii, S. I., Polyakov, B. B., & Ostrovskii, G. M. (2012). A new approach to the optimal design of industrial chemical-engineering apparatuses. *Theoretical Foundations of Chemical Engineering*, *46*(5), 437–445. doi:10.1134/S0040579512040112

Dyundyukov, V., & Tarassov, V. (2013). Goal-Resource Networks and Their Application to Agents Communication and Coordination in Virtual Enterprises. *Proceedings of the IFAC Conference on Manufacturing Modelling, Management and Control (MIM 2013).*

ECH-INE. (2007). *Uruguay Investment and Promotion Agency, 2007*. Retrieved from http://www.uruguayxxi.gub.uy/

ECH-INE. (2016). *Uruguay Investment and Promotion Agency, 2016*. Retrieved 18.12.2017 from http://www.uruguayxxi.gub.uy/

ECTS. (2009). ECTS User's Guide. Brussels, Belgium: European Commission /Luxembourg: Office for Official Publications of the European Communities. doi:10.2766/88064

Edmodo. (n.d.). Retrieved from https://www.edmodo.com/

Edström, K., & Kolmos, A. (2012). Comparing two approaches for engineering education development: PBL and CDIO. In *Proceedings of the 8th International CDIO Conference*. Brisbane: Queensland University of Technology. Retrieved from www.cdio.org/files/document/file/comparing_two_approaches_for_engineering_education_development_pbl_and_cdio.pdf

EERN. (2017). *UK and Ireland Engineering Education Research Network*. Retrieved from https://hefocus.raeng.org.uk/research-network/

Efimov A.I., & Novikov A.I. (2016). Algoritm Poetapnogo Utochneniya Proectivnogo Preobrazovaniya Dlya Sovmesheniya Izobrajenii [Algorithm of projective transformation staged refinement for image superimposition]. *Computer Optics, 40*(2), 258-265.

Efimov, A. I., Novikov, A. I., & Sablina, V. A. (2016). Image Superimposition Technique In Computer Vision Systems Using Contour Analysis. Bar, Montenegro: IEEE. doi:10.1109/MECO.2016.7525721

Ehrenberg, R. (2002). Reaching for the brass ring: The U.S. News & World Report Rankings and competition. *The Review of Higher Education*, *26*(2), 145–162. doi:10.1353/rhe.2002.0032

Electronic Educational System of Bauman Moscow State Technical University. (n.d.). Retrieved from: http://e-learning.bmstu.ru/new_face/

Electronic University of Bauman Moscow State Technical University. (n.d.). Retrieved from https://webvpn.bmstu.ru/+CSCO+0075676763663A2F2F72682E6F7A6667682E6568++/

Eliseev, V. L. (2016). *NNACS software project*. Retrieved June 18, 2016, from https://github.com/evlad/nnacs

Eliseev, V.L., & Filaretov, G.F. (2011) Programmniy paket dlya modelirovaniya i obucheniya metodam neirosetevogo upravleniya [Software for neural network training and modeling]. *Otkritoye obrazovaniye*, *86*(2), 98-101.

El-Khalili, N. H. (2013). Teaching agile software engineering using problem-based learning. *International Journal of Information and Communication Technology Education*, *9*(3), 1–12. doi:10.4018/jicte.2013070101

Ellcessor, E. (2016). *Media, Disability, and the Politics of Participation*. NYU Press.

ENIC-NARIC. (2017). Retrieved April 5, 2017 from http://www.enic-naric.net

Espinosa, L. J., Grandall, M., & Tukibayeva, M. (2014). *Rankings, Institutional Behavior, and College and University Choice Framing the National Dialogue on Obama's Ratings Plan.* Retrieved May 2, 2017, from www.acenet.edu/news-room/Documents/Rankings-Institutional-Behavior-and-College-and-University-Choice.pdf

EUA Council for Doctoral Education. (2017). Retrieved April 13, 2017 from http://www.eua.be/cde

EUbusiness. (2008). *A Guide to Languages in the European Union.* Retrieved from http://www.eubusiness.com/topics/Languages/eu-languages-guide/

Faletta, J.-P., Meier, J. A., & Balderas, J. U. (2016). High-Impact Practices: Integrating the First-Year Experience with Service-Learning and Study Abroad. In K. González & R. Frumkin (Eds.), *Handbook of Research on Effective Communication in Culturally Diverse Classrooms* (pp. 337–355). Hershey, PA: IGI Global. doi:10.4018/978-1-4666-9953-3.ch017

Federkeil, G. (2002). Some Aspects of Ranking Methodology - The CHE Ranking of German Universities. *Higher Education in Europe, 27*(4), 389–397. doi:10.1080/0379772022000071878

Fedorov, I. B., Norenkov, I. P., & Korshunov, S. V. (2006). Organizacionnaya podgotovka specialistov v oblasti komp'yuternyh nauk, tekhniki i tekhnologij [Organizational training of specialists in the field of computer science, technics and technology]. *Informacionnye tekhnologii - Information Technology, 12*(9), 73-77.

Fedorov, I. B., Norenkov, I. P., & Korshunov, S. V. (2006). Organizacionnaya podgotovka specialistov v oblasti komp'yuternyh nauki, tekhniki i tekhnologij, otrazhennaya v rossijskih i zarubezhnyh dokumentah [Organizational training of specialists in the field of computer science, technics and technology; as reflected in Russian and foreign documents]. *Informacionnye tekhnologii - Information Technology, 12*(9), 73-77.

Fedorov, I.B., & Medvedev, V.E. (2011). Engineering education: problems and tasks. [Inzhenernoye obrazovaniye: problemy i zadachi]. *High Education in Russia, 12.*

Feisel, L., & Rosa, A. (2005). The role of the laboratory in undergraduate engineering education. *Journal of Engineering Education, 94*(1), 121–130. doi:10.1002/j.2168-9830.2005.tb00833.x

Figuera, C., Morgado, E., Gutiérrez-Pérez, D., Alonso-Atienza, F., del Arco-Fernández-Cano, E., Caamaño, A. J., ... Requena-Carrión, J. (2011). A multidisciplinary problem based learning experience for telecommunications students. *International Journal of Human Capital and Information Technology Professionals, 2*(3), 15–28. doi:10.4018/jhcitp.2011070102

Filippovich, A. Y., & Filippovich, Y. N. (2015). Osnovnyie podhodyi k postroeniyu proektno-tehnologicheskoy magistraturyi [Basic approaches to the construction of a design and technological magistracy]. In *Aktualnyie problemyi realizatsii elektronnogo obucheniya i distantsionnyih obrazovatelnyih tehnologiy*. Moscow: Izdatel'stvo SGU.

Filosova, E. I., & Rykova, O. V. (2010). The use of ontologies in the object learning technologies. In *Proceedings of the Seventh International Scientific Conference "New Educational Technologies in High School*" (*vol. 2*, pp. 123-127). Ekaterinburg, Russia: Federal State Autonomous Educational Institution of Higher Education "Ural Federal University named after the first President of Russia B.N. Yeltsin".

Finansovyj slovar'. (2017). [Financial Terms Dictionary]. Retrieved from: http://dic.academic.ru/dic.nsf/fin_enc/13778

Fino, H., Solovyev, V., & Zinchenko, L. (2013). Challenges for students mobility between European and Russian Universities. *Proc. ICL.* 10.1109/ICL.2013.6644629

Fiorito, L. (2005). *How is English for Specific Purposes (ESP) different from English as a Second Language (ESL), also known as general English?* Retrieved May 20, 2016, from https://www.usingenglish.com/articles/teaching-english-for-specific-purposes-esp.html

Flórez-Puga, G., Gómez-Martín, M., Díaz-Agudo, B., & González-Calero, P. A. (2008). Dynamic expansion of behaviour trees. In *Proceedings of the 4th Artificial Intelligence and Interactive Digital Entertainment Conference, AIIDE 2008* (pp. 36-41). Stanford, CA: AIIDE.

Florman, S. C. (1997). Non-technical Studies for Engineers: The Challenge of Relevance. *European Journal of Engineering Education*, *22*(3), 249–258. doi:10.1080/03043799708923457

Flyvbjerg, B. (2010). Femmis for stålelser om case-studiet. In L. Tanggaard & S. Brinkmann (Eds.), *Kvalitative metoder*. København: Hans Reitzels Forlag.

Foresight, A. S. I. "Education 2035". (2017). *Presentacia forsaita "Obrazovanit 2035"*. etrieved from http://asi.ru/molprof/foresight/12254/ (in Russian)

Fortelius, C., & Akerman, M.-L. (2015). Project/problem based learning in the field of materials, food, and chemical engineering at Helsinki Metropolia University of Applied Sciences. *International Journal of Quality Assurance in Engineering and Technology Education*, *4*(4), 39–46. doi:10.4018/IJQAETE.2015100103

Fredriksson, J., Larsson, V., Olsson, C., Enqvist, O., & Kahl, F. (2016). Efficient algorithms for robust estimation of relative translation. Image and Vision Computing, 52, 114-124

Frick, J., Sautter, M., & Ovrebekk, S. (2000). Effective Online Learning – Both a Utilization of Technology and Methods. In F. Albalooshi (Ed.), *Virtual Education: Cases in Learning & Teaching Technologies* (pp. 179–184). Hershey, PA: IRM Press. doi:10.4018/978-1-93177-739-1.ch012

Froyd, J. E., Wankat, P. C., & Smith, K. A. (2012). Five major shifts in 100 years of engineering education. *Proc. IEEE*, 100. 10.1109/JPROC.2012.2190167

Frumina, E., & West, R. (2012). *Internationalization of Russian higher education: The English language dimension*. Moscow, Russia: British Council. Retrieved May 30, 2016, from https://www.teachingenglish.org.uk/sites/teacheng/files/RussianBaselineReport2012.pdf

Frumina, E. (2013). *Modernizing English Language Teaching: a case of a pilot project at NUST MISiS. NUST MISiS. The report by Elena Frumina, Advisor to Rector NUST MISiS, made at the CDIO Academy training program*. Moscow, Russia: SkolTech.

Fuller, M., Bradley, A., & Healey, M. (2004). Incorporating disabled students within an inclusive higher education environment. *Disability & Society*, *19*(5), 455–468. doi:10.1080/0968759042000235307

Furman, Y. A., Kreveckij, A.V., Peredreev, A.K., Rozhencov, A.A., Khvazov, R.G., Egoshina, I.L., & Leukhin, A.N. (2003). *Vvedenie v Konturnii Analiz. Prilojeniya k Obrabotke Izobrajenii I Signalov 2e izdanie* [Introduction into contour analysis. Applications to process images and signals (2nd ed.)]. Moscow, Russia: Fizmatlit.

Gablasova, D. (2015). Learning technical words through L1 and L2: Completeness and accuracy of word meanings. *Journal Elsevier: English for Specific Purposes*, *39*, 62–74. doi:10.1016/j.esp.2015.04.002

Galiamova, E. V., Karpenko, A. P., & Sokolov, N. K. (2009). The evaluation of conceptual knowledge of a subject taught in educational system. Science and Education. *Electron. Phys.*, *2009*(2). Available at: http://technomag.edu.ru/doc/115086.html

Galiullin, L. A., & Zubkov, E. V. (2011). Hybrid Neural Network for the Adjustment of Fuzzy System when Simulating Tests of Internal Combustion Engines. *Russian Engineering Research*, *31*(5), 439–443. doi:10.3103/S1068798X11050273

Gardel, A., Bravo, I., Revenga, P. A., Lázaro, J. L., & García, J. (2012). Implementation of industrial automation laboratories for e-learning. *International Journal of Electrical Engineering Education*, *49*(X), 402–418. doi:10.7227/IJEEE.49.4.4

Garrido, A., & Morales, L. (2014). E-Learning and Intelligent Planning: Improving Content Personalization. *Revista Iberoamericana de Tecnologias del Aprendizaje*, *9*(1), 1–7. doi:10.1109/RITA.2014.2301886

Gartner. (2016). *Gartner's 2016 Hype Cycle for Emerging Technologies identifies three key trends that organizations must track to gain competitive advantage*. Retrieved July 18, 2017 from http://www.gartner.com/newsroom/id/3412017

Gartner. (2017a). *Gartner says worldwide IT spending forecast to grow 1.4 percent in 2017.* Retrieved July 18, 2017 from http://www.gartner.com/ newsroom/id/3672818

Gartner. (2017b). *Magic Quadrant for Data Management Solutions for analytics*. Retrieved July 18, 2017 from https://www.gartner.com/doc/reprints?id=1-3TZLQ0P&ct=170221&st=sb%3f

Gavrilina, E., Zakharov, M., Karpenko, A., Smirnova, E., & Sokolov, A. (2016). Model of integral assessment quality of training graduates of higher engeneering education. *CEUR Workshop Proceedings. Selected Papers of the XI International Scientific-Practical Conference Modern Information Technologies and IT-Education (SITITO 2016).*

Geraschenko, A., & Efremov, A. (2014). The Targeted Training of the Students Enrolled in Aeronautical Programs. *Proceedings of the 10th International CDIO Conference.*

Gershunsky, B. S. (1998). *Filosofiya obrazovaniya dlya XXI veka: V poiskakh praktiko-orientirovannykh kontseptsiy* [Philosophy of education for the 21st century: In search of concepts]. Moscow: Sovershenstvo.

Gertz, M. W., Stewart, D. B., & Khosla, P. K. (1994). A human-machine interface of distributed virtual laboratories. *IEEE Robotics & Automation Magazine*, *1*(4), 5–12. doi:10.1109/100.388265

Ghinea, V. M. (2014). Modelling and Simulation of the Need for Harmonizing the European Higher Education Systems. In A. M. Dima (Ed.), *Handbook of Research on Trends in European Higher Education Convergence* (pp. 62–83). Hershey, PA: IGI Global. doi:10.4018/978-1-4666-5998-8.ch004

Gill, J., Ayre, M., & Mills, J. (2017). Revisioning the Engineering Profession: How to Make It Happen! In M. Gray & K. D. Thomas (Eds.), *Strategies for Increasing Diversity in Engineering Majors and Careers* (pp. 156–175). Hershey, PA: IGI Global. doi:10.4018/978-1-5225-2212-6.ch008

Glavexpertcenter. (2017). Retrieved April 5, 2017 from http://nic.gov.ru

Glebova, L. N., & Kuznetsova, M. D. (2012). Monitoring kachestva vysshego pedagogicheskogo obrazovaniya [Monitoring of quality of higher pedagogical education]. Logos.

Glew, B., & Elsworth, T. (2010). *Development of a work-based learning MSc course which incorporates the development and demonstration of professional engineering competence standards*. Engineering Education Conference 2010, Birmingham, UK.

Gomes, A. J., Mendes, A. J., & Marcelino, M. J. (2015). Computer Science Education Research: An Overview and Some Proposals. In R. Queiros (Ed.), *Innovative Teaching Strategies and New Leaning Paradigms in Computer Programming* (pp. 1–29). IGI Global; doi:10.4018/978-1-4666-7304-5.ch001

Gomes, L., & Bogosyan, S. (2009). Current trends in remote laboratories. *IEEE Transactions on Industrial Electronics*, *56*(12), 4744–4756. doi:10.1109/TIE.2009.2033293

Gonzalez, R. C., Woods, R. E., & Eddins, S. L. (2006). *Digital image. Processing environment MATLAB*. Richmond, TX: Ergodebooks.

Google Class. (n.d.). Retrieved from https://edu.google.com/intl/ru/products/productivity-tools/classroom/

Google. (2017a). *Google Cloud Platform documentation.* Retrieved July 18, 2017 from https://cloud.google.com/docs/

Google. (2017b). *Google Cloud Platform solutions.* Retrieved July 18, 2017 from https://cloud. google.com/solutions/

Gordon, B. M. (2011, June). *Capabilities of effective engineering leaders. Version 3.6.* MIT Engineering Leadership Program. Retrieved from gelp.mit.edu/sites/default/files/documents/leadershipcapabilities.pdf

Gorelov, V. I. (2007). *Management of the society development.* Moscow: Ekon-Inform Press.

Gornev, V. F. (1998). *Kompyuterno-orientirovannye obuchayuschie tekhnologii v inzhenernoy podgotovke* [The computer focused teaching technologies in engineering training]. Moscow: NIIVO.

GOST R 541-2010. (2010). *GOST R (Russian national standard) 541-2010: Industrial Automation Systems and Integration. Guidelines for Standards Implementation, Structure and Glossary.* Moscow, Russia: Standartinform. (in Russian)

GOST R 54147-2010. (2010). *GOST R (Russian national standard) 54147-2010: Strategic and Innovation Management. Terms and Definitions.* Moscow, Russia: Standartinform. (in Russian)

GOST R ISO 15704-2008. (2008). *Promyshlennye avtomatizirovannye sistemy. Trebovaniya k standartnym arhitekturam i metodologiyam predpriyatiya* [GOST R (Russian national standard) ISO 15704-2008: Industrial Automated Systems. Standard Architecture and Enterprise Methodology Requirements]. Moscow, Russia: Standartinform. (in Russian)

GOST R ISO 19439-2008. (2008). *Integraciya predpriyatiya. Osnova modelirovaniya predpriyatiya* [GOST R (Russian national standard) ISO 19439-2008: Enterprise Integration. The basis of Enterprise Modelling]. (in Russian)

Government Decree About measures of state support of leading universities of the Russian Federation. (2013, March 16). Retrieved from ConsultantPlus: http://www.consultant.ru/sys/english/

Grasso, D., & Burkins, M. B. (Eds.). (2010). Holistic Engineering Education. Beyond Technology. Springer Science+Business Media. doi:10.1007/978-1-4419-1393-7

Gray, M., & Lundy, C. (2017). Engineering Study Abroad: High Impact Strategy for Increasing Access. In M. Gray & K. D. Thomas (Eds.), *Strategies for Increasing Diversity in Engineering Majors and Careers* (pp. 42–59). Hershey, PA: IGI Global. doi:10.4018/978-1-5225-2212-6.ch003

Greshilova, A. V. (2014). The Content of meta-subject competencies of students in secondary vocational education. *Magister Dixit: Scientific and Pedagogical Magazine of Eastern Siberia, 1*(13). Available at: http://md.islu.ru/sites/md.islu.ru/files /rar/greshilova_statya_md_0.pdf

Griffith, A., & Rask, K. (2005). *The influence of the U.S. News and World Report collegiate rankings on the matriculation decision of high-ability students.* Retrieved May 2, 2017, from http://digitalcommons.ilr.cornell.edu/cgi/viewcontent.cgi?article=1029&context=cheri

Grigoryan, S., & Lebedeva, E. (2014). *Peculiarities of the introduction of the model of integrated mastery of content and foreign language teaching into the Russian educational system.* Science and Education: Economy and Economics. *Entrepreneurship Law and Management, 5,* 17–21.

Grigoryeva, K. (2016). *Formation of technical college students' foreign language competence in the field of professional communication based on CLIL technology* (Unpublished doctoral dissertation). Kazan State University, Kazan, Russia.

Grossmann, I. E. (2012). Advances in mathematical programming models for enterprise-wide optimization. *Computers & Chemical Engineering, 47,* 2–18. doi:10.1016/j.compchemeng.2012.06.038

Guess, A. R. (2011). *5 key issues any Big Data integration platform should address.* Retrieved July 18, 2017 from http://www. dataversity.net/5-key-issues-any-big-data-integration-platform-should-address/

Gugina, E. V., & Kuzenkov, O. A. (2014). Education standards of Lobachevsky State University of Nizhniy Novgorod. *Herald of Lobachevsky State University of Nizhniy Novgorod, 3-4*, 39-44. Retrieved from https://elibrary.ru/item.asp?id=22862964

Gurov, V. S., Koryachko, V. P., Taganov, A. I., Moiseenko, V. P., & Taganov, R. A. (2010). Opyt sozdaniya i primeneniya resursov elektronnoy informatsionno-obrazovatel'noy sredy po napravleniyu CALS i CASE tekhnologii [Experience of development and application of resources of the electron informational and educational media in the direction CALS and CASE-technologies]. In *Papers of the VII All-Russian scientific and methodical conference Telematics-2010*. Academic Press.

Gurov, V.S., Kolodko, G.N., Kostyashkin, L.N., Loginov, A.A., Muratov, E.R., Nikiforov, M.B., ... Yukin, S.A. (2016). *Obrabotca Izobrajenii v Aviacionnih Sistemah Technicescogo Zreniya* [Image processing in aircraft computer vision systems]. Moscow, Ruissia: Fizmatlit.

Gusev, V. A., Zatsepin, V. A., Nisman, O. Y., & Kogan, E. Y. (2014). Development of educational programs using automated information system. *Vestnik of Samara State Technical University. Psychological and Pedagogical Sciences Series*, *1*(21), 59–64.

Gynnild, V. (2014, August 25). *From 'quality assurance' to 'quality enhancement': Addressing the transition from 'teaching' to 'learning' in engineering education.* Paper presented at the iCEER2014 - International Conference on Engineering Education and Research, Hamilton, Ontario, Canada.

Hadzigeorgiou, Y., Fokialis, P., & Kabouropoulou, M. (2012). Thinking about Creativity in Science Education. *Creative Education*, *5*(05), 603–611. doi:10.4236/ce.2012.35089

Hall, A. D. (1962). *A Methodology for Systems Engineering*. Princeton, NJ: Van Nostrand.

Ham, F. M., & Brown, R. G. (1983). Observability, eigenvalues, and Kalman filtering. *IEEE Transactions on Aerospace and Electronic Systems*, *19*(2), 269–273. doi:10.1109/TAES.1983.309446

Harms, U. (2000). Virtual and remote labs in physics education. *Proceedings of the Second European Conference on Physics Teaching in Engineering Education*, 1-6.

Harmuth, H. (1977). *Sequence Theory: foundations and applications*. New York: Academic Press.

Harrison, R. T., & Leitch, C. (2010). Voodoo institution or entrepreneurial university? Spin-off companies, the entrepreneurial system and regional development in the UK. *Regional Studies*, *44*(9), 1241–1262. doi:10.1080/00343400903167912

Hast, A., & Marchetti, A. (2014). Invariant Interest Point Detection Based on Variations of the Spinor Tensor. *22nd International Conference in Central Europe on Computer Graphics, Visualization and Computer Vision (WSCG)*, 49-56.

Hauke, K. (2015). *Process approach to management knowledge objects managerial expertise for distance learning*. Paper presented at the Federated Conference on Computer Science and Information Systems, Łódź, Poland. 10.15439/2015F328

Hazelkorn, E. (2009). Attitudes to Rankings: Comparing German, Australian and Japanese Experiences. In *Addressing Critical Issues on Quality Assurance and University Rankings in Higher Education in the Asia Pacific.* Retrieved May 2, 2017, from http://arrow.dit.ie/cgi/viewcontent.cgi?article=1002&context=cserart

Hazelkorn, E. (2011). Globalization and the Reputation Race in *Rankings and the Reshaping of Higher Education: the Battle for World Class Excellence.* Palgrave MacMillan. Retrieved May 2, 2017, from http://arrow.dit.ie/cgi/viewcontent.cgi?article=1010&context=cserbk

Hazelkorn, E. (2014). The Effects of Rankings on Student Choices and Institutional Selection. In *Access and Expansion Post-Massification: Opportunities and Barriers to Further Growth in Higher Education Participation.* London: Routledge. Retrieved May 2, 2017, from http://arrow.dit.ie/cgi/viewcontent.cgi?article=1015&context=cserbk

HEFCE. (2017a) *NSS Results 2017.* Retrieved from http://www.hefce.ac.uk/lt/nss/results/2017/

HEFCE. (2017b). *Teaching Excellence Framework.* Retrieved from http://www.hefce.ac.uk/lt/tef/

Heinecke, W. F., Blasi, L., Milman, N., & Washington, L. (1999). *New Directions in the Evaluation of the Effectiveness of Educational Technology.* Paper presented at The Secretary's Conference on Educational Technology, Washington, DC. DOI: 10.1300/J025v18n02_07

Helbig, M., & Ulbricht, L. (2010). *Perfekte Passung: Finden die besten Hochschulen die besten Studenten? Gefälligkeitsübersetzung: Perfect fit: do the best universities find the best students?* Retrieved May 2, 2017, from www.wzb.eu/sites/default/files/u128/scan_perfekte_passung_0.pdf

HERI. (2007). *College Rankings & College Choice: How important are college rankings in students' college choice process?* Los Angeles: Higher Education Research Institute University of California. Retrieved May 2, 2017, from https://www.heri.ucla.edu/PDFs/pubs/briefs/brief-081707-CollegeRankings.pdf

Heyneman, S. P., & Lee, B. (2014). *The impact of international studies of academic achievement on policy and research. In Handbook of international large-scale assessment: Background, technical issues and methods of data analysis* (pp. 37–72). Boca Raton, FL: Taylor and Francis Group.

Heywood, J. (2016). *Categorizing the Work Done by Engineers: Implications for Assessment and Training. In The Assessment of Learning in Engineering Education: Practice and Policy.* Wiley-IEEE Press; doi:10.1002/9781119175575

Hido, S. (2016). *Complex neural networks made easy by Chainer.* O'Reilly Media, Inc. Retrieved July 18, 2017 from https://www.oreilly.com/learning/complex-neural-networks-made-easy-by-chainer

Hodge, S., & Harris, R. (2012). Discipline, governmentality and 25 years of competency-based training. *Studies in the Education of Adults, 44*(2), 155–170. doi:10.1080/02660830.2012.11661630

Hogenboom, F., Frasincar, F., Kaymak, U., & de Jong, F. (2011). An overview of event extraction from text. In *CEUR Conference Proceedings, Workshop on Detection, Representation, and Exploitation of Events in the Semantic Web (DeRiVE 2011) at Tenth International Semantic Web Conference (ISWC 2011)* (Vol. 779, pp. 48-57). Koblenz, Germany: CEUR-WS.org. Retrieved from http://ceur-ws.org/Vol-779/derive2011_submission_1.pdf

Holland, H. J. (1975). *Adaptation in natural and artificial systems.* Ann Arbor, MI: University of Michigan Press.

Holmegaard, H., Ulriksen, L., & Madsen, L. (2010). Why students choose (not) to study engineering. *Joint International IGIP-SEFI Annual Conference.* Retrieved from http://www.sefi.be/wp-content/papers2010/abstracts/203.pdf

Holz-Clause, M., Guntuku, D., Koundinya, V., Clause, R., & Singh, K. (2015). Current and future trends in higher education learning: Implications for curriculum design and delivery. In Handbook of Research on Enhancing Teacher Education with Advanced Instructional Technologies. IGI Global. doi:10.4018/978-1-4666-8162-0.ch015

Horie, M. (2002). The internationalization of higher education in Japan in the 1990s: Aeconsideration. *Higher Education, 43*(1), 65–84. doi:10.1023/A:1012920215615

HPI. (2017). *The Hasso Plattner Institute.* Retrieved July 18, 2017 from https://hpi.de/en/the-hpi/overview.html

Hung, N. M., & Quyen, N. V. (2009). Sales tax: Specific or ad valorem tax for a non-renewable resource? *Economics Letters, 102*(2), 132–134. doi:10.1016/j.econlet.2008.12.001

Hung, W. (2011). Theory to reality: A few issues in implementing problem-based learning. *Educational Technology Research and Development*, *59*(4), 529–552. doi:10.100711423-011-9198-1

Hutchinson, T., & Water, A. (1987). The development of ESP. In *English for Specific Purposes: A learning-centered approach* (pp. 9–15). Cambridge, UK: Cambridge University Press. doi:10.1017/CBO9780511733031.005

Hyper-V Overview. (2012). Retrieved April 20, 2017, from https://msdn.microsoft.com/en-us//library/hh831531(v=ws.11).aspx

IBM. (2017). *IBM Bluemix catalog*. Retrieved July 18, 2017 from https://console.ng.bluemix.net/catalog/

IEEE 1484.11.2 Standard for Learning Technology. (2003). *ECMA Script Application Programming Interface for Content to Runtime Services Communication.*

IET. (2017). *New Approaches to Engineering Higher Education.* Retrieved from http://www.theiet.org/policy/panels/education/22-05-17.cfm

Ilyasov, D. F. (2010). Organizacija obuchenija pedagogov v uchrezhdenii povyshenija kvalifikacii kadrov [Organization of teachers training in the institution of skills development]. *Tomsk State Pedagogical University Bulletin.*, *1*(91), 64–69.

IMechE. (2010). When STEM? Institution of Mechanical Engineers.

IMechE. (2011). *Meeting the Challenge: Demand and Supply of Engineers in the UK.* Institution of Mechanical Engineers.

Improving Diversity in STEM. (2016). *A report by the Campaign for Science and Engineering (CaSE).* Retrieved from http:// www.sciencecampaign.org.uk/resource/ImprovingDiversityinSTEM2014/

IMS Content Packaging Information Model, Version 1.1.4 Final Specification. (2004). IMS Global Learning Consortium, Inc.

Inclusion in Russian universities. (2016) Retrieved from http://www.istok-audio.com/special/articles_for_specialists_in_t/detail.php?id=2902/

Inozemtseva, K., Bondaletova, E., & Borisova, T. (2016). Evolution of ESP as a methodology for teaching a foreign language for professional purposes in non-linguistic universities in Russia. *Humanitarian Research, 2*. Retrieved from http://human.snauka.ru/2016/02/13994

Internet Assigned Numbers Authority. (2017). *Service name and transport protocol port number registry*. Retrieved April 4, 2017, from http://www.iana.org/assignments/service-names-port-numbers/service-names-port-numbers.xhtml

Intranet: Academic—Sistema avtomatizacii upravlenija uchebnym processom VUZa (n.d.). [The system of automation of management of educational process of the University]. Retrieved from http://softwareinc.ru/solutions/intranet-academic/

Ionesov, V. I., & Nesterenko, V. M. (2013). Human being in the system of spatiotemporal relations: Projections and challenges of sociocultural communications. *Herald of Samara State Technical University*, *2*(20), 142–148.

Iskrenko, E. V., & Poulton, T. A. (2008). Problemno-orientirovannoe obuchenie: osobennosti metodiki prepodavaniya v Velikobritanii [The specificity of problem based learning: the technique of operation in St. St. George University of London, Great Britain]. Nauchnyie vedomosti – Scientific statements, 10(50), 214-218.

Ivachnenko, A. G. (1971). Polynomial theory of complex systems. *IEEE Transactions on Systems, Man, and Cybernetics*, *1*(4), 364–378. doi:10.1109/TSMC.1971.4308320

Ivanov, O. B., & Ivanova, S. V. (2015). Special features of education space formation in economic and social conditions inherent in the post-industrial epoch. *Lifelong Learning Proceedings of the 13th International Conference*, 43-47.

Ivanova, S. V., & Ivanov, O. B. (2016). Society demands for the quality of education as a factor of modern education space forming. In S.V. Ivanova & E.V. Nikulchev (Eds.), *SHS Web of Conferences (Vol. 29). 2016 International Conference on Education Environment for the Information Age (EEIA-2016).* Moscow, Russia: SHS. Retrieved April 10, 2017, from: http://www.shs-conferences.org/articles/shsconf/abs/2016/07/contents/contents.html

Ivanov, I., Lukin, A. F., & Soloviev, G. I. (2002). *Elektrotehnika. Osnovnyie polozheniya, primeryi i zadachi* [Electrical engineering. The main provisions, examples and exercises]. Sankt-Peterburg, Russia: Lan Publishing house.

Jeanes, E., & Huzzard, T. (Eds.). (2014). *Critical Management Research: Reflections from the Field.* Los Angeles, CA: SAGE Publications. doi:10.4135/9781446288610

John, D., Katherine, A. R., Elizabeth, J. M., Mitchell, J. N., & Daniel, T. W. (Eds.). (2013). Improving Students' Learning With Effective Learning Techniques Promising Directions From Cognitive and Educational Psychology. *Psychological Science in the Public Interest, 14*, 1. PMID:26173287

Jones, T. H., & Rocco, P. (1999). Research framework and dimensions for evaluating the effectiveness of educational technology systems on learning outcomes. *Journal of Research on Computing in Education, 32*(1), 17–27. doi:10.1080/08886504.1999.10782266

Jovic, A., Brkic, K., & Bogunovic, N. (2014). An overview of free software tools for general data mining. In *Proceedings of the 37th International Convention on Information and Communication Technology, Electronics and Microelectronics (MIPRO) 2014.* IEEE Xplore. 10.1109/MIPRO.2014.6859735

Kalistratov, A.P., Zaikin, S.I., Kuzovlev, V.I., & Semkin, P.S. (2016). Ispolzovanie snapshotov pri organizacii uchebnoy sredy na baze virtualnyh mashin. *Science and Education: Scientific Publication*, (7), 209-215.

Kalman, R. E. (1961). A new approach to linear filtering and prediction problems. *Transactions of the ASME –. Journal of Basic Engineering, 82*(1), 35–45. doi:10.1115/1.3662552

Kalugyan, K. H. (Ed.). (1999). Test systems in high school as a learning management tool. Rostov-on-Don.

Kamens, D.H., & McNeely, C.L. (2009). Globalization and the growth of international educational testing and national assessment. *Comparative Education Review, 54*(1), 5–25.

Kamens, D. H., & McNeely, C. L. (2009). Globalization and the growth of international educational testing and national assessment. *Comparative Education Review, 54*(1), 5–25. doi:10.1086/648471

Kamensky, E. G. (2015). Context of NBIC-technologies development: Institutions, ideology and social myths. *Mediterranean Journal of Social Sciences, 6*(6), 181–185.

Karasik, V. I. (2004). *Yazykovoi krug: lichnost', kontsepty, diskurs* [Language Circle: Personality, Concepts, and Discourse]. Moscow, Russia: Gnozis.

Karasik, V. I. (2009). *Yazykovye klyuchi* [Language Keys]. Moscow, Russia: Gnozis.

Karpenko, A. P. (2010). Measures the importance of concepts in semantic networks of ontology knowledge base. *Electron. Phys., 2010*(7). Available at: http://technomag.edu.ru/doc/151142.html

Karpenko, A. P., & Sokolov, N. (2008a). To Complexity estimation of semantic network into a tutoring system. *Science and Education, 2008*(11). Available at: http://technomag.bmstu.ru/doc/106658.html

Karpenko, A. P., & Sokolov, N. (2008b). Expanded semantic network of a tutoring system and its complexity. *Science and Education, 2008*(12). Available at: http://technomag.bmstu.ru/doc/111716.html

Karpicke, J., Blun, D., & Janell, R. (2011). Retrieval Practice Produces More Learning than Elaborative Studying with Concept Mapping. *Science, 331*(6018), 772-775.

Kassinopoulos, M. (2004). Student and staff mobility in EHEA during the first years of Bologna Process. ICEE 04.

Kassinopoulos, M. (2007). The three cycle structure in European Higher Education Institutions. *Proc. ICEE 07.*

Kassinopoulos, M., & Dodridge, M. (2004). Academic and professional recognition of UK degrees in Cyprus. *Proc. ICEE 04.*

Kazinik, M. S. (2010). *Secrets of the geniuses*. Moscow: Legan Press.

Kelbley, J., & Sterling, M. (2010). *Windows Server 2008 R2 Hyper-V: Insider's Guide to Microsoft's Hypervisor*. Indianapolis, IN: Wiley Publishing, Inc. doi:10.1002/9781118257357

Kelly, J.-P., Botea, A., & Koenig, S. (2008). Offline planning with hierarchical task networks in video games. In *Proceedings of the 4th Artificial Intelligence and Interactive Digital Entertainment Conference, AIIDE 2008* (pp. 60-65). Stanford, CA: AIIDE.

Kilpatrick, W. H. (1918). The project method. *Teachers College Record, 19*, 319–335.

Kim, W. C., & Mauborgne, R. (2005). *Blue ocean strategy: How to create uncontested market space and make competition irrelevant*. Boston: Harvard Business School Press.

King, J., & Magoulas, R. (2017). *2017 European Data Science Salary Survey.* O'Reilly Media, Inc. Retrieved July 18, 2017 from http://www.oreilly.com/data/free/files/2017-european-data-science-salary-survey.pdf

King, W. R. (Ed.). (2009). Knowledge Management and Organizational Learning. Annals of Information Systems, 3-12. doi:10.1007/978-1-4419-0011-1

Kirikov, I. A., Kolesnikov, A. V., Listopad, S. V., & Rumovskaya, S. B. (2015). Melkozernistie gibridnie intellektualnie sistemy. Chast 1: Lingvisticheskiy podhod [Fine-Grained Hybrid Intelligent Systems. Part 1: Linguistic Approach]. *Informatika i ee Primeneniya, 9*(4), 98–105.

Kirikov, I. A., Kolesnikov, A. V., Listopad, S. V., & Rumovskaya, S. B. (2016). Melkozernistie gibridnie intellektualnie sistemy. Chast 2: Dvunapravlennaya gibridizatsia [Fine-Grained Hybrid Intelligent Systems. Part 2: Bidirectional Hybridization]. *Informatika i ee Primeneniya, 10*(1), 96–105.

Kirikov, I., Kolesnikov, A., Listopad, S., & Soldatov, S. (2015). Geterogennie intellektualnie kompyuternie sistemy podderzhki prinyatia resheniy: Modeli coordinatsii i soglasovannosti [Heterogeneous Intelligent Computer Decision Support Systems: Models of Coordination and Consistency]. *Sistemy i Sredstva Informatiki, 25*(2), 96–110.

Kirkup, G., Zalevski, A., Maruyama, T., & Batool, I. (2010). *Women and men in science, engineering and technology: the UK statistics guide 2010.* Bradford, UK: the UKRC. https://www.wisecampaign.org.uk/uploads/wise/files/archive/final_sept_15th_15.42_ukrc_statistics_guide_2010.pdf

Klementiev, I. E., & Ustinov, V. A. (2011). *Akademija Microsoft: Vvedenie v oblachnye vychislenija* [Microsoft Virtual Academy: Introduction to cloud computing]. Retrieved April 20, 2017, from http://www.intuit.ru/studies/courses/673/529/lecture/11915

Knoll, M. (1997). The project method: Its vocational education origin and international development. *Journal of Industrial Teacher Education, 34*, 59–80.

Knyaginin, V. N., Meshkov, N. A., & Utolin, K. V. (2016). Advanced education in the information society. *International Review of Management and Marketing, 6*(3), 89–99.

Kochetov, A. I., Krapuhina, N. V., & Pronichkin, S. V. (2009). Razrabotka sistem podderzhki prinjatija reshenij dlja upravlenija kachestvom dejatel'nosti vuza [Development of decision support systems for quality management of the university. Human Ecology]. *Jekologiya Cheloveka*, *9*, 39–45.

Kohlert, C., Kohlert, M., Chistyakova, T., & Ivanov, A. (2010). Counterfeit-proofing based on the principle of randomness. *Kunststoffe international, 7*, 32-35.

Kol'tsova, R. V. (2012, September 28). 65 let Likhoslavl'skomu zavodu «Svetotekhnika» [Likhoslavl Factory "Svetotechnika" celebrates 65 years]. *Svetskaya zhizn' - Social life*. Retrieved from http://www.bl-g.ru/upload/iblock/231/svetskaya_zhizn_20_prevew.pdf

Kolesenkov, A.N., & Taganov A.I. (2015). Kontseptsiya geoinformatsionnoy tekhnologii monitoringa obrazovatel'nykh programm onlayn-obucheniya [Conception of geoinformation technology for monitoring of online-learning education programs]. In *Open and online education*. Tomsk: Publishing House of Tomsk State University.

Kolesnikova, I. A. (2009). *Theory and practice of modular transformation of learning environment at educational institution*. Saint-Petersburg.A. I. Herzen.

Kondrat'ev, J. V., & Chemezov, I. S. (2015). The change of Russian higher education to CDIO standards: Content, prospect, problem. In Proceedings of Voronezh State University (vol. 2, pp. 41-50). Academic Press.

Kornienko, I. G., Chistyakova, T. B., & Novozhilova, I. V. (2014). E-learning study of hard alloy production process management. *Izvestiya MGTU MAMI*, *5*(3), 157–163.

Korzhov, V. (1997). *Multi-level client-server system*. Retrieved March 31, 2017, from http://www.osp.ru/nets/1997/06/142618

Kostrov, B.V., & Baranchikov, A.I. (2013) Teoriya i metody issledovaniya modeley i algoritmov predstavleniya dannykh dlya predmetnykh oblastey s ranzhiruyemymi atributami [Theory and methods of researches of models and algorithms for representation of data for subject areas with ranged attributes]. *RSREU Vestnik, 47*, 59-64.

Kotel'nikov, V. (1959). *The Theory of Optimum Noise Immunity*. New York: McGraw-Hill Book Co.

Kovalenko, I. V. (2012). Theoretical bases of lifelong education research as an innovation system. *Izvestija of the Tula state University, 1*, 281-287.

Kremlin. (2015). *The list of instructions following the results of a meeting of the State Council in April 28, 2015*. Retrieved 06/07/2017 from http://kremlin.ru/catalog/ keywords/40/events/49214

Kubryakova, Y. S. (Ed.). (1996). *Kratkii slovar' kognitivnykh terminov* [Concise Dictionary of Cognitive Terms]. Moscow, Russia: MSU Press.

Kuhar, A. (2011). *Paravirtualizatsiya* [Para-virtualization]. Retrieved April 20, 2017, from www.vmgu.ru/articles/Paravirtualizatsiya

Kulkarni, P. (2012). *Reinforcement and systemic machine learning for decision making*. Hoboken, NJ: John Wiley & Sons, Inc. doi:10.1002/9781118266502

Kumaran, G., & Allan, J. (2005). Using names and topics for new event detection. In *Proceedings of the Conference on Human Language Technology and Empirical Methods in Natural Language Processing* (pp. 121-128). Stroudsburg, PA: Association for Computational Linguistics. 10.3115/1220575.1220591

Kumpikaite, V. (2008). Human resource development in learning organization. *Journal of Business Economics and Management*, *9*(1), 25–31. doi:10.3846/1611-1699.2008.9.25-31

Kurovskaya, Y. G. (2014). Osnovy kognitivnoi lingvistiki cherez prizmu pedagogicheskoi nauki [Basics of cognitive linguistics through the prism of pedagogical science]. *Novoe v psikhologo-pedagogicheskikh issledovaniyakh, 3*, 90-97.

Kurovskaya, Y. G. (2016). Linguistics and Cognitive Linguistics as Tools of Pedagogical Discourse Analysis. In S.V. Ivanova & E.V. Nikulchev (Eds.), *SHS Web of Conferences (Vol. 29). 2016 International Conference on Education Environment for the Information Age (EEIA-2016).* Moscow, Russia: SHS. Retrieved April 10, 2017, from: http://www.shs-conferences.org/articles/shsconf/abs/2016/07/contents/contents.html

Kustov, V. N., & Novozhilova, I. V. (2012). Electronic training of specialists for the systems of professional development. *Programmnyie produktyi i sistemyi, 2*, 125-127.

Kuzenkov, O. A., & Tikhomirov, V. V. (2013). Using the methodology of "Tuning" in the development of national ICT competences framework. In *Modern Information Technology and IT Education.* Moscow, Russia: Fund for Promotion of Internet media, IT education, human development League Internet Media. Retrieved from https://elibrary.ru/contents.asp?issueid=1371690

Kuznetsova, T., & Kirsanova, G. (2015). *English for students of Optics.* Moscow, Russia: Bauman State Technical University Press.

Lakoff, G. (1987). *Women, fire, and dangerous things: What categories reveal about the mind.* Chicago: University of Chicago Press. doi:10.7208/chicago/9780226471013.001.0001

Lande, D. V., Braichevskii, S. M., Grigoryev, A. N., Darmokhval, A. T., & Radetzkii, A. B. (2007). Vyyavlenie novyh sobytiy iz potoka novostey [Detection of new events from news flow]. In *Kompyuternaya lingvistika i intellektualnye tekhnologii. Trudy mezhdunarodnoy konferentsii «Dialog–2007»* (pp. 349-352). Moscow, Russia: Nauka. Retrieved from http://dwl.kiev.ua/art/dia07/dia07.pdf

Langacker, R. W. (1987). Foundations of cognitive grammar: Vol. 1. *Theoretical prerequisites.* Redwood City, CA: Stanford University Press.

Langacker, R. W. (1991). Foundations of Cognitive Grammar: Vol. 2. *Descriptive Application.* Redwood City, CA: Stanford University Press.

Lanir, J., Maltz, M., & Rotman, S.R. (2007). Comparing multispectral image fusion methods for a target detection task. *Optical Engineering, 46*(6).

Lantsberg, A. V., Troitzch, K. G., & Buldakova, T. I. (2011). Development of the electronic service system of a municipal clinic (based on the analysis of foreign Web resources). *Automatic Documentation and Mathematical Linguistics, 45*(2), 74–80. doi:10.3103/S0005105511020075

Lantz, B. (2013). *Machine learning with R.* Retrieved from https://www.packtpub.com/big-data-and-business-intelligence/machine learning-r

Lapteva, T. (2012). *Some aspects of using the CLIL methodology for teaching foreign languages.* The Siberian State Geodesic Academy Journal.

Larose, D. T., & Larose, C. D. (2015). *Data mining and predictive analytics* (2nd ed.). Hoboken, NJ: John Wiley & Sons, Inc.

Lattanze, A. J. (2008). *Architecting Software Intensive Systems: A Practitioners Guide.* New York, NY: Auerbach Publications. doi:10.1201/9781420045703

Lawson, L. (2011). *Big Data platform should support data exploration.* ITBusinessEdge. Retrieved July 18, 2017 from http://www.itbusinessedge.com/cm/blogs/lawson/big-data-platform-should-support-data-exploration/?cs=48159

Lebedeva, E. (2014). Improvement of foreign language competence of law students. Science and education: Economy and Economics. *Entrepreneurship Law and Management*, *7*, 17–21.

Lee, Y.H., & Medioni, G. (2016). RGB-D camera based wearable navigation system for the visually impaired. In *Computer Vision and Image Understanding* (pp. 3-20). New York, NY: Elsevier.

Lee, A. (2014). Students with disabilities choosing science, technology, engineering and math (STEM) Majors in Postsecondary Institutions. *Journal of Postsecondary Education and Disability*, *27*(3), 268.

Leonenkov, A. V. (2005). Nechetkoye modelirovaniye v MATLAB i fuzzyTECH [Fuzzy modeling in MATLAB and fuzzyTECH]. St. Petersburg: Academic Press.

Lewis, D. (2016). *The top 10 companies working on education in virtual reality and augmented reality.* Touchstone Research, Inc. Retrieved June 2, 2017, from http://touchstoneresearch.com/the-top-10-companies-working-on-education-in-virtual-reality-and-augmented-reality/

Lia, B. N., Fua, B. B., & Dong, M. C. (2008). Development of a mobile pulsewaveform analyzer for cardiovascular health. *Computers in Biology and Medicine*, *38*(4), 438–445. doi:10.1016/j.compbiomed.2008.01.008 PMID:18328471

Li, B., Springer, J., Bebis, G., & Gunes, M. H. (2013). A survey of network flow applications. *Journal of Network and Computer Applications*, *36*(2), 567–581. doi:10.1016/j.jnca.2012.12.020

Lipaev, V. V. (2006). Programmnaya engeneria. Metodologicheskie osnovy [Software Engineering Methodological Foundations]. Moscow: TEIS.

Li, Z., Wang, B., Li, M., & Ma, W.-Y. (2005). A probabilistic model for retrospective news event detection. In *Proceedings of the 28th Annual International ACM SIGIR Conference on Research and Development in Information Retrieval* (pp. 106-113). New York, NY: ACM. 10.1145/1076034.1076055

Logvinov, S. I., Romanov, V.A. (2014). Primenenie informacionno-analiticheskih sistem v obrazovatel'nom processe vuza: zveno "Fakul'tet kafedra" [Application of information and analytical systems in the educational process of the university: Faculty Department]. *Sovremennye problemy nauki i obrazovanija, 1*, 12-19.

Lorenzo, N., & Gallon, R. (2015). Higher Education and Globalization. In F. M. Ribeiro, Y. Politis, & B. Culum (Eds.), *New Voices in Higher Education Research and Scholarship* (pp. 127–156). Hershey, PA: IGI Global. doi:10.4018/978-1-4666-7244-4.ch007

Lukatskii, M. A. (2013). O mezhdistsiplinarnoi issledovatel'skoi initsiative, ob"edinivshei pedagogiku i kognitivnuyu lingvistiku, i o perspektivakh razrabotki pedagogicheskoi semiologii [On Interdisciplinary Research Initiative, Uniting Pedagogy and Cognitive Linguistics, and on the Perspectives of Pedagogical Semiology]. *Otechestvennaya i zarubezhnaya pedagogika, 5*(14), 62–76.

Lukianova, N., & Fell, E. (2016). Support for students with disabilities in the UK universities. *SHS Web of Conferences*, 28. doi: 10.1051hsconf/20162801066

Lund, D. (2009). Rent taxation for nonrenewable resources. *Annual Review of Resource Economics*, *1*(1), 287–308. doi:10.1146/annurev.resource.050708.144216

Lunev, A., Fedotova, A., & Rybakov, A. (2014). Complex Strategy of CDIO Initiative Implementation in a Regional Russian University. *Proceedings of the 10th International CDIO Conference.*

Lunev, A., Zaripova, V., & Petrova, I. (2013). Implementation of CDIO Initiative Approach at a Russian Regional University. *Proceedings of the 9th International CDIO Conference.*

Lyubimov, D., & Palumbo, A. (2016). *Apache Mahout: Beyond MapReduce. Distributed algorithm design.* CreateSpace Independent Publishing Platform.

Macías, J. A. (2012). Enhancing project-based learning in software engineering lab teaching through an e-portfolio approach. *IEEE Transactions on Education, 4*(55), 502–507. doi:10.1109/TE.2012.2191787

Magazinnikova, A. L. (2015). Ob izmeneniyah kursa matematiki dlya studentov-radiotehnikov [The alteration of mathematics course for students-radio engineers]. *Izvestiya vyisshih uchebnyih zavedeniy. Fizika, 8/3*(58), 324-326.

Maher, M. L., Merrick, K., & Saunders, R. (2008). Achieving creative behavior using curious learning agents. In *AAAI Spring Symposium 2008 - Creative Intelligent Systems.* Stanford, CA: AAAI.

Makarenko, S. I. (2009). *Intellektual'nye informacionnye sistemy: uchebnoe posobie* [Intellectual informational systems. Tutorial]. Stavropol, Russia: SF MGGU im. M. A. Sholohova.

Malakhov, A. A., Smirnova, E. V., Vishnyakov, N. V., Kholomina, T. A., & Lunyakov, A. E. (2016). Criterion identification for student's meta-competences evaluation during laboratory work execution on an unique scientific equipment. *Science and Education,* (7): 51–72. Available at http://technomag.neicon.ru/doc/843958.html

Malhotra, K., Gardner, S., & Patz, R. (2007). Implementation of elliptic-curve cryptography on mobile healthcare devices. *IEEE International Conference on Networking, Sensing and Control,* 239-244. 10.1109/ICNSC.2007.372784

Manesh, H. F., & Schaefer, D. (2010). Virtual Learning Environments for Manufacturing. In W. Ritke-Jones (Ed.), *Virtual Environments for Corporate Education: Employee Learning and Solutions* (pp. 89–109). Hershey, PA: IGI Global. doi:10.4018/978-1-61520-619-3.ch006

Marchenko, N., Osipova, S., & Amautov, A. (2016). New Role of Employer in the Educational Process of Metallurgy Programme. *Proceedings of the 12th International CDIO Conference.*

Margaryan, T. (2014). Formy avtonomnogo obuchenia v tekhnicheskom vuze v gruppakh ESP (angliyskiy dlia specialnykh tseley [Autonomous learning in ESP groups at a technical university]. *Gumanitarny Vestnik MGTU im. N.E. Baumana - Journal of Humanities Bulletin of BMSTU, 2*(16). Retrieved May 27, 2016, from http://hmbul.ru/articles/166/166.pdf

Marginson, S. (2009). Global perspectives and strategies of Asia-Pacific research universities. *Evaluation in Higher Education, 3*(2), 1–43.

Marsh, D. (1994). Bilingual Education & Content and Language Integrated Learning. Paris, France: University of Sorbonne & International Association for Cross-cultural Communication, Language Teaching in the Member States of the European Union (Lingua).

Marsh, D. (2002). *CLIL/EMILE – The European Dimension: Actions, Trends and Foresight Potential Public Services. Contract DG EAC.* European Commission.

Martinez, F., Herrero, L. C., & de Pablo, S. (2011). Project-based learning and rubrics in the teaching of power supplies and photovoltaic electricity. *IEEE Transactions on Education, 1*(54), 87–96. doi:10.1109/TE.2010.2044506

Martynov, V. V., & Filosova, E. I. (2014). Methods for developing CSRP-system in education for human capacity SMART-region. In *Proceedings of the Workshop on Computer Science and Information Technologies (CSIT2014)* (*vol. 2*, pp. 180-184). Ufa State Aviation Technical University Press.

Martynov, V. V., & Zaytseva, A. I. (2015). Educational methodical complex development support system. In *Proceedings of the 2015 Workshop on Computer Science and Information Technologies (CSIT2015)* (*vol. 1*, pp. 183-185). Ufa State Aviation Technical University Press.

Martynov, V. V., Filosova, E. I., & Filosova, V. K. (2015). The use of ontologies in the task of drawing up training plans for the requirements of employers' information technology. In *Proceedings of the International Scientific-Practical Conference "Problems and Solutions"* (vol. 1, pp. 242-246). Publishing House "East Press".

Martynov, V. V., Filosova, E. I., & Guzairov, M. B. (2014). Improving the efficiency of training requirements for employers. In *Proceedings of the International Scientific-Practical Conference "Innovative Information Technologies"* (pp. 376-381). Moscow: National Research University Higher School of Economics.

Martynov, V. V., Filosova, E. I., Sharonova, J. V., & Shiryaev, O. V. (2015). The technology of ontological analysis in educational activities. In *Proceedings of the Workshop on Computer Science and Information Technologies (CSIT'2015) (vol. 1*, pp. 173-178). Ufa State Aviation Technical University Press.

Marzinotto, A., Colledanchise, M., Smith, C., & Ogren, P. (2014). Towards a unified behavior trees framework for robot control. *Proceedings of 2014 IEEE International Conference on Robotics and Automation (ICRA 2014)*, 5420-5427. 10.1109/ICRA.2014.6907656

Massachusetts Institute of Technology. (2017). *PhD program in social & engineering systems*. Retrieved July 18, 2017 from https://idss.mit.edu/academics/ses_doc/

MATLAB Neural Network Toolbox. (n.d.). Retrieved June 18, 2016, from http://www.mathworks.com/products/neural-network

Matzdorf, F., & Greenwood, J. (2015). *Student choice, league tables and university facilities*. Retrieved May 2, 2017, http://shura.shu.ac.uk/10423/1/EuroFM2015_StudentChoicesLeagueTables%26Facilities_final.pdf

Ma, Y., Fang, J., Wang, W., & Li, J. (2014). Decoupled observability analyses of error states in INS/GPS integration. *Journal of Navigation*, *67*(03), 473–494. doi:10.1017/S0373463313000829

McCulloch, W. S., & Pitts, W. (1943). A logical calculus of the ideas immanent in nervous activity. *The Bulletin of Mathematical Biophysics*, *5*(4), 115–133. doi:10.1007/BF02478259

McDermott, B. (2014). *Winners dream: A journey from corner store to corner office*. New York, NY: Simon & Schuster.

McDonough, P., Antonio, A., Walpole, M., & Perez, L. (1998). College Rankings: Democratized College Knowledge for Whom? *Research in Higher Education*, *39*(5), 513–537. doi:10.1023/A:1018797521946

Medvedev, V. E., & Dobriakov, A. A. (2013). Conceptual clauses and peculiarities of the cognitively-structured training of engineering elite. In *Proceedings of the International Conference on Interactive Collaborative Learning, (ICL)* (pp. 203-207). Academic Press.

Melnik, N. M. (2013a). Conception and technology of qualification upgrading for engineers in the innovative development of economics. *Herald of Samara State Technical University. Psychological-Pedagogical Sciences*, *1*(19), 86–93.

Melnik, N. M. (2013b). Innovative technology for global competition leader training. *Herald of Samara State Technical University. Psychological-Pedagogical Sciences*, *2*(20), 129–141.

Melnik, N. M., & Nesterenko, V. M. (2007). *Conception of the evolutional-activity professional education*. Moscow: VLADOS.

MEPhI MA educational programs. (2017). *Unikal'nyye obrazovatel'nyye programmy dlya nabora v 2017 godu* [Unique Educational Programs for the 2017 Recruitment]. Retrieved from https://admission.mephi.ru/admission-2017/magistracy/unique

Merritt, J., Lee, M. Y., Rillero, P., & Kinach, B. M. (2017). Problem-based learning in K-8 mathematics and science education: A literature review. *Interdisciplinary Journal of Problem-Based Learning, 11*(2). Retrieved online at 10.7771/1541-5015.1674

Mertins, K. V. (2012). Higher liberal education prediction by way of educational technologies. *European Researcher. Series A, 2*(17), 205–209.

MetaMath. (2016). Objectives. In *MetaMath project official web-site*. Retrieved from http://www.metamath.eu/objectives/

MEXT. (2012). *Selection for the FY2012 Re-Inventing Japan Project*. Retrieved from http://www. mext. go.jp/en/policy/education/highered/title02/detail02/sdetail02/sdetail02/1374092.htm

MEXT. (2014). *Selection for the FY2014 Top Global University Project*. Retrieved from http:// docplayer. net/16379288-Selection-for-the-fy-2014-top-global-university-project-we-hereby-nnounce-the-selectionof-universities-for-the-top-global-university-project.html

Microsoft. (2017a). *Microsoft Azure solutions*. Retrieved July 18, 2017 from https://azure.microsoft.com/en-us/solutions/

Microsoft. (2017b). *Introduction to R Server and open-source R capabilities on HDInsight*. Retrieved July 18, 2017 from https://docs.microsoft.com/en-us/azure/hdinsight/hdinsight-hadoop-r-server-overview

Microsoft. (2017c). *Graph Engine. Serving big graphs in real-time*. Retrieved July 18, 2017 from https://www.graphengine.io/

Microsoft. (2017d). *Microsoft Azure Machine Learning Studio*. Retrieved July 18, 2017 from https://studio.azureml.net/

Ministry of Education and Science of the Russian Federation. (2013). *The Russian Federation State Programme on the Development of Education for the period of 2013-2020*. Retrieved from http://government.ru/programs/202/events/

Ministry of Education and Science of the Russian Federation. (2014). *Federal Educational Standard (FES) for bachelor's studies in Physics*. Retrieved from http://www.edu.ru/db/mo/Data/d_14/m937.pdf/

Ministry of Education and Science. (2013). *Strategy for the development of the system of training of personnel and the development of applied qualifications in the Russian Federation 2013-2020*. Retrieved 06/13/2017 from http://profiok.com/files/minobr-education.pdf

Mirzoyan, L.A. (2013). *Primeneniye CDIO v Mosckovskom Aviatsionnom Institute* [Implementation of CDIO in Moscow Aviation Institute]. The report by Lolita A. Mirzoyan, Head of Department of Educational Programs NUT MAI, made at the CDIO Academy training program. Moscow, Russia: SkolTech.

Mnasser, A., Bouani, F., & Ksouri, M. (2016). Neural Networks Predictive Controller Using an Adaptive Control Rate. In I. Management Association (Ed.), Psychology and Mental Health: Concepts, Methodologies, Tools, and Applications (pp. 614-633). Hershey, PA: IGI Global. doi:10.4018/978-1-5225-0159-6.ch026

Moallem, M., Morge, S.P., Narayan, S., & Tagliarini, G.A. (2015). The power of computational modeling and simulation for learning STEM content in middle and high schools. *Improving K-12 STEM Education Outcomes through Technological Integration*, 135-171.

Moats, J. (2015). Influences on the acceptance of innovative technologies used in learning opportunities: A theoretical perspective. In F. M. Nafukho & B. J. Irby (Eds.), *Handbook of Research on Innovative Technology Integration in Higher Education* (pp. 262–281). Hershey, PA: IGI Global. doi:10.4018/978-1-4666-8170-5.ch013

Model View Culture. (2016). *Q&A with STEM Professionals with Disabilities*. Retrieved from http://www.modelview-culture.com/pieces/qa-with-stem-professionals-with-disabilities/

Moeini, S. (2014). *6 (very useful!) Approaches to identify research gaps and generate research questions.* Retrieved July 18, 2017 from https://www.linkedin.com/pulse/20140912150946-275561203-6-very-useful-approaches-to-identify-research-gaps-and-generate-research-questions/

Monsalve, J. C., Uribe, A., Cardona-Gil, J. A., Osorio, M., Hincapie, C. A., & Isaza, C. A. (2016). Development of an automatic control system employing CDIO standards and competence-based learning. In *Proceedings of the 15th International Conference on Information Technology Based Higher Education and Training (ITHET).* Istanbul: IEEE. 10.1109/ITHET.2016.7760755

Moodle Docs 2.8. (n.d.). Retrieved from https://docs.moodle.org/28/en/Main_page

Moon, N. W., Todd, L. R., Morton, D. L., & Ivey, E. (2012). *Accommodating Students with Disabilities in Science, Technology, Engineering, and Mathematics (STEM).* Findings from Research and Practice for Middle Grades through University, National Science Foundation. Retrieved from http://www.Educationhttp://advance.cc.lehigh.edu/sites/advance.cc.lehigh.edu/files/accommodating.pdf/

Moore, D., Gorra, A., Adams, M., Reaney, J., & Smith, H. (Eds.). (2011). Disabled Students in Education: Technology, Transition, and Inclusivity. IGI Global. doi:10.4018/978-1-61350-183-2

Moore, S., May, D., & Wold, K. (2014). Developing Cultural Competency in Engineering through Transnational Distance Learning. In Information Resources Management Association (Ed.), Cross-Cultural Interaction: Concepts, Methodologies, Tools, and Applications (pp. 1571-1589). Hershey, PA: IGI Global. doi:10.4018/978-1-4666-4979-8.ch089

Moore, A. W., & Zuev, D. (2005). Internet traffic classification using Bayesian analysis techniques. In *Proceedings of the 2005 ACM SIGMETRICS international conference on Measurement and modeling of computer systems.* Association for Computing Machinery. 10.1145/1064212.1064220

Moore, A., Crogan, M., & Zuev, D. (2005). *Discriminators for use in flow-based classification.* London, UK: Intel Research Tech.

Moorthy, A. K., Wang, Z., & Bovik, A. C. (2011). Visual perception and quality assessment, in Optical and Digital Image Processing. Wiley-VCH Verlag GmbH & Co. KGaA.

Moscow Centre for Complex Rehabilitation of the Deaf and Hard-of-Hearing (GUIMC). (n.d.). Retrieved from http://www.guimc.bmstu.ru/center/about_center/

Mosey, S., Guerrero, M., & Greenman, A. (2016). Technology entrepreneurship research opportunities: Insights from across Europe. *Journal of Technology Transfer,* 1-9. Retrieved 06/20/2017 from http://link.springer.com/article/10.1007/s10961-015-9462-3#copyrightInformation

Mueller, R., & Rockerbie, D. (2005). Determining demand for university education in Ontario by type of student. *Economics of Education Review, 24*(4), 469–483. doi:10.1016/j.econedurev.2004.09.002

Muratova, E., & Tayurskaya, M. (2014). Stakeholders' Evaluation of Learning Outcomes in Educational Programs Improvement. *Proceedings of the 10th International CDIO Conference.*

Muratova, E. I., Dvoretsky, S. I., & Voyakina, E. Y. (2013). *Organization of postgraduate students training in the technical field of sciences. International Conference on Interactive Collaborative Learning,* 46(5), 437-445. 10.1109/ICL.2013.6644622

Murphy, E., & Rodríguez-Manzanares, M. A. (2013). *Activity Theory Perspectives on Technology in Higher Education.* Portland, OR: Book News Inc.; doi:10.4018/978-1-4666-4590-5

Mustar, P., Wright, M., & Clarysse, B. (2008). University spin-off firms: Lessons from ten years of experience in Europe. *Science & Public Policy*, *35*(2), 67–80. doi:10.3152/030234208X282862

NAFI. (2016). *Small and medium businesses do not seek external financing.* Retrieved 06/13/2017 from http://nafi.ru/analytics/malyy-i-sredniy-biznes-ne-obrashchaetsya-za-vneshnim-finansirovaniem/

National Research Nuclear University. (2017). *Obrazovatelniye standarty vyshevo obrazovaniya Natsionalnovo issledovatelskovo yadernovo universiteta MIFI. Napravleniye podgotovki 14.03.02 Yaderniye fizika I tehnologii* [Educational Standard of Higher Education of National Research Nuclear University "MEPhI". Direction of training 14.03.02 Nuclear Physics and Technologies]. Retrieved from https://mephi.ru/obrdeyat/edst-vo.php

National Science Foundation, National Center for Science and Engineering Statistics. (2015) Retrieved from https://nsf.gov/statistics/2017/nsf17310/static/data/tab9-35.pdf

Naumov, I.S., &Vykhovanets, V.S. (2016). Using syntactic text analysis to estimate educational tasks' difficulty and complexity. *Automation and Remote Control, 77*(1), 159-178.

NEA. (2012). *Nuclear education and training: From concern to capability.* Organization for Economic Co-operation and Development. Nuclear Energy Agency. Retrieved July 18, 2017 from http://www.oecd-nea.org/ndd/pubs/2012/6979-nuclear-education.pdf

Neshchadim, I. (2013). Developing communicative competence in engineering students through Internet-based project work. In *Proceedings of the 1st International Conference on Teaching English for Specific Purposes in Serbia* (pp.372-376). Nish, Serbia: University of Nish.

Nesterenko, V. M. (2011). Parametric knowledge management in the professional problem solving process. *Herald of Samara State Technical University. Psychological-Pedagogical Sciences*, *1*(15), 89–99.

Nesterenko, V. M. (2015). Conceptual aspects of the solution system generation of actual professional problems. *Herald of Samara State Technical University*, *3*(27), 161–170.

Neusypin, K. A., Proletarsky, A. V., Shen, K., Liu, R., & Guo, R. (2014). Aircraft self-organization algorithm with redundant trend. *Journal of Nanjing University of Science and Technology*, *5*, 602–607.

Newton, J. (2002). Views from Below: Academics coping with quality. *Quality in Higher Education*, *8*(1), 39–61. doi:10.1080/13538320220127434

Ninomiya, A., & Urabe, M. (2011). Impact of PISA on education policy: The case of Japan. *Pacific-Asian Education, 23*(1), 23–30.

Novikov A.I. (2015). *Nachala Lineinoi Algebra I Analitichescaya Geometriya* [Foundations of linear algebra and analytic geometry]. Moscow, Russia: Fizmatlit.

Novikov, A. I., Sablina, V. A., Nikiforov, M. B., & Loginov, A. A. (2015). The Contour Analysis and Image-Superimposition Problem in Computer Vision Systems. Pattern Recognition and Image Analysis, 25(1), 73-80. doi:10.1134/S1054661815010149

Novikov, A.I., Sablina, V.A., Efimov, A.I., & Nikiforov, M.B. (2016). Contour Analysis in the tasks of real and virtual images superimposition. *Journal Coupled Systems and Multiscale Dynamics, 4*(4), 251-259.

Office of Educational Technology. (2016). *Openly Licensed Educational Resources*. Retrieved from https://tech.ed.gov/open/

Oliveira, L., Vasconcelos, N., Queiroz, F., Queiroz, J., & Hékis, H. (2011). Contribution of integrated management systems to university management: Case study of the federal university of Rio Grande Do Norte. *Journal of Social Sciences*, *7*(3), 415–422. doi:10.3844/jssp.2011.415.422

On measures to implement state social policy. (2012, May 7). Decree of the President of the Russian Federation vol. 577.

Onykij, B. N., Mishulina, O. A., & Scherbakov, I. B. (1996) Laboratornyj praktikum po kursu 'Teoriya nejronnyh setej' [Laboratory works for course 'Theory or neural networks']. In Fizicheskoe obrazovanie v vuzah, 2(1), 77-90.

OPORA ROSII. (2017). *Anti-crisis monitoring of small and medium-sized enterprises in Russia.* Retrieved 06/13/2017 from http://www.oporamo.ru/upload/users/b473c3114edc79b3618d0d7b f2d090b6.pdf

OPORA. (2017). *OPORA RSBI Index.* Retrieved 06/13/2017 from http://opora.ru/projects/indeks-opory-rsbi/?sphrase_id=1662

Oppenhein, A., & Schafer, R. (2012). *Digital Signal Processing*. Englewood Cliffs, NJ: Prentice-Hall.

Osmakov, V. S., & Pastukhov, V. A. (Eds.). (2015). Inzhiniring I promyshlennyi dizain [Engineering and Industrial Design]. Moscow, Russia: Onebook.ru. (in Russian)

Osovsky, C. (Ed.). (2002). Neural networks for information processing. Finance and Statistics, 235-238.

Palahicky, S. (2014). Utilizing learning management system (LMS) tools to achieve differentiated instruction. In J. Keengwe & J. J. Agamba (Eds.), *Models for improving and optimizing online and blended learning in higher education* (pp. 12–33). Hershey, PA: IGI Global.

Pang, A. (2016). *Worldwide enterprise applications market to hit $208B by 2020 as replatforming intensifies.* Apps Run The World. Retrieved July 18, 2017 from https://www.appsruntheworld.com/worldwide-enterprise-applications-market-to-hit-208b-by-2020-as-replatforming-intensifies/

Paradiso, R., Loriga, G., & Taccini, N. (2005). A wearable health care system based on knitted integrated sensors. *IEEE Transactions on Information Technology in Biomedicine*, *9*(3), 337–344. doi:10.1109/TITB.2005.854512 PMID:16167687

Peat, M., & Franklin, S. (2002). Supporting student learning: The use of computer–based formative assessment modules. *British Journal of Educational Technology*, *33*(5), 515–523. doi:10.1111/1467-8535.00288

Pegat, A. (2013) Nechetkoye modelirovaniye i upravleniye [Fuzzy simulation and management]. Binom.

Pentreath, N. (2015). *Machine learning with Spark.* Birmingham, UK: Packt Publishing Ltd.

Perfilieva, I., Yarushkina, N., Afanasieva, T., & Romanov, A. (2016). Web-Based System for Enterprise Performance Analysis on the Basis of Time Series Data Mining. *Proceedings of the First International Scientific Conference on Intelligent Information Technologies for Industry" (IITI'16), Rostov-on-Don, Russia.* 10.1007/978-3-319-33609-1_7

Perkins, J. (2013). *Review of Engineering Skills.* Retrieved from https://www.gov.uk/government/uploads/system/uploads/attachment_data/file/254885/bis-13-1269-professor-john-perkins-review-of-engineering-skills.pdf

Pesoshin, V.A., Zvezdin, V.V., Iljuhin, A.N., Saubanov, R.R., & Saubanov, R.R. (2016). Avtomatizirovannaya sistema testirovaniya kak instrument povisheniya kachestva otsenki znaniy. *Science and Education: Scientific Publication*, (7), 137-142.

Petrova, I., Zaripova, V., Ishkina, E., Militskaya, S., Malikov, A., Kurmishev, N., ... Zakharova, I. (2013). *Tuning Russia: Reference Points for the Design and Delivery of Degree Programmes in Information and Communication Technologies.* Bilbao, Spain: University of Deusto. Retrieved from http://www.deusto-publicaciones.es/index.php/main/libro/1041/eu

Petrov, D. N., Chistyakova, T. B., & Charyikov, N. A. (2014). Mathematical model for management training in fullerene synthesis processes. *Bulletin of the Saint Petersburg State Institute of Technology*, *52*(26), 72–79.

Petrovskaya, T. (2013). Using CDIO Concept to Develop Engineering Education in Tomsk Polytechnic University. *Proceedings of the 9th International CDIO Conference*.

Pinker, S. (2013a). *Substantsiya myshleniya: Yazyk kak okno v chelovecheskuyu prirodu* [The Stuff of Thought: Language as a Window into Human Nature]. Moscow, Russia: Knizhnyi dom LIBROKOM.

Pinker, S. (2013b). *Language, Cognition, and Human Nature: Selected Articles*. New York, NY: Oxford University Press. doi:10.1093/acprof:oso/9780199328741.001.0001

Pittel, K., & Bretschger, L. (2008). *Sectoral heterogeneity, resource depletion, and directed technical change: Theory and policy* (Working paper 08/96). CER-ETH – Center of Economic Research at ETH Zurich, Switzerland. Retrieved from https://www.ethz.ch/content/dam/ethz/special-interest/mtec/cer-eth/cer-eth-dam/documents/working-papers/wp_08_96.pdf

Pokrovsky, N. E. (Ed.). (2010). Virtualizacija mezhuniversitetskih i nauchnyh kommunikacij: metody, struktura, soobshhestva [Virtualization of inter-University and academic communication methods, structure, communities]. Moscow: SoPSo.

Polly, D., Martin, C., Wang, C., Lambert, R.G., & Pugalee, D. (2016). Supporting the enactment of standards-based mathematics pedagogies: The cases of the CoDE-I and APLUS projects. *Innovative Professional Development Methods and Strategies for STEM Education*, 137-148.

Poncela, A. (2013). A blended learning approach for an electronic instrumentation course. *International Journal of Electrical Engineering*, *50*, 1–18.

Ponkratov, V. V. (2014). Tax maneuver in Russian oil production industry. *Neftyanoe khozyaystvo – Oil Industry*, *9*, 58-61.

Ponkratov, V. V., & Pozdnyaev, A. S. (2016). The oil production taxation in Russia – consequences of tax maneuver. *Neftyanoe khozyaystvo – Oil Industry*, *3*, 24-27.

Prado, M., Roa, L., Reina-Tosina, J., Palma, A., & Milan, J. A. (2002). Virtual center for renal support: Technological approach to patient physiological image. *IEEE Transactions on Biomedical Engineering*, *49*(12), 1420–1430. doi:10.1109/TBME.2002.805454 PMID:12542237

Pressman, R. (2009). *Software Engineering: A Practitioner's Approach*. New York, NY: McGraw-Hill Education.

Primary Engineer. (2017). Retrieved from http://home.primaryengineer.com/

Prishchepa, S. V. (2017). Vyyavlenie sobytiya, ego subekta i obekta v tekstovyh dokumentah. [Identification of an event, its subject and object in text documents] In *Informatsionnye tekhnologii i bezopasnost. Materialy XVI Mezhdunarodnoy nauchno-prakticheskoy konferentsii ITB-2016* (pp. 121–129). Kiev, Ukraine: Institute for Information Recording of the NAS of Ukraine. Retrieved from http://dwl.kiev.ua/art/itb2016/ITB-2016-pdf.pdf

Proletarskij, A. V., & Neusypin, K. A. (2014). Osobennosti ispol'zovanija sovremennyh informacionnyh tehnologij v obrazovanii [Features of using modern information technologies in education]. *European Social Science Journal*, *1-1*(40), 63-65.

Proletarsky, A. V., Shen, K., & Neusypin, K. A. (2015). Intelligent control systems: Contemporary problems in theory and implementation in practice. In *Proceedings of the 2015 5th International Workshop on Computer Science and Engineering: Information Processing and Control Engineering*. Moscow: The Science and Engineering Institute.

Providing support across science, technology, engineering, mathematics, and medicine for disabled workers, current and aspiring disabled students and their teachers. (2016). Retrieved from http:// www.stemdisability.org.uk/

Prudnikova, N. (2013). ESP teaching at the institutions of higher education in modern Russia: problems and perspectives (pp. 390-396). In *BCES Conference Books: Education in one world: Perspectives from different nations* (vol. 11). Sofia, Bulgaria: University of Sofia. 10.2139srn.2232203

Pruss, Y. V. (2013, February). To the problem of exploration of technogenic complex of Kolyma. *Gornyi Zhurnal, 2*, 38–40.

QS World University Rankings® 2015/16. (2015, August 10). Retrieved from Web site World University Rankings: https://www.topuniversities.com

Rachford, J. (2017). Qualified People – It is a Safety and Skill Set Thing. *Transactions on Industry Applications, 53*(1), 700–708. doi:10.1109/TIA.2016.2604223

Radinsky, K., & Horvitz, E. (2013). Mining the web to predict future events. In *Proceedings of the Sixth ACM International Conference on Web Search and Data Mining* (pp. 255-264). New York, NY: ACM. 10.1145/2433396.2433431

RAEng. (2007). *Educating Engineers for the 21st Century.* London: Royal Academy of Engineering.

Raimond, Y., & Abdallah, S. (2007). *The event ontology.* Retrieved April 21, 2017, from http://motools.sourceforge.net/event/event.html

Rajala, S. (2012). Beyond 2020: preparing engineers for the Future. *Proc. IEEE,* 100. 10.1109/JPROC.2012.2190169

Rani, R. M., & Ismail, W. R. (2012). Operator allocation in labor-intensive manufacturing system. In *ICSSBE 2012 - Proceedings, 2012 International Conference on Statistics in Science, Business and Engineering: "Empowering Decision Making with Statistical Sciences"* (pp. 433-436). Academic Press. 10.1109/ICSSBE.2012.6396602

Rauhvargers, A. (2008). Implementation of the Lisbon Recognition Convention in the countries participating in the Bologna Process. In A. Rauhvargers & S. Bergan (Eds.), New Challenges in Recognition, Council of Europe Higher Education Series No. 10 (pp. 13–27). Academic Press.

Rauhvargers, A. (2011). *Global university rankings and their impact.* Retrieved May 2, 2017, from http://www.eua.be/Libraries/publications-homepage-list/Global_University_Rankings_and_Their_Impact.pdf?sfvrsn=4

Reaney, J., Gorra, A., & Hassan, H. (2011). Support for disabled students in higher education: A move towards inclusion. In D. Moore, A. Gorra, M. Adams, J. Reaney, & H. Smith (Eds.), *Disabled Students in Education: Technology, Transition, and Inclusivity* (pp. 192–208). IGI Global. doi:10.4018/978-1-61350-183-2.ch010

Rebrin, O., Sholina, I., & Berestova, S. (2014). Interdisciplinary Project for Bachelor Engineering Program. *Proceedings of the 10th International CDIO Conference.*

Rechistov, G., & Plotkin, A. (2014). Computer Engineering Educational Projects of MIPT-Intel laboratory in the Context of CDIO. *Proceedings of the 10th International CDIO Conference.*

Regueras, L. M., Verdú, E., Verdú, M. J., & de Castro, J. P. (2011). Design of a competitive and collaborative learning strategy in a communication networks course. *IEEE Transactions on Education, 2*(54), 302–307. doi:10.1109/TE.2010.2053933

Reinig, G., Winter, P., Linge, V., & Nägler, K. (1998). Training simulators: Engineering and use. *Chemical Engineering & Technology, 21*(9), 711–716. doi:10.1002/(SICI)1521-4125(199809)21:9<711::AID-CEAT711>3.0.CO;2-H

Ritter, F. E., & Schooler, L. J. (2001). The learning curve. In *International encyclopedia of the social and behavioral sciences.* Amsterdam: Pergamon. doi:10.1016/B0-08-043076-7/01480-7

Roach, N. (2016). Open Source in Higher Education: Top 10 Universities [Blog post]. Retrieved March 26, 2017, from https://www.axelerant.com/blog/open-source-in-higher-education

Roco, M. C., & Bainbridge, W. S. (Eds.). (2003). *Converging technologies for improving human performance: Nanotechnology, biotechnology, information technology and cognitive science*. Kluwer Academic Publishers. doi:10.1007/978-94-017-0359-8

Rodriguez, S. B., Fuertes, M. C., Piera, A. F., Garcia, I. P., & Solsona, F. J. A. (2012). Lessons learned in the use of WIRIS quizzes to upgrade Moodle to solve electrical circuits. *IEEE Transactions on Education*, *3*(55), 412–417. doi:10.1109/TE.2011.2181381

Romanova, A. O. (2011). Virtualizacija v vysokoproizvoditel'nyh vychislitel'nyh sistemah [Virtualization in high-performance computing systems]. *Science and Education: Scientific Publication of BMSTU*, *3*(93), 1–23.

Rosch, E., Varela, F. J., & Thompson, E. (1991). *The Embodied Mind: Cognitive Science and Human Experience*. Cambridge, MA: MIT Press.

Rose Technologies. (2016). *IBM Big Data Platform.* Retrieved July 18, 2017 from http://www.rosebt.com/blog/ibm-big-data-platform

Rosenblatt, F. (1957). *The Perceptron: a perceiving and recognizing automaton*. New York: Cornell Aeronautical Laboratory.

Russell, L. A. (1981). *Creating the corporate future: Plan or be planned for*. Wiley.

Russell, S. J., & Norvig, P. (2010). *Artificial Intelligence: A modern approach* (3rd ed.). Upper Saddle River, NJ: Prentice Hall.

Russian Government Decree. (2015). *Rasporyazhenie Pravitel'stva RF ot 06.01.2015 N°7-r «Ob utverzhdenii perechnya special'nostej i napravlenij podgotovki vysshego obrazovaniya, sootvetstvuyu-shchih prioritetnym napravleniyam modernizacii i tekhnologicheskogo razvitiya rossijskoj ehkonomiki»* [Russian Government Decree of 6 January 2015 #7-p "On Approval of the Higher Education Specializations and Programmes List Which Corresponds to the Priority Areas of Modernization and Technological Development of the Russian Economy"]. Author. (in Russian)

Russian Government. (2000). *About the Federal program of state support of small business in Russian Federation for 2000-2001 years*. Resolution of 14 February 2000 #121.

Russian Government. (2016). *Universities as centers of innovation creation space*. Retrieved 06/13/2017 from http://government.ru/media/files/OnTUmegFLNj5Uqtac57y1WG1EtMG9ABe.pdf

Russian state tertiary educational standards. (2009). Retrieved May 20, 2016, from http://fgosvo.ru/uploadfiles/fgos/15/20111115151321.pdf

RVC. (2015). *The development of innovation ecosystems universities and research centers.* Retrieved from https://www.rusventure.ru/ru/programm/analytics/docs /Innovation_ecosystem_analytical_report.pdf

RVC. (2016). *The National Report on Innovations in the Russian Federation 2016.* Retrieved 06/13/2017 from http://www.rvc.ru/upload/RVK_innovation_2016_v.pdf

Rybushkina, S., & Sidorenko, T. (2015). Modular approach to teaching ESP in engineering programs in Russia. In *Proceedings of 2015 International Conference on Interactive Collaborative Learning (ICL)* (pp. 105-108). Florence, Italy: IEEE. 10.1109/ICL.2015.7317988

Rykov V.P. (2013) *Automated technology modular principle of learning and self-organizing artificial neural networks.* Vestnik Tambov University. Series: Natural and Technical Sciences. N°4-1. 1428-1430.

Ryneveld, L. V. (2016). Introducing educational technology into the higher education environment: A professional development framework. In K. Dikilitaş (Ed.), *Innovative Professional Development Methods and Strategies for STEM Education* (pp. 126–136). Hershey, PA: IGI Global. doi:10.4018/978-1-4666-9471-2.ch008

Saaty, T. L. (1980). *The Analytic Hierarchy Process*. New York, NY: McGraw-Hill.

Sadler, D. R. (2013). The futility of attempting to codify academic achievement standards. *Higher Education, 7*. doi:0.100710734-013-9649-1

Saleh, H. M. (2016). *Program of discipline "Software Engineering Methodology"*. Retrieved from https://www.hse.ru/data/2016/10/04/1117071415/program-1513156455-EHWApXuS2m.pdf

Samohvalov, E., Revunkov, G., & Gapanyuk, Yu. (2015). Metagraphs for Describing Semantics and Pragmatics of Information Systems. *Herald of Bauman Moscow State Technical University, 1*(100), 83–99.

Sandvine. (2016). *Global Internet phenomena, Africa, Asia-Pacific, and Middle East*. Retrieved from https://www.sandvine.com/downloads/general/global-internet-phenomena/2016/global-internet-phenomena-apac-mea.pdf

Sangaran, S., Raju, R., Haron, S., & Yassin, I. M. (2017). Simulation in development of skilled and competent operators. *Journal of Engineering and Applied Sciences (Asian Research Publishing Network), 12*(10), 3299–3303.

Santos, R. (2008). *Intrusion prevention with L7-Filter.* Retrieved from InfoSec Reading Room WebSite: https://www.sans.org/reading-room/whitepapers/intrusion/intrusion-prevention-l7-filter-32868

SAP. (2017). *SAP HANA Platform.* Retrieved July 18, 2017 from https://help.sap.com/viewer/product/SAP_HANA_PLATFORM/2.0.00/en-US

Saprykin, D.L. (2012). Inzenernoe obrazovanie v Rossii: istoriya, kontseptsiya, perspektiva [Engineering education in Russia: history, conception, future trends]. *Vyssheye obrazovaniye v Rossii., 1*, 125-137.

Sarkisyan, S. A. (1977). Teoriya prognozirovaniya i prinyatiya resheniy [Theory of prediction and decision making]. Higher school.

Sauthwik, F. (2013). Opinion: Academy suppresses creativity. *Journal of Science and Practice Problems of Governance, 8*, 62-65.

Scharnhorst, A., & Ebeling, W. (2016). Evolutionary Search Agents in Complex Landscapes – a New Model for the Role of Competence and Meta-competence (EVOLINO and other simulation tools). *The Virtual Knowledge Studio*. Available at: http://virtualknowledgestudio.nl/documents /_andreascharnhorst/arxiv_final.pdf

Schedrovitsky, G. P. (2005). *The structure of sign: senses, meanings, knowledge* (Vol. 1). Moscow: Vost. Lit.

Schmidt, H. G., Rotgans, J. I., & Yew, E. H. J. (2011). The process of problem-based learning: What works and why. *Medical Education, 45*(8), 792–806. doi:10.1111/j.1365-2923.2011.04035.x PMID:21752076

Schmidt, H., Hatebur, D., & Heisel, M. (2015). Developing Secure Software Using UML Patterns. In V. Diaz, J. Lovelle, & C. Garcia-Bustelo (Eds.), *Handbook of Research on Innovations in Systems and Software Engineering* (pp. 32–70). IGI Global; doi:10.4018/978-1-4666-6359-6.ch002

Schoology. (n.d.). Retrieved from https://www.schoology.com/higher-ed

Scopus. (2015, August 10). Retrieved from Web site of Scientific Periodicals: https://www.scopus.com

SCORM 2004 4th Edition Run-Time Environment Version 1.0. (2009). Advanced distributed learning.

SEFI. (2013). *European Society for Engineering Education (SEFI), Brussels.* A Framework for Mathematics Curricula in Engineering Education. A Report of the Mathematics Working Group. Brussels, Belgium: European Society for Engineering Education (SEFI). Retrieved from http://www.sefi.be/?page_id=125

Senge, P. (2003). *The Fifth Discipline. The Art and Practice of the Learning Organization.* Moscow: Olimp-Biznes.

Sergeeva, M. G. (2016). Kriterial'nye pokazateli sformirovannosti sotsial'noy kompetentnosti obuchayushchikhsya [Criteria indicators for students' formation of social expertise]. *Professional'noe Obrazovanie i Obshchestvo, 1*(17), 30–37.

Sevo, R. (2011). *Basics about disabilities and science and engineering education.* Atlanta, GA: Center for Assistive Technology and Environmental Access, Georgia Institute of Technology. Retrieved from https://books.google.co.uk/books?hl=en&lr=&id=OfvKAQAAQBAJ&oi=fnd&pg=PT7&dq=,+R.+(2011).+Basics+about+disabilities+and+science+and+engineering+education&ots=VPfyugh3u-&sig=a-2GAj2JRgLst6EgY8UwwrtAP7g#v=onepage&q=doi&f=false

Shabalina, O., Mozelius, P., Vorobkalov, P., Malliarakis, C., & Tomos, F. (2015). Creativity in digital pedagogy and game-based learning techniques; theoretical aspects, techniques and case studies. *Proceedings of IISA 2015 - 6th International Conference on Information, Intelligence, Systems and Applications.* 10.1109/IISA.2015.7387963

Shakespeare, T. (2010). The social model of disability. In L. J. Davis (Ed.), *The Disability Studies Reader* (3rd ed.; pp. 266–273). New York: Routledge.

Shakhnov, V., Vlasov, A., & Zinchenko, L. (2012). O metodicheskom obespechenii ingenernogo obrazovaniya v sovremennuh usloviyah [Features of Engineering Education]. Vysshee obrazovanie - Higher Education in Russia, 3, 104-108.

Shakhnov, V., Vlasov, A., Rezchikova, E., & Zinchenko, L. (2013). Visual learning environment in electronic engineering education. In *Proceedings International Conference on Interactive Collaborative Learning (ICL)* (pp. 389-398). Academic Press. 10.1109/ICL.2013.6644605

Shakhnov, V.A., Zhuravleva, L.V., & Vlasov, A.I. (2013). Nanoengineering Ontology. *International Research Journal, 12*(19), 50-67.

Shale, J. (2015). *Households Below Average Income 2013-14.* DWP. Retrieved from https://www.gov.uk/government/uploads/system/uploads/attachment_data/file/437246/households-below-average-income-1994-95-to-2013-14.pdf

Shallish, L. (2015). Just how much diversity will the law permit?: The Americans with Disabilities Act, diversity and disability in higher education. *Disability Studies Quarterly, 35*(3). doi:10.18061/dsq.v35i3.4942

Shaw, M. C. (2001). *Engineering Problem Solving: A Classical Perspective.* Norwich, NY: William Andrew.

Shpak, M., Smirnova, E., Karpenko, A., & Proletarsky, A. (2016). Mathematical Models of Learning Materials Estimation Based on Subject Ontology. *Proceedings of the First International Scientific Conference on Intelligent Information Technologies for Industry (IITI'16).* 10.1007/978-3-319-33609-1_24

Siegel, D. S., & Wright, M. (2015). University technology transfer offices, licensing, and start-ups. In Chicago handbook of university technology transfer and academic entrepreneurship. Chicago, IL: University of Chicago Press.

Siegwart, R., Büchi, R., & Buehler, Ph. (1998). Mechatronics Education at ETH Zurich based on 'Hands on Experience. In J. Adolfsson & J. Karlsén (Eds.), *Mechatronics 98* (pp. 667–672). Elsevier Science Ltd. doi:10.1016/B978-008043339-4/50108-5

Simmons, O., & Gregory, T. (2003). Grounded Action: Achieving Optimal and Sustainable Change. *Forum Qualitative Social Research, 4*(3).

Simonyants, R. P. (2014). *Problems of engineering education and their decision involving industry.* Moscow, Russia: BMSTU. (in Russian)

Sinicyna, E. Y., Kurdyukova, G. N., & Abramova, E. Y. (2013). Razvitie paradigmy kreativnogo obrazovaniya posredstvom obucheniya cherez issledovaniya [Creative education paradigm development via 'Learning through Research']. *Vestnik MJEI - MPEI Vestnik, 6*, 217-219.

Skvortsova, V. S. (2014). Kontseptsiya obuchayushcheysya organizatsii i yeye primeneniye v praktike menedzhmenta [The concept of the learning organization and its application in the management practice]. *Ekonomika i menedzhment innovatsionnykh tekhnologiy - Economy and management of innovative technologies, 3,* 1. Retrieved from http://ekonomika.snauka.ru/2014/03/3844

Sloan, J. D. (2001). *Network troubleshooting tools.* Sebastopol, CA: O'Reilly & Associates, Inc.

Smirnova, E., Suzev, V., Proletarsky, A., & Gurenko, V. (2017) Signal's Simulation Methodologies used in Scientific and Educational tasks of real-time Information System's Modelling. In *Procceedings of 11th International Technology, Education and Development Conference INTED2017* (vol. 11, pp. 9140 – 9143). Valencia, Spain: IATED Academy.

Soloviev, M. A., & Zamyatina, O. M. (2013). Elite technical education system in Tomsk polytechnic university. *Tomskij politehnik, 18*, 96-103.

Sorina, G. V. (2016). Cross-cultural communications in national educational spaces in a global world. In S.V. Ivanova & E.V. Nikulchev (Eds.), *SHS Web of Conferences (Vol. 29). 2016 International Conference on Education Environment for the Information Age (EEIA-2016).* Moscow, Russia: SHS. Retrieved April 10, 2017, from: http://www.shs-conferences.org/articles/shsconf/abs/2016/07/contents/contents.html

Spencer, J., & Jordan, R. K. (1999). Learner centred approaches in medical education. *British Medical Journal, 318*(7193), 1280–1283. doi:10.1136/bmj.318.7193.1280 PMID:10231266

Sphinx Technologies Inc. (2010, August). *How Sphinx relevance ranking works.* Retrieved April 21, 2017, from http://sphinxsearch.com/blog/2010/08/17/how-sphinx-relevance-ranking-works/

Spillane, J. P., & Hopkins, M. (2013). Organizing for instruction in education systems and school organizations: How the subject matters. *Journal of Curriculum Studies, 45*(6), 721–747. doi:10.1080/00220272.2013.810783

Srisawasdi, N. (2015). Motivating inquiry-based learning through a combination of physical and virtual computer-based laboratory experiments in high school science. *Improving K-12 STEM Education Outcomes through Technological Integration*, 108-134.

St. Amant, K., & Flammia, M. (2016). *E-Learning and Technical Communication for International Audiences. In Teaching and Training for Global Engineering: Perspectives on Culture and Professional Communication Practices.* Wiley-IEEE Press; doi:10.1002/9781119084297

Stagner, H. (2009). *Pro Hyper–V.* New York, NY: Apress. doi:10.1007/978-1-4302-1909-5

STEM Disability. (2016). Retrieved from http://www.stemdisability.org.uk/events/stemm-dac-conference-2016/

Stepanov, Y. S. (1997). *Konstanty. Slovar' russkoj kul'tury. Opyt issledovaniya.* [Constants. Dictionary of Russian culture. Study experience]. Moscow, Russia: Shkola Yazyki russkoj kul'tury.

Stepanov, O. A., & Toropov, A. B. (2010). A comparison of linear and nonlinear optimal estimators in nonlinear navigation problem. *Gyroscopy and Navigation, 1*(3), 183–190. doi:10.1134/S2075108710030053

Stepanov, O. A., & Vasiliev, V. A. (2010). Comparison of the Bayesian and neural network algorithms in nonlinear navigation estimation problems. In *Proceedings of the 2010 International Joint Conference on Neural Networks*. Barcelona: IEEE Computational Intelligence Society. 10.1109/IJCNN.2010.5596834

Steuer, L., Bouffier, A., Gaedicke, S., & Leicht-Scholten, C. (2017). Diversifying Engineering Education: A Transdisciplinary Approach From RWTH Aachen University. In M. Gray & K. D. Thomas (Eds.), *Strategies for Increasing Diversity in Engineering Majors and Careers* (pp. 201–235). Hershey, PA: IGI Global. doi:10.4018/978-1-5225-2212-6.ch010

Stojanovic, R., Mitropulos, P., Koulamas, C., Koubias, S., & Papadopoulos, G. (2001). Real-Time Vision-Based System for Textile Fabric Inspection. Real-Time Imaging, 7(6), 507-518. doi:10.1006/rtim.2001.0231

Strategy Partners Group & Minpromtorg RF (Ministry of Industry and Trade of the Russian Federation). (2015). *Forsajt-issledovanie v oblasti perspektivnyh professional'nyh kompetencij specialistov v oblasti inzhiniringa i promyshlennogo dizajna* [Foresight-research in promising engineering and industrial design professional skills]. Authors. (in Russian)

Strevens, P. (1988). ESP after twenty years: A re-appraisal. In M. Tickoo (Ed.), *ESP: State of the Art* (pp. 1–13). Singapore: SEAMEO Regional Centre.

Studying/Thinking about study. (2016). Retrieved from http://www.stemdisability.org.uk/resources/students/

Sukhobokov, A. V. (2016). Perspektivy razvitiia system obrabotki informatsii i upravleniia v sleduiuschem desiatiletii (2020 – 2030) [Prospects of data processing and management systems development in the next decade (2020 – 2030)]. *Informatsionno-izmeritelnyie i upravliaiuschie sistemy, 2016*(12), 69-78.

Sun, H.-C. (2003). Conceptual clarifications for 'organizational learning', 'learning organization' and 'a learning organization'. *Human Resource Development International*, *6*(2), 153–166. doi:10.1080/13678860110086465

Surin, V. I., Biryukov, A. P., & Dubkov, B. A. (2013). Ispol'zovaniye CAD-sistem v uchebno-tekhnicheskikh uchebnykh kursakh dlya studentov Natsional'nogo issledovatel'skogo yadernogo universiteta MIFI [The use of CAD-systems in educational engineering training courses for the students of the National Research Nuclear University MEPhI]. In *V trudakh Mezhdunarodnoy konferentsii «Sistemy proyektirovaniya, predplanirovaniya i upravleniya zhiznennymi tsiklami promyshlennogo produkta CAD / CAM / PDM 2013»*. Moscow, Russia: V.A. Trapeznikov Institute of Control Sciences of Russian Academy of Sciences.

Svyatkina, M., Tarassov, V., & Dolgiy, A. (2016). Logical-Algebraic Methods in Constructing Cognitive Sensors for Railway Infrastructure Intelligent Monitoring System. *Proceedings of the First International Scientific Conference on Intelligent Information Technologies for Industry (IITI'16)*. 10.1007/978-3-319-33609-1_17

Swales, J. (1971). Writing Scientific English. London, UK: Thomas Nelson and Sons.

Swart, A. J. (2015). Student usage of a learning management system at an open distance learning institute: A case study in electrical engineering. *International Journal of Electrical Engineering Education*, *52*(X), 142–154. doi:10.1177/0020720915575925

Syuzev, V.V. (2014). *Introduction to digital signal processing*. Moscow: RTSoft. (in Russian)

Syuzev, V. V. (2015). Fast generalized Hartley transform in single-base number systems. [in Russian]. *Bauman University Journal of Machinary Devices*, *6*, 63–81.

Taganov, A. I., & Taganov, R. A. (2006). *Metodicheskiye osnovy sozdaniya informatsionnykh sistem sfery obrazovaniya* [Methodical fundamentals for development of information systems for education: Tutorial.]. Ryazan: RSREU.

Talalakina, E. V., Brown, N. A., & Eggington, W. (2014). *Mastering English through global debate*. Washington, DC: Georgetown University Press.

Talmy, L. (2000a). Toward a Cognitive Semantics: Vol. 1. *Concept structuring systems*. Cambridge: MIT Press.

Talmy, L. (2000b). Toward a Cognitive Semantics: Vol. 2. *Typology and process in concept structuring*. Cambridge: MIT Press.

Tan, C. (2012). The culture of education policymaking: Curriculum reform in Shanghai. *Critical Studies in Education, 53*(2), 153–167.

Tax Code of the Russian Federation. (2017). Russian Federation.

Teaching, E. S. P. (2012). *Best Practices*. Moscow, Russia: REPETITOR Multi Media.

Tech. (2014). *World Industry Report: Global Engineering Services Market*. London, UK: Technavio.

Technical Chamber of Cyprus. (2014). Retrieved April 18, 2017 from http://www.etek.org.cy

Teichler, U. (2013). Opportunities and problems of comparative higher education research: The daily life of research. *Higher Education, 11*. doi:10.100710734-013-9682-0

Teradata. (2017). *Teradata Integrated Big Data Platform*. Retrieved July 18, 2017 from http:// teradata.ru/products-and-services/integrated-big-data-platform

Terehov, S. I. (1998). *Lekcii po teorii i prilozhenijam iskusstvennyh nejronnyh setej* [Lectures for theory and applications of artificial neural networks]. Retrieved June 19, 2016, from http://alife.narod.ru/lectures/neural/Neu_index.htm

The Art of the Possible. (2011). Using Technology to Make Teaching More Inclusive Simon Ball (JISC TechDis. In D. Moore, A. Gorra, M. Adams, J. Reaney, & H. Smith (Eds.), *Disabled Students in Education: Technology, Transition, and Inclusivity*. IGI Global. doi:10.4018/978-1-61350-183-2.ch001

The California State University. (n.d.). *Course Redesign with Technology*. Retrieved June 2, 2017, from http://courseredesign.csuprojects.org/wp/

The EAR Manual. (2017). Retrieved April 7, 2017 from http://www.eurorecognition.eu/emanual

The European Qualifications Framework. (2017). Retrieved April 9, 2017 from http://ec.europa.eu/ploteus

The Global Innovation Index. (2014). *The human factor in innovation*. Retrieved from https://www.globalinnovationindex.org/userfiles/file/reportpdf/GII-2014-v5.pdf

The Resolution Foundation. (2016). *Employing new tactics: the changing distribution of work across British households*. Retrieved from http://www.resolutionfoundation.org/wp-content/uploads/2016/01/Employing-new-tactics.pdf

The United Nations Educational, Scientific and Cultural Organization. (1989). *The modular approach in technical education*. Paris, France: UNESCO.

Thelwall, M. (2000). Computer-based assessment: A versatile educational tool. *Computers & Education, 34*(1), 37–49. doi:10.1016/S0360-1315(99)00037-8

Thindwa, H. (2015). The Role of Technology in Improving Quality of Teaching in Higher Education: An International Perspective. In F. M. Nafukho & B. J. Irby (Eds.), *Handbook of Research on Innovative Technology Integration in Higher Education* (pp. 54–73). Hershey, PA: IGI Global. doi:10.4018/978-1-4666-8170-5.ch003

Thomas, L. (2012). *What Works – Final Report*. Retrieved from http://www.phf.org.uk/publications/works-student-retention-success-final-report/

Titov, I., & Smirnova, E. (2013). Network architectures of remote laboratories proposal of a new solution and comparative analysis with existing ones. *International Journal of Online Engineering*, *9*(6), 41–44. doi:10.3991/ijoe.v9i6.3142

Tribou, M. J., Wang, D. W. L., & Waslander, S. L. (2016). Degenerate motions in multicamera cluster SLAM with non-overlapping fields of view. Image and Vision Computing, 50(C), 27-41.

Troshin, A., Alexeev, A., Yaminsky, I., Smagulova, D., & Burlutsky, D. (2015, February). Staffing of nanotechnology. *Nanoindustry*, *55*(1), 14–21. doi:10.22184/1993-8578.2015.55.1.14.21

Trukhin A.V. (2002). Ob ispol'zovanii virtual'nykh laboratoriy v obrazovanii [About use of virtual laboratories in education.] *Otkrytoe i distantsionnoe obrazovanie,* (4), 67-69.

Tsibizova, T. Yu., & Neusypin, K. A. (2012). Some restructuring aspects of the modern educational-scientific centers control system. *Automation and modern technologies, 1*, 30-34.

Tsibizova, T.Yu. (2011). Podgotovka vysokokvalifitsirovannykh spetsialistov v sisteme nepreryvnogo professional'nogo obrazovaniya (na primere MGTU im. N.E. Baumana) [Training of highly-knowledgeable specialists in a system of non-stop professional development (per the example of Bauman Moscow State Technical University)]. *European Social Science Journal*, *2*(5), 154-159.

Tsibizova, T. Yu. (2012). Profil'noe obuchenie kak komponent sistemy nepreryvnogo professional'nogo obrazovaniya [Profile education as a part of the system of continued professional education] [Profile School]. *Profil'naya Shkola*, *4*, 9–13.

Tulloch, M. (2010). *Understanding Microsoft Virtualization Solutions, From the Desktop to the Datacenter* (2nd ed.). Redmond, WA: Microsoft Press.

Turi. (2017). *GraphLab Create.* Retrieved July 18, 2017 from https://turi.com/learn/

Turro, L. J., & Farell, M. P. (2013). Higher Education and IT. In P. Pablos & R. Tenisson (Eds.), *Strategic Role of Tertiary Education and Technologies for Sustainable Competitive Advantage* (pp. 120–127). Hershey, PA: IGI Global. doi:10.4018/978-1-4666-4233-1.ch004

U.S. Department of State. (2016). *Virtual Student Foreign Service*. Retrieved June 2, 2017, from http://www.state.gov/vsfs/

Ugin, E. G., & Demihof, K. E. (2005). *Founders of the scientific school of Bauman Moscow State Technical University.* Moscow: BMSTU Press.

UK Government. (2010). *Securing a Sustainable Future for Higher Education.* Retrieved from https://www.gov.uk/government/publications/the-browne-report-higher-education-funding-and-student-finance

UK Government. (2016). *Success as a Knowledge Economy: teaching excellence, social mobility and student choice.* Retrieved from https://www.gov.uk/government/publications/higher-education-success-as-a-knowledge-economy-white-paper

UK Government. (2017). *Teaching Excellence and Student Outcomes Framework Specification.* Retrieved from https://www.gov.uk/government/publications/teaching-excellence-and-student-outcomes-framework-specification

Universities and Colleges Act. (2005). Retrieved from http://lovdata.no/

University Information System of Mendel University in Brno. (n.d.). Retrieved from http://is.mendelu. cz/

University Management System UMS. (n.d.). Retrieved from http://ampletrails.com/university-management-system-ums

University of Minnesota. (2017). *Doctoral students. General degree information.* Retrieved July 18, 2017 from https://www.cs.umn.edu/academics/graduate/phd

University. (2017). In *Duden Online-Wörterbuch* [Duden Online Dictionary] (27th ed.). Berlin, Germany: Bibliographisches Institut GmbH, Dudenverlag. Retrieved July 1, 2017, from: http://www.duden.de/rechtschreibung/Universitaet#b2-Bedeutung-1

Valtanen, J., Berki, E., Georgiadou, E., Hatzipanagos, S., Ross, M., Stamelos, I. G., & Staples, G. (2012). Features for suitable problems: IT professionals' and IT students' opinions. *International Journal of Human Capital and Information Technology Professionals*, *3*(3), 27–41. doi:10.4018/jhcitp.2012070103

Van Ryneveld. (2015). Introducing Educational Technology into the Higher Education Environment: A Professional Development Framework. In Innovative Professional Development Methods and Strategies for STEM Education (pp. 126-136). Hasan Kalyoncu University.

Vaqqas, M. (2016). *RESTful web services: A tutorial*. Retrieved from http://www.drdobbs. com/web-development/restful-web-services-a-tutorial/240169069

Velan, P., Cermak, M., Celeda, P., & Drasar, M. (2015). A survey of methods for encrypted traffic classification and analysis. *International Journal of Network Management*, *25*(5), 355–374. doi:10.1002/nem.1901

Verhovskaya, O. R., Dorohina, M. V., & Sergeeva, A. V. (2013). *Global entrepreneurship monitor*. Retrieved 06/20/2017 from http://www.gsom.spbu.ru/images/cms/data/faculty/gem_2013_final20_all.pdf

Veshneva, I., Singatulin, R., Bolshakov, A., Chistyakova, T., & Melnikov, L. (2015). Model of formation of the feedback channel within ergatic systems for monitoring of quality of processes of formation of personnel competences. *International Journal of Qualitative Research*, *9*(3), 495–512.

Virtual laboratory. (2006). In *EduTechWiki*. Retrieved June 2, 2017, from http://edutechwiki.unige.ch/en/Virtual_laboratory

Vishnevskaya, T. I., & Romanova, T. N. (2012). *Techonologiya programmirovaniya: Metodicheskie ukazaniya k laboratornomu praktikumu – Ch. 3* [Programming Technology: Methodical instructions for laboratory works – P.3]. Moscow: Moscow State Technical University Publishing House. Retrieved from http://www.catalog.inforeg.ru/Inet/GetEzineByID/293408

Vishnevskaya, T. I., & Romanova, T. N. (2017). *Metodologiya programmnoy ingenerii: Metodicheskie ukazaniya k laboratornomu praktikumu* [Methodology of software engineering: Methodical instructions for laboratory works]. Moscow: Moscow State Technical University Publishing House. Retrieved from http://ebooks.bmstu.ru/catalog/199/book1531.html

Vladimirov, A. I. (2011). *Ob injenerno-tekhnicheskom obrazovaniy* [About engineering-technical education]. Moscow, Russia: Nedra Publishing House.

Vlasov, A. I., & Yudin, A. V. (2011). Distributed control system in mobile robot application: General approach, realization, and usage. *Communications in Computer and Information Science*, *156*, 180–192. doi:10.1007/978-3-642-27272-1_16

Vlasov, A. I., Zhuravleva, L. V., Sharipov, N. R., & Sharipova, A. F. (2012). Architecture of adaptive multiservice information and training systems. *Scientific Review (Singapore)*, *6*, 152–154.

Wang, C., Zhang, H., & Ye, Z. (2015). A peer to peer traffic identification method based on wavelet and particle swarm optimization algorithm. *International Journal of Wavelets, Multresolution, and Information Processing*, *13*(6), 1550048. doi:10.1142/S0219691315500484

Wang, L., & Xia, Y. (2015). Observability analysis of Mars entry integrated navigation. *Advances in Space Research*, *56*(5), 952–963. doi:10.1016/j.asr.2015.05.016

Wen, L., Tuffley, D., & Dromey, R. G. (2014). Formalizing the transition from requirements' change to design change using an evolutionary traceability model. *Innovations in Systems and Software Engineering*, *10*(3), 181–202. doi:10.100711334-014-0230-6

Wheeler, M. W., & Bailer, A. J. (2009). Benchmark dose estimation incorporating multiple data sources. *Risk Analysis*, *29*(2), 249–256. doi:10.1111/j.1539-6924.2008.01144.x PMID:19000080

Whiddett, S., & Hollyforde, S. (2003). *A practical guide to competencies: How to enhance individual and organisational performance* (2nd ed.). Edinburgh, UK: Chartered Institute of Personnel and Development.

White, C. M., Daniel, E. J., & Teague, K. A. (2012). A real-time network analysis tool to aid in characterizing VoIP system performance. *International Journal of Electrical Engineering Education*, *42*(2), 119–131. doi:10.7227/IJEEE.42.2.1

Wikipedia. (2018). *Uruguay*. Retrieved 18.12.2017 from https://en.wikipedia.org/wiki/Uruguay

Wilson, S. (2013). Open source in higher education: How far have we come? [Blog post] Retrieved March 28, 2013, from https://www.theguardian.com/higher-education-network/blog/2013/mar/28/open-source-universities-development-jisc

Wilson, S. (2014). *Tackling the challenges of open source adoption in education*. Retrieved from https://opensource.com/education/14/5/choosing-open-source-education

Wilson, P. (2016). *Design Recipes for FPGAs: Using Verilog and VHDL* (2nd ed.). London: Elsevier.

Winters, J., & Wang, Y. (2003). Wearable sensors and telerehabilitation. *IEEE Engineering in Medicine and Biology Magazine*, *22*(3), 56–65. doi:10.1109/MEMB.2003.1213627 PMID:12845820

Winters, M. E., Tissari, H., & Kathryn, A. (2010). *Historical Cognitive Linguistics*. Berlin, New York: Walter de Gruyter. doi:10.1515/9783110226447

Wisdom, S. (2015). *Handbook of Research on Advancing Critical Thinking in Higher Education*. Hershey, PA: IGI Global; doi:10.4018/978-1-4666-8411-9

Witten, I. H., Frank, E., & Hall, M. A. (2011). *Data mining: Practical machine learning tools and techniques* (3rd ed.). San Francisco, CA: Elsevier Inc.

Women, Minorities and Persons with Disabilities in Science and Engineering. (2016). Retrieved from http:// www.nsf.gov/statistics/2015/nsf15311/digest/nsf15311-digest.pdf/

World University Rankings 2015-2016. (2015, August 10). Retrieved from Web site The Times Higher Education World University Rankings: https://www.timeshighereducation.com

Wortham, J. (2013, Jan 30). Virtually there: Working remotely. *The New York Times*. Retrieved June 2, 2017, from http://www.nytimes.com/2013/02/03/education/edlife/virtual-internships.html

Wright, M., Clarysse, B., Mustar, P., & Lockett, A. (Eds.). (2007). *Academic entrepreneurship in Europe*. Cheltenham, UK: Edward Elgar. doi:10.4337/9781847205575

Wu, D., Lu, J., & Zhang, G. (2015). A Fuzzy Tree Matching-Based Personalized E-Learning Recommender System. *Transactions on Fuzzy Systems*, *23*(6), 2412–2426. doi:10.1109/TFUZZ.2015.2426201

Xin, T., Li, F., & Li, L. (2007). An International Comparison of Elementary Education Quality Assessment. *Journal of Beijing Normal University*, 44–47.

Yakovlev, D., Pryakhin, A., Korolev, S., Shaltaeva, Y., Samotaev, N., Yushkov, E., & Avanesyan, A. (2015). Engineering competitive education using modern network technologies in the NRNU MEPhI. In *Proceedings of the 2015 IEEE Workshop on Environmental, Energy, and Structural Monitoring Systems*. Trento, Italy: IEEE. 10.1109/EESMS.2015.7175849

Yang, Y., Pierce, T., & Carbonell, J. (1998). A study on retrospective and on-line event detection. In *Proceedings of the 21st Annual International ACM SIGIR Conference on Research and Development in Information Retrieval* (pp. 28-36). New York, NY: ACM. 10.1145/290941.290953

Yang, Y., Zhang, J., Carbonell, J., & Jin, C. (2002). Topic-conditioned novelty detection. In *Proceedings of the Eighth ACM SIGKDD International Conference on Knowledge Discovery and Data Mining* (pp. 688-693). New York, NY: ACM. 10.1145/775047.775150

Yarushkina, N. (2004). *Prikladnie intellektualnie sistemy, osnovannie na myagkih vichisleniyah* [Applied intelligent systems based on soft computing]. Ulianovsk: UlGTU.

Yarushkina, N., Moshkin, V., Andreev, I., Klein, V., & Beksaeva, E. (2016). Hybridization of Fuzzy Inference and Self-learning Fuzzy Ontology-Based Semantic Data Analysis. *Proceedings of the First International Scientific Conference on Intelligent Information Technologies for Industry (IITI'16).* 10.1007/978-3-319-33609-1_25

Yentes, J. M., Hunt, N., Schmid, K. K., Kaipust, J. P., McGrath, D., & Stergiou, N. (2013). The appropriate use of approximate entropy and sample entropy with short data sets. *Annals of Biomedical Engineering, 41*(2), 349–365. doi:10.100710439-012-0668-3 PMID:23064819

Yerevan Communique. (2015). *Paper presented at Ministerial Conference, Yerevan*. Retrieved May 2, 2017, from www.ehea.info/Uploads/SubmitedFiles/5_2015/112705.pdf

Yonezawa, A. (2011). The "global 30" and the consequences of selecting "world-class universities" in Japan. In N. C. Liu, Q. Wang, & Y. Cheng (Eds.), *Paths to a world-class university: Lessons from practices and experiences*. Rotterdam: Sense Publishers. doi:10.1007/978-94-6091-355-6_3

Yonezawa, A., Akiba, K., & Hirouchi, D. (2009). Japanese university Leaders' perceptions of Internationalization: The Role of Government in Review and Support. *Journal of Studies in International Education, 13*(2), 125–142. doi:10.1177/1028315308330847

Yudin, A., Kolesnikov, M., Vlasov, A., & Salmina, M. (2017). Project oriented approach in educational robotics: From robotic competition to practical appliance. In M. Merdan, W. Lepuschitz, G. Koppensteiner, & R. Balogh (Eds.), *Robotics in education. Advances in intelligent systems and computing* (Vol. 457, pp. 83–94). Springer. doi:10.1007/978-3-319-42975-5_8

Yu, Z., Cui, P.-y., & Zhu, S.-y. (2015). Observability-based beacon configuration optimisation for mars entry navigation. *Journal of Guidance, Control, and Dynamics, 38*(4), 643–650. doi:10.2514/1.G000014

Zakharia, M. (2014). *Spark's role in the Big Data ecosystem.* Spark Summit, San Francisco, CA. Retrieved July 18, 2017 from https://spark-summit.org/2014/wp-content/ uploads/2014/07/Sparks-Role-in-the-Big-Data-Ecosystem-Matei-Zaharia1.pdf

Zakharia, M. (2015). *Introduction to Spark.* Databricks Intern Event. Retrieved July 18, 2017 from https://www.slideshare.net/databricks/introduction-to-spark-intern-event-presentation

Zakharova, I. V., Dudakov, S. M., & Yazenin, A. V. (2016). On the development of an exemplary curriculum for UGNS (combined group of specialties) "Computer and Information Sciences" in accordance with professional standards. *Herald of Tver State University. Series: Pedagogy and Psychology, 2*, 84-100. Retrieved from https://elibrary.ru/item.asp?id=26555960

Zakharova, I. V., Dudakov, S. M., Yazenin, A. V., & Soldatenko, I. S. (2015). On methodological aspects of the development of exemplary educational programs of higher education. *Educational Technology & Society*, *18*(3), 330-354. Retrieved from https://elibrary.ru/item.asp?id=24102764

Zakharova, I. V., Kuzenkov, O. A., & Soldatenko, I. S. (2014). The use of modern educational technologies for the improvement of mathematical education within the engineering courses in Russian universities. *Modern Information Technology and IT Education*, *10*, 159-171. Retrieved from https://elibrary.ru/item.asp?id=23020629

Zamyatina, O., & Mozgaleva, P. (2014). Competence component of the project-oriented training of elite engineering specialists. In *Proceedings* of *the 2014 IEEE Global Engineering Education Conference*. New York, NY: IEEE. 10.1109/EDUCON.2014.6826077

Zander, S., & Schmoll, C. (2009). *Calculating flow statistics using Netmate*. Retrieved April 12, 2017, from https://dan.arndt.ca/nims/calculating-flow-statistics-using-netmate/

Zaripova, R. (2015). *CLIL-based method of teaching in a foreign language at university* (Unpublished doctoral dissertation). Kazan Federal University, Kazan, Russia.

Zelencova, N.F., Zelencova, E.V., & Zelencov, V.V. (2014). Problemy professional'nogo stanovleniya i vosproizvodstva nauchnykh i nauchno-pedagogicheskikh kadrov v nepreryvnoy sisteme profil'nogo inzhenerno-tekhnicheskogo obrazovaniya [Problems of professional growth and scientific-pedagogical staff reproduction in a system of continued profile engineering education]. *Nauka i Obrazovanie: Nauchnoe Izdanie MGTU im. N.E. Baumana*, *11*, 827-839.

Zhao, Q., Mitra, P., & Chen, B. (2007). Temporal and information flow based event detection from social text streams. In *Proceedings of the Twenty-Second AAAI Conference on Artificial Intelligence. AAAI-07* (Vol. 2, pp. 1501–1506). Palo Alto, CA: AAAI Press. Retrieved from http://www.aaai.org/Papers/AAAI/2007/AAAI07-238.pdf

Zhu. (2015). Organisational culture and technology-enhanced innovation in higher education. *Technology, Pedagogy and Education*, *24*(1), 65-79. doi:10.1080/1475939X.2013.822414

Zhukov, A. A., & Korotaev, A. G. (2015). Metodicheskoe i informatsionnoe obespechenie kursa "Osnovyi rabotyi v SDO Moodle" [Methodical and information support of the course "Basics LMS Moodle"]. *Innovatsii na osnove informatsionnyih i kommunikatsionnyih tehnologiy – Innovations based on information and communication technologies*, *1*, 46-49.

Zhulego, V., & Kunina, G. (2012). Monitoring of effectiveness of scientific organizations dealing with research in the field of nanotechnology. *Nanotechnology. Ecology. Production*, *5*, 36–39.

Zhulego, V., & Kunina, G. (2012, March). Determination of centers of excellence among organizations, leading research and development in nanotechnology. *Nanotechnologies in Russia*, *7*(3-4), 12–17.

Zhuravleva, L. V. (2012). Formalization of lecture material. In *Proceedings Tenth International Symposium "Intellectual systems"* (pp. 576-580). Academic Press.

Zhuravleva, L., & Shakhnov, V. (1998). Features of business games in engineering education. In Proc. Padagogische Probleme in der lngenieurausbildung (pp.287-288). Academic Press.

Zhuravleva, L., & Shakhnov, V. (1995). Fachspile als Bestandteil der Ingenierpadagogik. Ingenieurausbildung and Strukturveranderungen am Arbeitsplatz des des Ausgehenden 20. Jahrhunders. In *Proceedings 24. Internationalen Symposiums. Ingenieurpadagogik*, *95*, 244–247.

Zinchenko, G. P. (1988). *Continuous education – command of time*. Moscow: Knowledge Press.

Zotov, L. V. (2012). *Teorija fil'tracii i obrabotka vremennyh rjadov, kurs lekcij* [Theory of filtering and time series processing]. Moscow, Russia: MSU Physical Faculty Publishing.

About the Contributors

Elena Smirnova, PhD, is a practical software engineer and an innovative teacher in the field of information systems, specialising in mainframes. She is a Deputy Head of the Computer Systems and Networks Department, Bauman Moscow State Technical University (BMSTU), Director of the IBM Mainframe Center of Excellence, university's represantative in the Worldwide CDIO Initiative. She has combined research experience at a senior level with the development of advanced educational methods. Her work has included the creation and management of information systems; preparation of grant applications; the development of sustainable links between academia and industry and extensive international proficiency, including participation in a UK funded 'Know-How' project and European Union TEMPUS and ERASMUS+ grant applications.

Robin Clark is a Professor of Practice and Associate Director of Postgraduate Programmes for WMG (Warwick Manufacturing Group, University of Warwick) with responsibility for the promotion of innovative learning and teaching practice, learning and teaching quality and engineering education research and scholarship. As a National Teaching Fellow, Robin has been recognised as one of the top HE learning and teaching practitioners in the UK. He has served a 4 year term on the Governing Board of the global Research in Engineering Education Network (REEN) and in 2014 was elected to the Board of Directors of SEFI. Robin has a Bachelors and PhD in Mechanical Engineering and an MBA. Prior to joining WMG, he spent 14 years working in industry as a mechanical engineer and engineering manager both in the UK and USA and then 14 years at Aston University. Robin is on the editorial board of EJEE and is Chair of the UK and Ireland Engineering Education Research Network. Robin is a Chartered Engineer, Chartered Manager and Principal Fellow of the Higher Education Academy.

* * *

Irina Abrashkina graduated from the specialty "Information and measuring technology and technology" at Tomsk Politechnical University (diploma with honors). For contribution to the development of Tomsk Polytechnic University was awarded a bronze medal of TPU. After graduation she was assigned to the Department of Information and Measuring Technology of the Electro-Physical Faculty as an assistant. In October 2001, she was appointed to the post of deputy dean for the commercial department of the Electro-physical Faculty in which she stayed until April 2003. In 2006, she was transferred to the position of the engineer of the Department of Information and Measuring Equipment, and was appointed responsible for the development and support of the cathedral site. In the same year she was appointed Deputy Head of the Department of Information and Measuring Equipment, in whose position she was until January 2011. From March to December 2011 she held the position of the leading expert of the Department of Automation of the Management of the Training Process of TPU. Acted as the Head of the Department. Since 1.01.12 is the head of the department. In February 2015, the Department of Automation of Educational Management was renamed the Department of Planning, Organization and Management of the Academic Process. For his contribution to the development of TPU, she was added to the Gallery of Honor of TPU and was awarded with the II degree medal "For Participation in the Development of TPU".

Anatoly Alexandrovich Aleksandrov is the Rector of Moscow State Technical University named after Bauman, the laureate of RF government prize in the field of science and technology, Professor, Doctor of Technical Sciences.

Konstantin Alekseev graduated from Moscow engineering physics Institute in 1976 with a degree "engineer-physicist". In 1976 – 1986 works at research and development of world's first mixing D2 – CO2 laser. From 2007 to 2014 – rector and Pro-rector of Federal state budget educational institution of further professional education "State Institute of new learning forms". Design, development and testing of the cross-University network system in interdisciplinary training and professional development for nanotechnology industry(www.nano-obr.ru), on the basis of scientific-educational centers of the national nanotechnology network's (NNN) educational segment aimed to implement the Federal target program (FTP) "Development of infrastructure of nanoindustry in Russian Federation 2008-2011".

Alexander Alimov is currently a PhD student at Volgograd State Technical University. The topic of his thesis is management of multi-task intellectual agents behavior in real time systems.

Natalia Alyavdina is an Associate Professor of the Linguistic Department at Bauman Moscow State Technical University. She graduated from Moscow State Linguistic University in 1976. She writes and presents widely on issues of ESP teaching, course designing, blended learning.

Vadim Anatolyevich Shakhnov was born in 1941, graduated from BMSTU in 1966. Dr.Sc., professor, corresponding member of RAS, Head of the Department of Design and Technology of Electronic Devices, BMSTU. The author more than 200 scientific operations in the field of micro and a nanoelectronics, radiotronics, production technologies of computer, microprocessor means and systems, reliabilities of an electronic equipment. A honorable title - "European engineer - educationist" laureate, a member IGIP.

Ark Mikhailovich Andreev, PhD, graduated from Bauman Moscow Higher Technical School in 1967. He is currently an Associate Professor at the Computer Systems, Complexes and Networks Department of Bauman Moscow State Technical University. His research interests include informatics and control systems.

Jane Andrews is a Sociologist and a Senior Lecturer in the School of Engineering and Applied Science at Aston University, UK. With 15 years of experience teaching, she has played an active role in leading learning and teaching innovation in engineering education. For over 7 years she has been a key figure in the international and national field of Engineering Education Research and Practice. Jane has a PhD and is a Senior Fellow of the Higher Education Academy.

Vitaliy Baklikov was born in 1983. In 2007, he graduated from the University of Maryland College Park with a degree in Computer Science. Over the recent years, Vitaliy became an accomplished professional with an extensive background in enterprise architecture. His competence lies within Big Data space with a strong emphasis on Hadoop stack and parallel computations. His interests extend to data science, machine learning, and visualization. Vitaliy works as a freelancer advising major banks in London on Big Data strategy, BI, and Hadoop adoption. He is also a co-founder of several startups in the space and sits at executive management level. In 2016 -2017, Vitaliy is a guest Lecturer at Bauman Moscow State Technical University. He is passionate about training new talent and mentoring young specialists.

Gennady Baryshev is vice-head of the Council of Young Scientists and Specialists of MEPhI. Graduated from MEPhI. Area of research: solid state physics, non-destructive testing.

Aleksandr Berestov is a Vice-Head of the Department of Engineering Science and Technology of MEPhI. Graduated from MEPhI. Area of research: solid state physics, non-destructive testing.

Dmitry Valeryevich Berezkin, PhD, graduated from Bauman Moscow State Technical University (BMSTU) in 1990. He is currently an Assistant Professor at the Computer Systems, Complexes and Networks Department of BMSTU. His research interests include design of databases and knowledge bases, computational linguistics, systems analysis and design of complex systems.

Aleksandr Biryukov is a PhD student of the Department of Engineering Science and Technology of MEPhI. Graduated from MEPhI. Area of research: non-destructive testing.

Tatiana Bloshenko is a PhD in Economics. Area of scientific interests: state tax policy; natural rent and instruments of its withdrawal. Author more than 50 articles and 8 monographs.

Sergey Borisov, born in 1953. Graduated from Bauman Moscow State Technical University in 1978, specialization - machines and technology of highly efficient treatment processes. Works at Bauman Moscow State Technical University, Head of the Department of Innovative Entrepreneurship, Head of the School of technology entrepreneurship of BMSTU, Ph.D. Small and medium business, technology entrepreneurship. Deputy chairman of the government commission on competition and development of small and medium business; Member of the government expert council of the Russian Federation; Member of the russian council on foreign affairs; Chairman of the board of trustees of 'OPORA ROSSII'

Artem Borodkin is Associate Professor in National Research University "Moscow Power Engineering Institute".

Yuri Bozhko is a Vice-Head of the Counsil of Young Scientists and Specialists of MEPhI. Graduated from MEPhI. Area of research: solid state physics, non-destructive testing.

Miguel Brechner got a degree in Electronics at Kings College London and a Masters degree in Telecommunications at Imperial College, London. He was a successful executive and businessman in the IT field in Uruguay. In 2005 he resigned to all executive positions in the private sector and redirected his efforts to innovation and technology in the public sector, especially Education. Since 2006, he is the

President of Centro Ceibal, an institution created by himself. Centro Ceibal is changing the landscape of education in Uruguay. More than 720.000 students and teachers from public education have their own computers, 99% of them having connectivity in their education facility. Plan Ceibal is based on a profound equal opportunities vision that emphasizes social inclusion as a means of changing learning, education and the technology infrastructure of the country.

Tatyana I. Buldakova was born in 1956. In 1980 she graduated from Bauman Moscow State Technical University. Specialty in diploma – Computer Systems. Professor of Faculty "Informatics and Control Systems" BMSTU. Doctor of Technical Sciences, Professor. Field of research - System Analysis, Identification of Complex Systems, Remote Monitoring Systems, Information Security.

Valery Chernenkiy was born in 1941 and graduated from the Bauman Moscow State Technical University in 1964. Valery Chernenkiy completed his Doctor of Science (Habilitation) in Computer Science in 2000 and is now Head of the department of Information Processing and Control Systems.

Alexander Chernikov is the Head of Educational &Research Laboratory "Computer Information Systems and Technologies", the Dean of the Faculty of International Educational Programmes, the Head of the Centre of International Strategic Partnerships of Bauman Moscow State Technical University (BMSTU). The Honorary Doctor of Sciences of De Montfort University, UK. The Honorary Fellow of Glyndwr University, UK.

Alexander Chibisov is a Programmer of the Research & Educational Centre "Electronic University" of Bauman Moscow State Technical University.

Tamara Chistyakova is Chairwoman of Computer-Aided Design and Control Department, Professor, Dr. Sc. (Engineering), Saint-Petersburg State Institute of Technology, Saint-Petersburg, Russia Head of the leading scientific and pedagogical school of St. Petersburg in the field of information and telecommunication systems and technologies.

Alexander I. Chuchalin was born in 1952. In 1974 he graduated from Tomsk Polytechnic Institute as an Electrical Engineer. Currently Alexander I. Chuchalin is Doctor of Sciences, Professor of the Department of Engineering Pedagogy at Tomsk Polytechnic University. Prof. Chuchalin is focused on pedagogy of Higher Education, quality management in HEI, engineering programs design and accreditation. He published more than 350 papers, 5 monographs and 8 textbooks (idorcid/org/0000-0003-2807-0185). Prof. Chuchalin is a member of the WFEO Capacity Building Committee, member of the ENAEE Promotion Committee, member of the Consultative Committee of Aalborg Centre for PBL in Engineering Science and Sustainability under the auspices of UNESCO and member of the Editorial Boards of the International Journal of Quality Assurance in Engineering and Technology Education, European Journal of Engineering Education and Higher Education in Russia. Prof. Chuchalin received numerous awards and prizes and is a laureate of the Russian Federation Government Prize in Education in 2011.

Olga Alexandrovna Dotsenko graduated from Tomsk State University in 1990 with a degree in Radio physics, was admitted to the Radio Electronics Department of the Faculty of Radio physics and has been working there for the time being. In 2007 she earned the Ph.D. degree. Her research activities deal with electromagnetic wave propagation in magnetic and dielectric materials, the analysis of measurement methods for the microwave characterization of materials, modeling of magnetic and dielectric materials, and active methods of teaching.

Michael V. Dubkov, born in 1972. Graduated from Ryazan State Radio Engineering University. Head of the Department of Physics, Pro-rector RSREU, assistant professor, PhD, field of interest - target training.

Vladimir Eliseev is Associate Professor in National Research University "Moscow Power Engineering Institute".

Yuriy Fedorenko is a post-graduate student of Bauman Moscow State Technical University.

Elena Ivanovna Filosova has graduated from "The software of computer facilities and the automated systems", the Ufa State Aviation Technical University (USATU). She received her PhD in technical sciences in 2008. The theme of her work was "Management of the Multilevel Adaptive Training Systems as Elements of Formation of Quality of Educational Process". Filosova E.I. works as the associate professor in the department of Economic Informatics, UGATU, Ufa. She has more than 23 years of experience in scientific and pedagogical activity. Area of scientific interests: design of information systems of educational activity, ontology, management in social and economic systems. She published 2 monographs, 14 educational and methodical works and more than 100 scientific and methodical articles.

Yuriy Gapanyuk was born in 1974. Graduated from the Bauman Moscow State Technical University in 1998. PhD in Computer Science in 2006. Working in the IT industry since 1995.

Juan Gonzalez Gusev was born in 1979. Bachelor in Electronics and Telecommunications from the National Polytechnic School of Quito, Ecuador in 2004. MSc in Automated Systems from the Bauman Moscow State Technical University in 2015. PhD candidate in Computer Science from the same University since 2015. Has experience in telecommunication and IT area (mobile service companies and telecommunications regulatory organizations) since 2003.

Maxim Mikhailovich Grekhov, born in 1980, graduated from National Research Nuclear University MEPhI in 2003, PhD thesis in 2009 in the specialty "condensed matter physics". Work in NRNU MEPhI as head of the scientific and technical Department, the ORCID 0000-0003-3194-4269. Research interests: analytics developments and technologies of nanoindustry, educational resources on nanotechnology, monitoring of University activities in different scientific areas.

Vladimir Gurenko is Assistant Professor of the Department of Computer Systems and Networks of the Bauman Moscow State Technical University, Deputy Dean of the Faculty of Informatics and Control Systems. He is directly involved in the educational process for training specialists in the field of information technology and computer science. His specialization is in information technology, programming, digital signal processing.

Vidar Gynnild's main theme was the basic need for academics to take a less transmissive approach to teaching. He started by reviewing some different definitions of information literacy: which sounded good, but he was aware that his fellow academics were not terribly engaged in it and did not necessarily see it as their job to teach. He identified two paradigms. The first was the "Instruction paradigm" e.g. that the mission was to deliver information, the teaching/learning structures consisted of covering the required material, funding was by hours of instruction and faculty were primarily lecturers. However in the newer paradigm, you put learning first. In this you aim to produce learning, structures consists of producing specified learning results, funding goes by learning outcomes, and faculty are primarily designers of learning methods and environments. This means you have to start with the intended learning outcomes. He mentioned what he called "frame factors" that shape human agency (e.g. mentoring, learning activities, room facilities, social factors) which need to be thought about, and which may be of more or less important. Like Ralph Catts, he identified the importance of assessment in this process, and how students tend to ask "How does this count towards my grade?". He finished by emphasising how there needed to be collaboration between faculty and librarians to embed information literacy in the curriculum. He identified some key issues such as academic integrity that prompt such collaboration. He thought that academics such as he could be champions in helping this come about. They can work together to transform a teaching paradigm into a learning paradigm.

Gerard L. Hanley, Ph.D., is the Assistant Vice Chancellor for Academic Technology Services and the Executive Director of MERLOT (Multimedia Educational Resource for Learning and Online Teaching, www.merlot.org) for the California State University, Office of the Chancellor. At the CSU, Gerry oversees the management of systemwide academic technology initiatives including digital library services, course redesign with technology, affordable learning solutions, accessible technology, and Cal State Online, serving CSU's 23 campuses and 475,000 students. At MERLOT, he directs the development and sustainability of the international consortium and technology strategy to provide open educational services to improve teaching and learning. He is the Director of Skills Commons, the US Department of Labor's open repository of educational resources for workforce development. The CSU Long Beach Center for Usability in Design and Accessibility is also under his direction. Gerry's previous positions include professor of psychology at the CSU Long Beach campus and a wide range of program management positions in industry. EDUCATION: State University of New York at Stony Brook: Degree - Ph.D., Field Of Study: Experimental / Cognitive Psychology, Dates attended 1978 – 1984; Bachelor's Degree, Field Of Study Psychology, Dates attended 1974 – 1978.

Samih Jammoul is a PhD student in BMSTU since 2015. Master degree in telecommunication networks Telecom-Bretagne, France.

Ludmila Vasilievna Juravleva was born in 1949. The Associate Professor "Design and Technology of Electronic Devices" department of BMSTU, PhD. The expert in the field of active design methods of training, business games and information pedagogical technologies, the author of a number of classical textbooks in the field of engineering pedagogics.

Alexey Kalistratov, born on 1993. Received bachelor's degree at 2015 at BMSTU. Now is working at his magister degree at BMSTU. Maintainer of laboratory class at BMSTU. Also works at Fujitsu GDC as Field Support Engineer. Specializes at virtualization and data transfer networks. orcid.org/0000-0002-9394-0557

Maxim Mikhailovich Kapilevich, 1963 year of birth. Tomsk Medical Institute, 1986. Tomsk Polytechnic University, Head of the Department of Physical Culture, Doctor of Medicine, Professor at Tomsk State University, a Head of the Department of Sport Tourism, Physiology and Medicine. The field of scientific interests is the physical activity of the disabled, and physical fitness. orcid.org/0000-0002-2316-576X.

Marios Kasinopoulos is an Electrical Engineer member of the academic staff of the IPS. He first graduated from the University of Grenoble in France where he obtained the degree Maitrise in "Electrical Engineering" and then he continued his studies at the University Paul Sabatier in Toulouse (France) where he got his Ph.D in "Automatique". He has done also studies in Economics and he is a holder of the degrees Licence "Science Economiques" and Maitrise "Economie d'Entreprise". He worked for four years at the CNRS in France and then at the Higher Technical Institute (HTI) in Nicosia Cyprus in the position of Senior Lecturer. Then he worked as member of the academic staff for seven years in the Cyprus University of Technology. Since 2014 he is a member of the academic staff of IPS. He was involved in various research projects in the fields of Control Engineering, Telemedicine, Internet Addiction and Engineering Education. He has a long experience in the Bologna Process and in Quality Assurance (QA) of Higher Education. He is member of the Cyprus team of Bologna Experts and member as International Expert on QA in various countries like France, Belgium, Lithuania and Estonia.

Tatyana Kholomina, Professor, Doctor of Physical and Mathematical Sciences. The Head of the micro- and nanoelectronics department of Ryazan State Radio Engineering University (RSREU). Fields of research: new conceptions in engineering education, investigation of relaxation processes in semiconductor barrier structures.

Alena Kireyeva, Doctor of Economic. Area of scientific interests: tax system, tax policy, international taxation, integration processes. Author more than 100 publications.

Galina V. Kirsanova was born in 1954, graduated from Moscow State Linguistic University in 1977. Since then has worked at Bauman University Linguistic Department as a teacher of English, senior lecturer, assistant professor. Has held the position of Dean of the Linguistics Faculty for three years. Has participated in conferences both in Russia and abroad including USA, Great Britain, China, Check Republic. Has over twenty papers mostly devoted to teaching English to deaf and hard of hearing engineering students. Interested in innovative methods of teaching foreign languages including CLIL.

Nurlan Kiyasov graduated with honors from MISiS, where he studied metallurgy and economics of production; worked as a melter, engineer at ferroalloy plants. Received a Master's degree ("Micro and nano-technologies and their commercialization"), as well as an MBA ("The commercialization of scientific developments") at Cambridge University. Founder of the project "National Open School", initiator of the National portal "Open Education", founder and program director of the International Conference on the new educational technologies “EdCrunch”. Member of the working group of the Council for open education at the Russian Ministry of Education. His areas of interest include new technologies in education, state innovation system infrastructure, management in higher education, fundraising and endowment funds.

Tatiana Kochetkova graduated from Tomsk State University in 1987 with a degree in Radio physics and Electronics, was admitted to the Department of Radio Electronics of the Radio physics Department and has been working there for the time being. In 2003, she earned Ph.D. Scientific interests lie in the field of measuring the dielectric properties of materials, remote sensing, active methods of teaching.

Boris Vasilevich Kostrov, born in 1948. Graduated from Ryazan State Radio Engineering University in 1971. Head of the electronic computers department. Professor, Doctor of Technical Science. Honorary Worker of Russian higher education.

Ilya Andreyevich Kozlov graduated from the Computer Systems, Complexes and Networks Department of Bauman Moscow State Technical University (BMSTU) in 2013. He is currently a Junior Research Assistant of the Research Division of the Scientific-Educational Complex “Informatics and Control Systems” at BMSTU. His research interests include data mining and computational linguistics.

Kirill Kucherov is a Software Engineer student of 4th year in Bauman Moscow State Technical University.

Julia Kurovskaja. Date of Birth: 12.02.1973. Education: State Linguistics University of Nizhny Novgorod (LUNN), 1997, Specialist Degree: German Teacher, Spanish Teacher. Irkutsk State Linguistic University, 2009, Specialist Degree: Linguist, English Teacher. Ph. D. (Philology), associate professor. Name of employer, position: Bauman Moscow State Technical University, associate professor. Institute for Development Strategy of Education of the Russian Academy of Education, Senior Researcher. Interests: Education, Engineering Education, Pedagogical Semiology, Educational Discourse, Cognitive Linguistics, German.

Oleg Kuzenkov is Vice-director of Institute of Information Technologies, Mathematics and Mechanics at Lobachevsky University of Nizhni Novgorod (UNN) for methodical and educational work (since 2002). Dr. Kuzenkov graduated from UNN in 1984, diploma in applied mathematics. Candidate of physical and mathematical sciences (PhD) in 1989, associate professor at the department of Differential Equations, Mathematical and Numerical Analysis since 1996. Published more than 90 scientific publications. Participated in several Tempus projects in the area of modern educational technologies: Tuning, INARM, MetaMath (Grant 543851-TEMPUS-1-2013-1-DE-TEMPUS-JPCR). Research interests: differential equation, optimal control, theory of measure, mathematical modelling biological evolution, modernization of educational progammes in the area of ICT.

Viatcheslav Kuzovlev, born in 1948. Graduated from BMSTU at 1984. PhD. Works as Docent at IU5 department of BMSTU. Specializes at reliability and validity.

Dmitry Lakhvich was born in 1990. He received two degrees from Bauman Moscow Technical University (BMSTU): – 2013 - Specialist degree, Faculty of Special Machinery, specializing in "Space rocket engineering technologies"; – 2015 - Master degree from Bauman Moscow State Technical University, specializing in "Computer Science and Computer Engineering". Now Dmitry is Data Engineer at telecom company Maxima Telecom and extramural post-graduate student. Areas of research are Big Data technologies, Distributed computing systems, BI systems, Transfer pricing, Business Process Optimization, Machine Learning. He also works as a teaching assistant at BMSTU (prepares and conducts labs). He is interested in helping new specialist to become adept with new technologies and trends. ORCID ID: 0000-0002-2267-4204.

Vladimir Lazarev was born in 1986. He received the B.S. and M.S. (with distinction) degrees in optoelectronics from Bauman Moscow State Technical University (BMSTU), Moscow, Russia, in 2007 and 2009, respectively. He was a Research Engineer with the All-Russian Scientific and Research Institute for Optical and Physical Measurements, Moscow, Russia, in 2007–2012. He joined the Science and Education Center for Photonics and IR-Technology, BMSTU, in 2012 as a Senior Researcher. In 2013, he presented his Ph.D. thesis. Since 2017 he is a head of Stabilized Laser Systems Laboratory at BMSTU. He has authored 30 papers and 10 patents for inventions. His research interests include fiber optics, laser physics, ultrafast lasers, mid-IR laser sources, and optical frequency standards. Dr. Lazarev is a member of OSA, SPIE and IEEE.

Elena Leontyevna graduated from Tomsk State Pedagogical University in 1986 with a degree in "Foreign languages". In 2006 she earned the Ph.D. degree. In 2016 she graduated from Autonomous University of Barcelona with master degree. Scientific interests lie in the field of intercultural business communication, management in education, active methods of teaching.

Alexander Loginov, born in 1950. Graduated from Ryazan State Radio Engineering University in 1973. PhD, assistant professor. Field of interest as combined multispectral vision systems for aviation application, Information and communication capabilities of modern technologies in the teaching of special disciplines by order of enterprise.

Andrey Malakhov has scientific interests and results in the areas of scientific priorities of the higher school, electronic educational resources of new generation, interactive educational and scientific complexes to perform work in remote access mode, implementation of interactive forms of learning in cross-university networked system.

Anna Maltseva is the author of more than 200 scientific works on issues of innovation activity, innovation infrastructure, regional economy, including 11 monographs, the head of 5 and the executor of more than 20 scientific and research projects. In 2016 she was responsible for the implementation of the State Contract for the project "Development and implementation of methods to improve the effectiveness of the practice-oriented scientific and technical clubs for creative development of students and pupils" of the Federal Target Program for the Development of Education for 2016-2020. The project was devoted to the analysis of the activity of practice-oriented scientific and technical clubs for the creative development of students and pupils in engineering areas of training, functioning in the scientific and educational environment of educational institutions of higher education and the development of scientific and methodological foundations for increasing the effectiveness of their activities. Sphere of activity: scientific and research and analytical work in the field of studying the issues of the development of innovative systems on the micro- and meso levels, projects of creation of innovation infrastructure's objects, the economy of science and education, organizational and methodological support for the implementation of innovative activity in higher education.

Tatiana D. Margaryan graduated from Moscow State University of Education in 1990, since then she has been an instructor at Bauman Moscow State Technical University. In 2012 she received her PhD in Historical sciences. Currently, T. Margaryan is an Associate Professor of the Linguistic Department at Bauman Moscow State Technical University and a member of the coordination board of the National Association of Teachers of English in Russia (NATE Russia). She writes and presents widely on issues of ESP teaching, course designing, blended learning.

Vitaly Martynov was born on July 22, 1959. He graduated the Mari Polytechnic Institute with honors (specialty "Technology of mechanical engineering, metal-cutting machines and tools") in 1981 and then he continued to work in it as an assistant. In 1988 he graduated the full-time postgraduate study at the Moscow Aviation Institute. After graduation he worked as an assistant, senior teacher, docent, professor, the head of the Department of Economic Informatics, Ufa State Aviation Technical University (UGATU). He received his Ph.D. (specialty - Engineering Graphics and Applied Geometry) in 1989, Kiev Civil Engineering Institute and in 2000 he defended his doctoral thesis (specialty - Automated control systems), UGATU. 1996-1999 - doctoral studies of UGATU. In 2004 he was awarded the academic title of professor, the department of computational mathematics and cybernetics. Area of scientific interests include enterprise architecture, information systems in educational and socio-economic fields, applied geometry. The main lecture courses are "Informatics", "Databases", "Management of the Information Center of Information Technologies", "Modern IT" for bachelors and masters. Under his leadership and direct participation were carried out more than 15 research projects and contractual work on various programs and subjects in the IT field. Manages postgraduate students. 7 of his students successfully defended their Ph.D. theses. In 2012 he was awarded the title of "Honored Worker of Higher Professional Education of the Russian Federation". The list of works contains more than 300 titles: 5 monographs, about 80 articles in peer-reviewed editions, 20 teaching aids, articles, reports at conferences of various levels. ORCID - 0000-0002-8562-9267.

David Moffat's degrees in mathematics and computer science were from Cambridge and Canterbury universities. After a time in the software industry, he went to research AI at Edinburgh, and psychology at Amsterdam and Cardiff, then lecturing computing at GCU. He teaches game design and affective computing; and researches AI and emotion, creativity in education, serious games, and cognitive science.

Evgeniy Muratov, born in 1978. Graduated from Ryazan State Radio Engineering University in 2004. Assistant professor of electronic computers department, PhD, field of interest - technical vision systems.

Vladimir M. Nesterenko, Doctor of Sciences in Pedagogy (Habilitation), is currently a professor at Psyhology and Pedagogy Department of State Technical University, Russia. He is graduated from the Faculty of Electrical Engineering of Samara State Technical University, Russia, in 1969. He obtained his PhD degree in Electrical Engineering (1974) and DSc degree in Pedagogy (2001) also from State

Technical University, Russia. In 1974 he was appointed as an associate professor, then promoted to a professor position in 2001. In 2000-2014 he was a head of Psyhology and Pedagogy Department, and a dean of Faculty of Humanitarian Education in 2006-2011. His main research area is structural and technological innovations in Higher Education.

Konstantin Neusypin, Doctor of Technical Sciences, Professor, Laureate of the Russian Federation Government Prize, Director-General of Scientific and Educational Research Center for Intelligent Systems at Bauman Moscow State Technical University. His research interests include Kalman filtering, intelligent control systems, and observability analysis of dynamic systems.

Michael Nikiforov, born in 1946. Graduated from Ryazan State Radio Engineering University in 1968. Specialty of the electronic computers. Vice head of electronic computers department. Director of REC "Special computer", PhD, assistant professor. Corresponding Member of the Academy of Informatization and Education of the Russian Federation. Field of interest - target training, distance learning.

Anatoly Novikov, born in 1950. Graduated from Gorky State University. N.I. Lobachevsky 1972. Specialty applied mathematics. Works at Ryazan State Radio Engineering University in the department of Higher Mathematics. PhD, assistant professor. Field of interest as methods of teaching mathematics in secondary and higher schools, technology of mathematical training of engineering personnel.

Dorjsuren Odselmaa graduated from Mongolian University Science and Technology with a Bachelor's degree in Software Engineering in 2015. Currently, she is the student in Faculty of Computer Science and Control Systems at Bauman Moscow State University, pursuing master's degree in CS.

Anna Pilyugina, born in 1981. Graduated from Bauman Moscow State Technical University in 2004, specialization - business economics and management. Works at Bauman Moscow State Technical University, Deputy head of the Department of Innovative Entrepreneurship, associate professor, Ph.D. Technology entrepreneurship, economic-mathematical modeling. Expert of the Federal Education and Science Supervisory Service.

Valeriy Platonov is an Advisor on education to Vice-president of Rybakov Charity foundation, Moscow (rybakovfond.ru), Program committee of International conference on new educational technologies EdCrunch, Chairman of the first ever world competition of open online courses – Edcrunch Award OOC, Executive director of ANEO "Higher University of Sciences and Technologies" under the Russian Academy of sciences Institutions, Senior scientist at Prokhorov General Physics Institute, Moscow, CEO of Courseon.org CEO of private company "AGAT Concern R&D Centre" Ltd, based in Skolkovo foundation residency Education, May 2015 - December 2015, Moscow University for Industry and Finance "Synergy", Intangible assets and IP valuation - professional vocational program. In 1978 - Defended Thesis for a candidate of mathematical and physical sciences (Ph.D. in mathematical physics). 1975-1978 - post graduate student of Mathematical Faculty, Physics Department of Lomonosov Moscow State University. In 1975 - Honours diploma in mathematical physics from Physical Department of Lomonosov Moscow State University 1969-1975 - student of the Physical Department of Lomonosov Moscow State University.

Vadim Ponkratov is a PhD in Economics. Area of scientific interests: state financial policy; management of the oil and gas income of the budget, strategy of formation and use of means of sovereign funds; natural rent and instruments of her withdrawal; instruments of the state financial stimulation of economic development. He took part (as the head, the executive) in 50 research works on the aforesaid subject for the benefit of public authorities. Member of the committee of Chamber of Commerce and Industry of the Russian Federation on the energy strategy and development of fuel and energy complex. The member of Advisory council of Chamber of Commerce and Industry of the Russian Federation on improvement of the tax law and law-enforcement practice. Author more than 100 articles and 15 monographs.

Andrey Pozdnyaev is a Doctor of Economics. Area of scientific interests: state tax policy; management of the tax income of the budget. Author more than 20 articles.

Andrey Viktorovich Proletarsky, Doctor of Technical Sciences, Dean, Professor, Laureate of the Russian Federation Government Prize. His research interests include unmanned aerial vehicles control, information and communication technologies.

Georgiy Revunkov, born in 1948. Graduated from the Bauman Moscow State Technical University in 1971. PhD in Computer Science in 1976.

Ilya Rodko is a PhD student of the Department of Engineering Science and Technology of MEPhI, PhD. Graduated from MEPhI. Area of research: non-destructive testing.

Tatiana Nikolaevna Romanova obtained her Ph.D., physical and mathematical sciences, from Lomonosov Moscow State University, Russian Federation in 1984. In 1998 she started working as Associate Professor at the Faculty of Computer Science and Control Systems at The Bauman Moscow State Technical University. She writes and presents widely on issues of mathematical modeling and methodology of software engineering.

Peter Sakál (15.12.1951) is a professor at the Institute of Industrial Engineering and Management, Faculty of Materials Science and Technology in Trnava Science of the Slovak Technical University in Trnava, Slovak University of Technology in Bratislava. After graduation Master Degree (1976), he became a lecturer to Department of Mechanical and Electrotechnical Production Management, Faculty of Mechanical Engineering, Slovak University of Technology in Bratislava, where he taught subject Engineering Statistics lectured and practiced subjects Control Theory and subject Operational and System Analysis, which he supported by edition of three textbooks co-authorship. Since 1988, he lectured and exercised from the subject Economy Development of Company, for which he wrote co-authorship Exercise Guidance and he lectured and exercised subject Business Games. He was on an internship 2 times at the Moscow Institute of Management, also one time on Belarus Polytechnic Institute in Minsk, Kiev Polytechnic Institute in Kiev. In 2006, he was on research internship at the University of Iowa (USA). In 2006 and in 2010 he was on research internship at UGATU Ufa, ISEI UNC RAN a MEI Moscow. He also worked as an external teacher of the Swiss foundation TRANSFER Slovakia. Since 1993 until 1996 he served as Executive Director of the Ukrainian-Slovak enterprise EMIT in Kiev. But he still maintained contact with faculty (e.g. in 1994 on soil at Faculty of Materials Science and Technology organized international contracting and sales shows Ukraine '94). In 2000, after returning to the university he issued as a co-author of college textbook, Digital Control Systems. Further, he lectured and led the exercise of subject Operational analysis, for which wrote the co-authorship university textbooks of the same name in it based edition of Theory and Practice of Management. At present he lectures and practices the subject Strategic Management, for which he wrote the textbook co-authorship same name and monograph Strategic Management in Practice of Manager. In 2009, he issued the monograph Logistics of Operative Enterprise as head of collective authors'. He has trained about 60 bachelor students, 270 diploma students, 20 PhD students and now he trains 2 PhD students. He deals with the organization Student Scientific Conferences and supervised theses scientific theses some years. He is a member of the Scientific Board of the The Faculty of Materials Science and Technology in Trnava Slovak University of Technology in Bratislava, member and chairman of the Commission for state exams master studies and Bachelor studies, member of the Commission for the dissertation examinations in the study field of Enterprise Management, opponent and member of the Commission of Student Scientific Conferences. He is a member of the Presidency of Slovak Association of Process Management and he is member of Editorial Board the Journal process manager (Procesný manažér), the Journal Managers Forum (Fórum manažéra), the Journal Business Performance (Výkonnosť podniku), the Journal Problemy Razvituija Territoriji. He is the guarantor of the study program Personnel Policy in Industrial Plant at bachelor, master and PhD. level study. He was principal investigator of field and departmental research projects and many contracts, he leads 2 institutional projects. In 2004 he successfully completed as Project Manager solving scientific project VEGA No. 1/9099/02 "Enviromentally oriented management, marketing and logistics in strategic business units". In the years 2004-2006 he participated in solving the research project No. 019/2001: "Transforming Industry in Slovakia Through Participatory Ergonomic" (financially supported by common Slovakia-American Fund for Scientific and Technological Cooperation). In the years 2009-2014 he was the principal investigator of the project No. LPP-0384-09: "Koncept HCS modelu 3E vs. koncept Corporate Social Responsibility (CSR)", which it supported by the Slovak Research and Development Agency. In the years 2012-2014 he was the principal investigator of the project KEGA No. 037STU-4/2012: "Implementation of the subject 'Corporate Social Responsibility Entrepreneurship' into the study programme Industrial management in the second degree at MTF STU Trnava". He has extensive cooperation with Russian universities (MEI TU Moscow, UGATU Ufa, TU Iževsk) a RAV (ISEI UNC RAN, Vologda, Apatity/Murmansk). He is co-author of two improvement proposals. He was appointed as professor on November 29th 2005 in the field of Industrial Engineering and Management.

Pyotr Semkin was born on 1949. Graduated from BMSTU at 1986. Works as Associate Professor at BMSTU. Specializes at network software and operational systems.

Evgeniya Serebraykova graduated from the Tomsk University of Control Systems and Radioelectronics with a degree in Industrial Electronics. The second diploma in the specialty of "Financial Management". From 2004 to 2011 the main place of work is LLC Tomsk Electronic Company. In 2012, its own company, OOO IQ Group, was founded. The company is engaged in positioning and promotion. Since 2015 - the beginning of activity in the post of the head of the elite education department of Tomsk Polytechnic University.

Olga Shabalina is an Associate Professor at Volgograd State Technical University, in the Department of Computer Aided Design. Her research interests are in education research, including how best to teach engineering subjects, using technology enhanced learning in general. She is particularly interested in the application of game-based learning to the higher education curriculum. Her contributions to the field include research on the development and measurement of creativity in undergraduate students, and the use of games to stimulate creativity. She has chaired tracks in these areas at the ECGBL in recent years.

Kai Shen, Ph.D., is an assistant professor at Beijing Institute of Technology, Beijing, P.R. China. He graduated from Nanjing University of Sciences and Technology, Nanjing, P.R. China, in 2012. He graduated with honors from Bauman Moscow State Technical University, and received the M.S. degree (in 2015) and Ph.D. degree (in 2018) in control science and engineering, Bauman Moscow State Technical University, Moscow, Russia. His research interests include inertial navigation and intelligent navigation, Kalman filter and multisensor information fusion, system analysis based on the degree of observability, controllability and identifiability.

Marwa Ahmed Shouman is a Lecturer in the Department of Computer Science and Engineering at Menofiya University, Egypt.

Aleksey Skuratov graduated with honors from the Automatics and Computer Engineering Department of the Moscow Institute of Electronic Engineering in 1983, specializing in "Electronic computers", qualification "Engineer-system technician". In 1989, at the Moscow Institute of Electronics and Mathematics, he defended Ph.D. thesis in the field of automation systems for the design of microprocessor systems (specialty 05.13.12). He passed an internship at the Department of Real-Time Computer Systems of The University of Eskilstuna-Vasteras, Sweden, February 1 - June 15, 1993. In 2007, at the Moscow Institute of Electronics and Mathematics, he defended Doctoral thesis (Dr. of sciences) "Statistical monitoring and analysis of telecommunications networks" (specialty 05.13.13). 01.12.2011 received the title of Professor at the Department of Technical and Information Means of Control Systems of the Moscow Institute of Radio Engineering, Electronics and Automation. During the September 20-26, 2012, an internship was organized by the Ufa State Aviation Technical University, the Karlsruhe Institute of Technology, the COMMIT Project Partners GmbH in the program "Advanced problems of Informations Technologies". Laureate of two Prizes of the Government of the Russian Federation in the field of education for 2001 and 2011. Honored Worker of Science and Technology of the Russian Federation. Seniority: From 1983 to 1994 he worked at the Moscow State Institute of Electronics and Mathematics as an engineer, assistant, senior lecturer, associate professor, deputy dean of the Faculty of Automation and Computer Engineering. From 1994 to 1995 he worked in the Russian Research Institute of Information Systems as the head of the department "Videoconferences and satellite telecommunications". From 1996 to September 2013 he worked in the State Research Institute of Information Technologies and Telecommunications" ("Informika"). Since June 1997 as Deputy Director. From 01.10.2013 - senior researcher in Directorate of Scientific and Technical Programs.

Ilia Soldatenko Ph.D. in Physics and Mathematics. Vice-dean of Applied Mathematics and Cybernetics faculty, associate professor at Information Technology department, head of IT Department at Tver State University. Specializes in teaching programming, theory of possibility and fuzzy logic, soft computing, web-technologies and computer architecture, theoretical foundations of computer science.

Research interests: possibilistic optimization, fuzzy sets, intelligent optimization techniques, artificial intelligence, intelligent control systems, machine learning, development of automation systems, informatization of education and science.

Mikhail Solovyev graduated from the Faculty of Automation and Power Engineering of Tomsk Polytechnic University (specialty "Engineering and Electrophysics of High Voltages"). In Tomsk Polytechnic University from the moment of its termination: 1994-1998 - Junior Researcher of the High Voltage Research Institute under TPU (TPU graduate student); 1998 - 2002 - Assistant to the First Vice-Rector; (Senior lecturer of the Chair of High Voltage Technical and Electrophysics, Since 2000 - Associate Professor of the Department of Electric Power Systems and High-Voltage Technology(to the present)); 2002 - 2004 - Deputy Head of the Training and Methodology Department; 2004 - 2007 - Deputy Vice Rector for Academic Affairs; 2007 - 2009 - Head of the educational-methodical department; 2009 - present - Head of educational and methodical management. From 1994 to 1997 he studied at the post-graduate course of TPU on the specialty 05.14.12 "Technique of high stresses" under the scientific supervision of Professor V.I. Kurtz. The direction of the thesis is the study of the processes of electrical impulse destruction of dielectric materials. He defended his thesis on November 5, 1997. In 2001 he received a second higher education at the Faculty of Engineering and Economics of TPU (specialty "Economics and Management at the enterprise").

Radovan Stojanović received his BS and MS in Electrical Engineering from the University of Montenegro and a PhD in Electrical Engineering and Computer Technology from the University of Patras, Greece in 1990, 1994 and 2001 respectively. From 1990 until 1997 he worked as a teaching assistant and lecturer at the Faculty of Electrical Engineering, University of Montenegro, and then moved to the University of Patras where he stayed until 2003. At the University of Montenegro he was appointed to the positions of assistant, associate and full professor in Electronics and ICT, in 2002, 2008 and 2013 respectively. He spent periods of time undertaking research at numerous EU and oversees universities and institutes. He is an author/coauthor of more than 200 publications in international journals and conference proceedings and coordinator or sub-coordinator in more than 30 projects including (4 TEMPUS/ERASMUS, 2 NATO, 1 FP6, 1 FP7, 1 WUS and 1 World Bank, 4 Bilateral, 5 National…). He is an expert of EACEA (Education, Audiovisual and Culture Executive Agency), EURAMET (European Association of Metrology Institutes) and several governmental institutions in the Western Balkan Region. He is a member of the Board of Montenegrin Academy of Science for Natural and Technical Sciences, Think Tank Team of Montenegro, Board of Doctoral Studies at the University of Montenegro, a representative of Montenegro in H2020-ICT Committee as well as a president and establisher of the Montenegrin Association of New Technologies (MANT) and the founder and Chairman of the Mediterranean conference on embedded systems (MECO) and co-chairmen of EUROMICRO/IEEE Workshop on Embedded and Cyber-Physical Systems (ECYPS). He coauthored a book "Geospatial Visual Analytics" with R. DeAmici and G. Conti. in Springer Publication and an Chapter in a book "Effective Surveillance for Homeland Security: Balancing Technology and Social Issues" in edition of F. Flammini, R. Setola and G. Franceschetti published, by Chapman and Hall/CRC. He was the first director and one of the establishers of the Forum of Montenegrin University Teachers and Researchers (FUPI). He is also a member of board of directors of EUROMICRO (European Association for Micro Electronics). He has participated in scientific committees of more than 30 international and national academic conferences and serves as an editor or reviewer for several SCI/SCIE journals. He was a member of the Senate of the University of Montenegro. He is the founder of Montenegro's Centre for Applied Electronics- 2003, Laboratory for Medical Electronics (MEDEL)- 2016 as well as the creator of several new courses in the field of advanced electronics and ICT. He leads an active social life and as an independent scholar he has played a role in democratic movements in Montenegro in the last 20 years. He speaks English, Greek and basic Russian. More about Radovan Stojanović you can find at http://www.apeg.ac.me/rstojanovic.htm.

Andrey Sukhobokov was born in 1957. In 1980, he graduated from Tashkent State Technical University as "Engineer – Mathematician". He has 40 years of experience in IT systems design, development, implementation, support and operations. Andrey managed and participated in various roles in 70+ projects of integrating management systems, reorganizing management systems of enterprises, development of software for scientific projects. Currently, he is the CEO of technological startup Optimal Management, LLC which is using heavy mathematical models and Big Data tools for management of enterprises. In 2016 -2017, he is a guest Lecturer in Bauman Moscow State Technical University and a mentor for talented students. He is interested in training strong specialists and engaging them on new projects.

Artem Sukhobokov was born in 1983. Graduated from 3 leading Russian Universities: – 2004 – Specialist degree from Lomonosov Moscow State University (MSU), Mathematician, system programmer specializing in "Applied mathematics and informatics"; – 2006 – MSc from Bauman Moscow State Technical University, specializing in "Computer Science and Computer Engineering"; – 2011 – Finance University under the Government of the Russian Federation, Economist specializing in "Finance and Credit". In 2009, received PhD degree from Bauman Moscow State Technical University. Areas of research are Applications Architecture, Controlling methods, ERP systems, Adoption of mathematical models to SAP solutions. Currently, Upstream Oil&Gas Teamlead of SAP DBS CIS (Moscow). Artem is SAP certified Enterprise Architect, Solution Architect, Consultant, and Developer. TOGAF 9 certified Enterprise Architect. PMI certified Project Management Professional. In parallel to SAP, Artem is currently a professor (assistant) at BMSTU. A mentor to talented students, he is passionate about training strong specialists and engaging with them on new projects. ORCID ID: 0000-0002-1370-6905.

Sergey I. Suyatinov was born in 1953. In 1980 he graduated from Bauman Moscow State Technical University. Specialty in diploma – Automatic Control System. Associate Professor of Faculty "Informatics and Control Systems" BMSTU. Ph.D, Associate Professor. Field of research – Nonlinear System Modeling, Biotechnical Systems, Remote Monitoring Systems.

Vladimir Suzev is Doctor of Technical Science, Professor of the Computer Systems and Network Department, Bauman Moscow State Technical University, the Laureate of the Russian Federation Government Award 2009, honored Worker of Higher Education in Russian Federation. The author of more than 150 scientific works in the field of signal's digital processing methods specialised in real-time control systems.

Valery Terekhov, born in 1962. Graduated from the Lomonosov Moscow State University in1985. PhD in Computer Science in 2005. Working in the IT industry since 1986.

Ilya Tikhonov was born in 1993 and graduated from Russian State Technological University in 2013 with a degree in "Computer Science and Engineering". In 2015, he received Masters degree from Bauman Moscow State Technical University with degree of "Computer Science and Computer Engineering". Currently, Ilya is a PwC Consultant in financial consulting department and at the same time - extramural post-graduate student. Areas of his research are Big Data technologies, ERP systems, BI systems, Transfer pricing, Business Process Optimization, RPA technology. In parallel to PwC job, he works as an assistant at BMSTU (prepares and conducts labs). He is interested in training strong specialists and engaging them on new projects. ORCID ID: 0000-0001-8955-9908.

Tatiana Tsibizova, born in 1966, graduated from Bauman Moscow State Technical University, specialty Automatic Control Systems., in 1990, Lecturer, Professor, Head of Department of educational technology. Author of more than 100 scientific and educational materials in the field of automatic control systems, intelligent systems, and also in the theory of continuous education.

Tatiana Ivanovna Vishnevskaya obtained her Ph.D., physical and mathematical sciences, from Institute of Problems of Chemical Physics of Russian Academy of Sciences, Chernogolovka, Moscow region, Russian Federation in 1990. In 1999 she started working as Associate Professor in the Department of Computer Science and Control Systems at The Bauman Moscow State Technical University. She writes and presents widely on issues of mathematical modeling and methodology of software engineering.

Nikolay Vishnyakov was born in 1961. Associate Professor at the Department of micro - and nanoelectronics, Head of the Regional Centre of Probe Microscopy for the collective usage of Ryazan State Radio Engineering University.

Andrey Vlasov was born in 1968, graduated from BMSTU in 1994. The Deputy Head Design and Technology of Electronic Devices Department, BMSTU on scientific work, PhD. Specializes in area of the system analysis and information technologies. It is awarded by a medal the Ministry of Education and Science of the Russian Federation for the leadership in research work of students. ORCID: 0000-0001-5581-4982.

Vita Vlasova, born in 1991. Graduated from Bauman Moscow State Technical University in 2014, Master in Organization and management of science-intensive industries. Works at Bauman Moscow State Technical University. Assistant professor. Commercialization of technology innovations, management of science-technical results, economics of intellectual property.

Alexander Vasilievich Yazenin Doctor of Science in Physics and Mathematics, professor. Dean of Applied Mathematics and Cybernetics department at Tver State University. Honored Worker of Higher School of the Russian Federation. Specializes in teaching probability theory, fuzzy logic. Research interests: possibilistic-probabilistic optimization, theory of possibility, portfolio optimization.

Ravil Zagidullin is an Assistant Professor of the Department "Radioelectronic Devices" of Bauman Moscow State Technical University.

Sergey Zaikin, born on 1991. Received bachelor's degree at 2015 at BMSTU. Now is working at his magister degree at BMSTU. Also works at Moscow Exchange as System Engineer. Specializes at virtualization.

Irina Zakharova, Ph.D. in Physics and Mathematics. Vice-dean of Applied Mathematics and Cybernetics faculty, Associate professor at Mathematical Statistics and Systems Analysis department at Tver State University. Specializes in teaching probability theory, mathematical statistics.

Elena Zakieva was born on May 10, 1971 in Ufa, Russia. In 1993 she graduated from Ufa State Aviation Technical University (USATU) with a degree in Automated Systems for Information Processing and Management. She received her PhD in technical sciences in 2000. From 2012 to 2015 she studied in the "Management in social and economic systems" doctorate. Currently she works as an docent at the Department of Technical Cybernetics, USATU. Her research interests include system analysis, modeling and management of socio-economic systems, quality of life.

Alena Zaytseva received her bachelor degree in Business In-formatics at USATU, Ufa, 2011. In 2013 and 2014 she graduated from USATU with honors and received masters degrees in "Business Informatics" and "Management". In 2016 she finished her PhD studies and now is preparing her PhD thesis. She has 4 years of experience of scientific and pedagogical activity at the post of the assistant at the Department of Economic Informatics. She published 17 articles at conferences of various levels. Area of scientific interests: design of information systems of educational activity, ontology, management in social and economic systems.

Andrey Zhukov graduated from Tomsk State University in 1985 with a degree in Radio physics, was admitted to the Department of Radio Electronics of the Radio physics Department and has been working there for the time being. In 2001 she earned the Ph.D. degree. Scientific interests lie in the field of e-learning, programming, active methods of teaching.

Lyudmila A. Zinchenko obtained Engineer degree in 1987 from the Taganrog State RadioTechnical University, Taganrog, Russia. She obtained PhD in 1993 from the Southern Russian Technical University, Novocherkassk, Russia.She obtained Doctorate degree in 2000 from the High Attestation Committee, Russia. She works as a Professor of the Bauman Moscow State Technical University. Her scientific activities are nanoengineering, VLSI design, and informatics.

Index

D

E

F

G

H

I

J

K

L

M

N

O

P

Q

R

S

T

U

V

W

www.ingramcontent.com/pod-product-compliance
Lightning Source LLC
LaVergne TN
LVHW060008031026
849161LV00025B/2508
* 9 7 8 1 6 6 8 4 2 9 3 1 0 *